Instructor's Manual for

FUNDAMENTALS OF PHYSICS

Third Edition

Third Edition Extended

David Halliday
University of Pittsburgh

Robert Resnick
Rensselaer Polytechnic Institute

Prepared by

J. Richard Christman
United States Coast Guard Academy

Stanley A. Williams
Iowa State University

With the Assistance of

Walter Eppenstein
Rensselaer Polytechnic Institute

Edward Derringh
Wentworth Institute of Technology

Van E. Neie
Purdue University

JOHN WILEY & SONS   New York · Chichester · Brisbane · Toronto · Singapore

ISBN 0-471-81997-2

Printed in the United States of America

10 9 8 7 6 5 4 3 2 1

PREFACE

This manual contains material we hope will be useful in the design of an introductory physics course, based on the text FUNDAMENTALS OF PHYSICS by Halliday and Resnick. It may be used with either the extended or regular versions. We include material to help instructors choose topics, using both time and subject matter as criteria. We also provide lecture notes, outlining the important topics of each chapter, and suggest demonstration experiments, laboratory and computer exercises, films and videocassettes, and discussion questions. Many of the diagrams in the text have been reproduced in a large format for use as transparency masters. Answers are given to many of the suggested end-of-chapter questions.

In designing your course, you may also wish to consult the STUDY GUIDE by S.A. Williams, K. Brownstein, and T. Marcella. It gives students step-by-step help and is designed to accompany the text. Also consider IMPROVE YOUR PHYSICS GRADE by R. Aarons. Many stumbling blocks encountered by students are discussed in PRELUDE TO PHYSICS by Clifford Swartz. All three of these books are published by John Wiley & Sons. Consider making them available to your students and perhaps assigning one or more of them.

The principal authors are grateful to Edward Derringh, who helped with the answers to questions; to Walter Eppenstein, who helped with demonstration and laboratory experiments; to Benjamin Chi who helped with the transparency masters; and to Van Nie, who wrote the film and videocassette reviews. We appreciate the help and effort of Robert A. McConnin, former Physics Editor at Wiley, and Catherine Feduska, the present Physics Editor. We are also grateful to William Kellogg, who supervised production.

A special acknowledgement goes to our wives, Shirley Williams and Mary Ellen Christman, who supplied much encouragement and cheerfuly tolerated the many hours we worked on this manual.

J. Richard Christman   Stanley A. Williams
U.S. Coast Guard Academy   Iowa State University
New London, Connecticut   06320   Ames, Iowa   50011

CONTENTS

SECTION ONE
PATHWAYS AND TIME ESTIMATES

In this section of the manual we indicate ways in which the order of presentation of topics can be altered. We also give our best estimate of the time required to adequately cover each chapter of the text.

ORDERING OF TOPICS

Although the order used by the authors of the text is widely accepted, many instructors reorder the material to meet specific needs of their courses. Great care must been taken if this is done since some discussions in later chapters presume coverage of prior material.

Some minor changes that are possible are mentioned in the Lecture Notes. These are chiefly in the nature of postponements. For example, the scalar product, discussed in Chapter 3, can easily be postponed until the discussion of work in Chapter 7. Similarly, the vector product can be postponed until the discussion of torque in Chapter 12. In addition, some topics are labeled optional or supplementary. The material that follows does not depend of them and, if desired, they may safely be omitted from the course.

Major changes in the order are more difficult. For example, it would be extremely difficult to precede dynamics with statics as is sometimes done with other texts. To do so, one would need to discuss torque, introduced in Chapter 11, and explain its relation to angular acceleration. This involves considerable effort and is of questionable value.

Major changes which can easily be carried out are listed on the following pages. Other possibilities exist, but they should be undertaken only after careful study of the text. It is probably best to use the text as it is for a year or so before attempting such changes.

I.   Mechanics
     The central concepts of classical mechanics are covered in Chapters 1 through 12. Of these only Chapter 10 (Collisions) can be omitted or significantly shortened with safety. The order of the chapters must

be retained.

Chapters 13 through 18 consist chiefly of applications. Many courses omit one or more of Chapters 13 (Equilibrium and Elasticity), 15 (Gravity), 16 (Fluids), and 18 (Waves - II). There is some peril in these omissions, however. Newton's law of gravity is used later to introduce Coulomb's law and the proof that the electrostatic force is conservative relies on the analogy. Fluid flux is used to introduce electric flux in Chapter 25 (Gauss' law). The concepts of pressure and density are explained in Chapter 16 and used again in the thermodynamics chapters. If any of these chapters are omitted, you should be prepared to make up for the loss of material, either by presenting alternative introductions or defining pressure and density when they are first used in your course.

Chapter 13 (Equilibrium and Elasticity) can be safely omitted. If it is, a brief description of the equilibrium conditions might be included in the discussion of Chapter 12. The few problems in later chapters that depend on material in this chapter can be passed over.

Chapter 14 (Oscillations) and Chapter 17 (Waves - I) are important parts of an introductory course and should be covered except when time constraints are severe. Chapter 14 is required for Chapter 17 and both are required for Chapter 18 (Waves - II). Chapter 14 is also required for Chapters 35 (Electromagnetic Oscillations) and 36 (Alternating Currents) and Chapter 17 is required for Chapters 38 (Electromagnetic Waves), 40 (Interference), 41 (Diffraction), and 44 (Waves and Particles).

Chapter 17 may be covered immediately following Chapter 14 and both may be covered just prior to their first use (either Chapter 18, or 35).

II. Thermodynamics
     Chapters 19 through 22 cover the ideas of thermodynamics. Most two
     term courses and some three term courses omit these chapters
     entirely. If they are covered, they can be placed as a unit almost
     anywhere after the mechanics chapters. The idea of temperature is
     used in Chapter 28 (Current and Resistance) and in the modern physics

chapters. If Chapter 19 is not covered prior to Chapter 28, you should plan to discuss the ideas of temperature and the Kelvin scale in connection with that chapter.

III. Electromagnetism

The fundamentals of electricity and magnetism are covered in Chapters 23 through 38. The first four sections of Chapter 34 (Magnetism and Matter) are basic and should be covered. The rest of the chapter may be omitted, if desired.

Chapters 35 (Electromagnetic Oscillations) and 36 (Alternating Currents) may also be omitted but Chapter 35 must be covered if Chapter 36 is included.

IV. Optics

Chapters 39 through 41 are the optics chapters. Some courses include Chapter 39 (Geometrical Optics) but exclude Chapters 40 (Interference) and 41 (diffraction). Chapter 41, however, cannot be covered without Chapter 40.

V. Modern Physics

Modern physics topics are scattered throughout the text. For example, Chapter 8 (Conservation of Energy) includes the quantization of energy and the equivalence of mass and energy, Chapter 12 (Rolling, Torque, and Angular Momentum) includes the quantization of angular momentum, Chapter 23 (Electric Charge) includes the quantization of charge. Although we have catagorized these topics as supplementary in the Lecture Notes, they constitute important extensions of the usual course in classical physics. Instructors can include many important modern ideas simply by covering these sections. By studying modern topics when they occur in the text, students are not only introduced to them but also can begin to see their relationship to classical ideas.

In addition, Chapters 42 through 49 provide a more formal development of some of the important ideas and applications of modern physics. All of these chapters may omitted, of course. Many instructors, however, use some or all of this material to round out their courses. Chapters 43 through 49 are included in the extended version of the text only.

Strictly speaking, Chapter 42 (Relativity) is not required for the chapters that follow. The results of relativity theory that are needed, chiefly the definitions of energy and momentum and the mass-energy relationship, appear in earlier chapters. If you do not desire more than minimum coverage, Chapter 42 can be omitted. However, relativity is treated as a coherent theory in the chapter and flesh is added to the bare bones discussions of earlier chapters. The chapter may be used as a capstone to the course or as an introduction to the modern physics that follows.

Chapters 43 (Quantum Physics - I), 44 (Quantum Physics - II), and 45 (All About Atoms), if included in the course, must follow in the order written. Selected parts can be omitted, if desired. See the Lecture Notes for details. If you include these chapters be sure earlier parts of the course include discussions of uniform circular motion, angular momentum, Coulomb's law, electrostatic potential energy, magnetic dipole moment, electromagnetic waves, and diffraction. $E = mc^2$ and $E^2 = (pc)^2 + (mc^2)^2$, from relativity theory, are used in discussions of the Compton effect. If Chapter 42 is not covered, supply some supplementary material.

Chapter 47 (Conduction of Electricity in Solids) can safely be omitted, if desired. If it is included be sure to cover the discussion of energy quantization in Chapter 8. It is also helpful, though not necessary, to cover Chapter 45 first. The ideas of temperature and the Kelvin scale are also used but, with a little supplementary material, this chapter can be covered even if the thermodynamic chapters are not.

Chapter 48 (Energy from the Nucleus) requires Chapter 47 (Nuclear Physics) for background material, but both chapters may be omitted or only Chapter 47 included. The idea of a matter wave is used, so Chapter 44 should be covered first. $E = mc^2$ and $E^2 = (pc)^2 + (mc^2)^2$ from relativity theory are also used. Either Chapter 42 should be covered or time should be alloted to cover these ideas briefly. The discussion of thermonuclear fusion uses some of the ideas of kinetic theory, chiefly the distribution of molecular speeds. Either Chapter

21 should be covered first or the students should be supplied with a small amount of supplementary material.

Chapter 49 (Quarks, Leptons, and the Big Bang) includes an introduction to high energy particle physics. Some knowledge of the Pauli exclusion principle and spin angular momentum (both discussed in Chapter 45) is required. Some knowledge of the strong and weak nuclear forces (discussed in Chapter 47) is also required. In addition, beta decay (discussed in Chapter 47) is used several times as an illustrative example. Nevertheless, the chapter can be made to stand alone with the addition of only a small amount of supplementary material.

SUGGESTED COURSES

A bare bones two semester course (about 90 meetings) can be constructed around Chapters 1 through 12, 14, 17, 23 through 33, the first four sections of 34, and Chapters 37 and 38. The course can be adjusted to the proper length by the inclusion or omission of supplementary material and optional topics. Some time might be available for a few of the modern physics topics. If 4 to 8 additional meetings are available each term, Chapter 15 or 16 (or perhaps both) can be inserted after Chapter 12 and Chapter 39 (and perhaps 40 and 41) can be inserted after Chapter 38. The inclusion of these chapters must be balanced with the desire to include more of the supplementary modern topics.

A three term course (about 135 hours) can be constructed by adding the thermodynamics chapters (19 through 22) and some or all of the modern physics chapters (42 through 49) to those mentioned above. If the needs of the class dictate a section on alternating current, two of the modern physics chapters can be replaced by Chapters 35 and 36.

ESTIMATES OF TIME

The following chart gives estimates of the time required to cover <u>all</u> of each chapter, in units of 50 minute periods. The second column of the chart contains estimates of the number of lecture periods needed and includes the time needed to perform demonstrations and discuss the main points of the chapter. The third column contains estimates of the number of recitation periods required and includes the time needed to go over problem solutions, answers to end-of-chapter questions, and points raised by students. If your course is organized differently you may wish to add the two numbers to obtain the total estimated time for each chapter.

The times given in the chart are based on our collective experience with the text and, where we differ, an average is given. Use the chart as a rough guide when planning the syllabus for a semester, quarter, or year course. If you omit parts of chapters, reduce the estimated time accordingly.

Unit: 50 minute period

Text Chapter	Number of Lectures	Number of Recitations	Text Chapter	Number of Lectures	Number of Recitations
1	0.3	0.2	26	1.8	1.8
2	2.0	2.0	27	1.5	2.0
3	1.0	1.0	28	1.0	1.0
4	2.0	2.5	29	2.0	2.2
5	2.0	2.0	30	2.0	1.8
6	2.0	2.0	31	2.0	1.2
7	1.8	1.5	32	1.8	1.0
8	2.0	2.0	33	2.0	2.0
9	2.0	1.5	34	0.8	0.3
10	2.0	1.6	35	1.0	1.0
11	2.0	1.5	36	2.0	2.0
12	2.0	2.0	37	1.0	0.5
13	1.0	2.0	38	2.5	2.0
14	2.5	1.8	39	3.0	3.0
15	2.3	2.0	40	2.0	2.0
16	2.0	2.0	41	2.0	2.0
17	2.5	2.0	42	2.5	2.0
18	2.5	2.0	43	2.0	2.0
19	0.7	1.0	44	1.5	1.5
20	2.0	2.0	45	2.2	2.0
21	2.0	1.4	46	2.0	2.0
22	1.5	1.6	47	1.8	2.0
23	1.0	1.0	48	2.0	2.0
24	1.6	1.3	49	2.0	2.0
25	1.8	1.8			

SECTION TWO
LECTURE NOTES

Lecture notes for each chapter of the text are grouped under the headings
BASIC TOPICS, SUPPLEMENTARY TOPICS, and SUGGESTIONS.

BASIC TOPICS contains main points of each chapter, in outline form. In
addition, one or two demonstrations are recommended to show the main theme
of each chapter. You may wish to pattern your lectures after the notes,
suitably modified, or simply use them as a check on the completeness of
your own notes.

SUPPLEMENTARY TOPICS consists of a list of topics which, although covered
in the text, may be safely omitted in order to save time. Many of these
are pertinent to laboratory experiments and might be assigned in
conjunction with the laboratory or as outside reading. Many of them deal
with modern topics and you may wish to include them to broaden the course.

The SUGGESTIONS sections recommend end-of-chapter questions and problems,
films, film loops, video cassettes, computer projects, alternate
demonstrations, and other material which might be useful for the course.
Many of the questions concentrate on points which seem to give students
trouble and it is worthwhile dealing with some of them before students
tackle a problem assignment. Some questions and problems can be
incorporated into the lectures, some can be assigned and used to generate
discussion by students in small recitation sections.

Films and Videocassettes

All of the films and loops are short, well done, and highly pertinent to
the chapter. It is not possible to review all available films and there
are undoubtedly many other fine films which are not listed. Films and
loops can be incorporated into the lectures, shown during laboratory
periods, or set up in special rooms for more informal viewing.

Two excellent sets of videocassettes, THE MECHANICAL UNIVERSE, can be
obtained from The Mechanical Universe, Caltech, MS 1-70, Pasadena, CA
91125. The set consists of 52 half hour segments dealing with nearly all
important concepts of introductory physics. Historical information and
animated graphics are used to present the concepts in an imaginative and

engaging fashion. Some physics departments run appropriate segments throughout the course in special viewing rooms. Accompanying textbooks, teacher manuals and study guides are also available.

Computers

Computers have now made significant contributions to the teaching of physics. They are widely used in lectures to provide animated illustrations, with parameters under the control of the lecturer; they also provide tutorials and drills which students can work through on their own. The Physics Courseware Laboratory at the University of North Carolina (Raleigh NC 27695-8202) maintains an up-to-date catalog of both commercial and public domain software, last published in The Physics Teacher of May, 1987. Personnel of the laboratory write reviews, which appear in The Physics Teacher.

Specialized programs are listed in appropriate SUGGESTION sections of the Lecture Notes in this manual. In addition, there are several packages which cover large portions of an introductory course. Three of them are:

> The Physics I and Physics II Series (Control Data Company, 3601 West 77th Street, Bloomington, MN 55435). Excellent problem solving tutorials for the IBM PC. Sixteen modules cover important topics in mechanics and 12 cover important topics in electrostatics, magnetostatics, and Faraday's law.

> PHYSICS I (Microphysics Programs, 1737 West 2nd Street, Brooklyn, NY 11223). A set of programs that generate problems covering particle dynamics and some aspects of thermodynamics. Different versions are available for IBM PC, Apple II, TRS-80 (models III and IV), Commodore Pet and 64.

> PHYSICS SIMULATIONS (Kinko's Service Corporation, 4141 State Street, Santa Barbara, CA 93110). A great many demonstrations of important topics in introductory physics. Individual programs are listed under appropriate chapters in the Lecture

Notes. These programs are for the Apple Macintosh.

Sensei Physics (Broderbund Sftware, P.O. Box 12947,
San Rafael, CA 94913-2947). Tutorial reviews with
animated graphics of most of the major topics of
introductory physics. Over 300 problems for student
practice. These programs are for the Apple Macintosh.

You might consider setting aside a room or portion of a lab, equip it with
several computers, and make tutorial, drill, and simulation programs
available to students. If you have sufficient hardware (and software) you
might base some assignments on computer materials.

Computers might also be used by students to perform calculations. Properly
selected problems can add greatly to the students' understanding of
physics. Problems involving the investigation of some physical system of
interest might be assigned as individual projects or might be carried out
by a laboratory class.

Commercial spreadsheet programs, of the type used by business, can
facilitate problem solving. For a detailed account of how they are used
and a collection of informative problems, see Wondering About Physics ...
Using Spreadsheets to Find Out by D.I. Dykstra and R.G. Fuller (John Wiley
& Sons, 1988). Commercial problem solving programs such as Eureka: The
Solver (Borland International, 4585 Scotts Valley Drive, Scotts Valley, CA
95066; IBM PC) and TK Solver! (Universal Technical Systems Inc., 1220 Rock
Street, Rockford, IL 61101; IBM PC, Macintosh) can easily be used by
students to solve problems and graph results. In many cases, data
generated by spreadsheets can also be imported to graph drawing programs.
All these programs allow students to set up a problem generically, then
view solutions for various values of input parameters. For example, the
range or maximum height of a projectile can be found as a function of
initial speed or firing angle, even if air resistance is taken into
account. Eureka is packaged with the text at modest additional cost. A
large number of suitable problems and projects can be found in the two
volume calculator supplement Physics Problems for Programmable Calculators
by J. Richard Christman (John Wiley & Sons, 1981 and 1982, now out of
print). Some projects are suggested in the Lecture Notes.

Demonstrations

Notes for most of the chapters are developed around demonstration experiments. Generally speaking, these use relatively inexpensive, readily available equipment, yet clearly demonstrate the main ideas of the chapter. The choice of demonstrations, however, is highly personal and you may wish to substitute others for those suggested here or you may wish to present the same ideas using blackboard diagrams. Several excellent books give many other examples of demonstration experiments. The following are available from the American Association of Physics Teachers, 5110 Roanoke Place, College Park, MD 20740:

Resource Letter PhD-1: Physics Demonstrations, J.A. Davis and B.G. Eaton, 6 pages (1979). Contains 103 references to books, monographs, indexes, and conference proceedings dealing with physics demonstrations.

A Demonstration Handbook for Physics, G.D. Freier and F.J. Anderson, 320 pages (1981). Contains 807 demonstrations, including many which use everyday materials and which can be constructed with minimal expense. Line drawings illustrate every demonstration.

Physics Demonstration Experiments at William Jewell College, Wallace A. Hilton, 112 pages (1982). Contains descriptions and photographs of over 300 demonstrations.

The following is available from Robert E. Krieger Publishing Company, Malabar, FL 32950:

Physics Demonstration Experiments, H.F. Meiners, ed. An excellent source of ideas, information, and construction details on over 100 experiments, with over 2000 line drawings and photographs. It also contains some excellent articles on the philosophical aspects of lecture demonstrations, the use of shadow projectors, TV, films, overhead projectors, and stroboscopes.

Appropriate demonstrations described in Freier and Anderson and in Hilton are listed in the SUGGESTIONS sections of the notes. Neither of these books give any construction details, but more information about most demonstrations can be obtained from the book edited by Meiners.

Laboratories

Hands-on experience with actual equipment is another important element of an introductory physics course. There are many different views as to the objectives of the physics laboratory and the final decision on the types of experiments to be used has to be made by the individual instructor or department. This decision is usually based on financial and personnel considerations as well as on the pedagogical objectives of the laboratory.

Existing laboratories vary widely. Some use strictly cookbook type experiments while others allow the students to experiment freely, with practically no instructions. The equipment ranges from very simple apparatus to rather complex and sophisticated equipment. Physical phenomena may be observed directly or simulated on a computer. Data may be taken by the students or fed directly into a microprocessor.

Many physics departments have written their own notes or laboratory manuals and relatively few physics laboratory texts are on the market. Two such books, both available from John Wiley & Sons, are

> Laboratory Physics, H.F. Meiners, W. Eppenstein,
> R.A. Oliva, and T. Shannon, 2nd ed. (1987).

> Laboratory Experiments in College Physics, C.H.
> Bernard and C.D. Epp, 6th ed. (1980).

Experiments from these books are listed in the SUGGESTIONS section of the Lecture Notes. MEOS is used to designate the Meiners, Eppenstein, Oliva, Shannon book while BE is used to designate the Bernard and Epp book. Both books contain excellent introductory sections explaining laboratory procedures to students. MEOS also contains a large amount of material on the use of microprocessors in the lab.

Chapter 1   MEASUREMENT

BASIC TOPICS

I.   Base and derived units.
    A.   Explain that standards are associated with base units and that
          measurement of a physical quantity takes place by means of a
          comparison with a standard. Show a 1 kg mass and a meter stick.
          Show the simple well-known procedure for measuring length with a
          meter stick.
    B.   Explain that derived units are combinations of base units.
          Explain that the speed of light is now a defined unit and the
          meter is a derived unit. Discuss an experiment in which the time
          taken for light to travel a certain distance is measured.
          Example: the reflection of a light signal from the moon. Use a
          clock and a meter stick to find your walking speed in m/s.
    C.   This is a good place to review area, volume, and mass density.
          Use simple geometric figures (circle, rectangle, triangle, cube,
          sphere, cylinder, etc.) as examples.

II.   Systems of units.
    A.   Give the 1971 SI base units (Table 1). Stress that the first
          three will be used extensively.
    B.   Point out the SI prefixes (Table 2). The important ones for this
          course are mega, kilo, centi, milli, micro, nano, and pico.
          Stress the simplicity of the powers of ten notation.
    C.   Most of the students' experience is with the British system.
          Relate the inch to the centimeter and the slug to the kilogram.
          Discuss unit conversion. Use speed as an example: convert 50 mph
          and 3 mph to km/h and m/s. Point out the conversion tables in
          Appendix F.

III. Properties of standards.
    A.   Use questions 1, 2, 6, 7, and 19 to discuss choices of standards.
    B.   Discuss secondary standards such as the meter stick used earlier.

IV.   Measurements.
    A.   Stress the wide range of magnitudes measured. See Tables 3, 4, 5,
          and 6.

B. Discuss indirect measurements. See questions 10, 11, and 20.

V. Skills.

A. Discuss unit arithmetic, unit conversion, and powers of ten arithmetic. Show how Appendix F is used.

B. Assign problems for drill according to the needs of the class.

SUGGESTIONS

1. Demonstrations

Examples of "standards" and measuring instruments: Freier and Anderson Mal – 3; Hilton Ml.

2. Film

Powers of Ten, 16 mm or 3/4" videocassette, b/w, 25 min. Pyramid Films, P.O. Box 1048, Santa Monica, CA 90406. This very popular film is an excellent way to introduce the students to the relative sizes of things in the universe.

3. Laboratory

MEOS Experiment 7-1: Measurement of Length, Area, and Volume. Gives students experience using the vernier caliper, micrometer, and polar planimeter. Good introduction to the determination of error limits (random and least count) and calculation of errors in derived quantities (volume and area).

BE Experiments 1 and 2: Determination of Length, Mass, and Density and Determination of π and Density by Measurements and Graphical Analysis. Roughly the same as the MEOS experiment, but a laboratory balance is added to the group of instruments and the polar planimeter is not included. Graphs of mass vs. radius and radius squared for a collection of disks made of the same material are used to establish the quadratic dependence of mass on radius.

MEOS Experiment 7-3: The Simple Pendulum and BE Experiment 3: The Period of a Pendulum – An Application of the Experimental Method. Students time simple pendula of different lengths, then use the data and graphs (including a logarithmic plot) to determine the relationship between length and period. They calculate the acceleration due to gravity. This is an exercise in finding functional relationships and does not require knowledge of dynamics.

Chapter 2   MOTION IN A STRAIGHT LINE

BASIC TOPICS

I.   Position.
     A.   Move a toy cart along a table top with constant velocity. Select
          an origin, place a meter stick and clock on the table, and
          demonstrate how x(t) is measured in principle. Emphasize that x
          is always measured from the origin; it is not the cart's
          displacement during any time interval. Draw a graph of x(t) and
          point out it is a straight line. Show what the graph looks like
          if the cart is not moving.
     B.   Go over the various graphs in Fig. 2. Point out that the slopes
          are steeper for faster objects. Use an overhead transparency to
          project the figure, draw a vertical line corresponding to an
          arbitrary time, and show that a steeper slope means a greater
          distance is covered in the same time.

II. Velocity.
     A.   Define average velocity over an interval. Stress the meaning of
          the sign. Go over Sample Problems 1 and 2.
     B.   Define instantaneous velocity. To demonstrate the limiting
          process, go over Table 1. Define instantaneous speed as the
          magnitude of the velocity. Average speed over an interval is the
          total path length divided by the time. It is not necessarily the
          magnitude of the average velocity.
     C.   Note that many calculus texts use a prime to denote a derivative.
          They also define the derivative of x with respect to time, for
          example, by the limit of $[x(t+\Delta t) - x(t)]/\Delta t$ rather than by the
          limit of $\Delta x/\Delta t$. Mention the different notations in class so
          students can relate their physics and calculus texts.

III. Acceleration.
     A.   Define average and instantaneous acceleration. Use Fig. 6 to
          discuss acceleration as the slope of a v(t) graph.
     B.   Interpret the sign of the acceleration. Give examples of objects
          with acceleration in the same direction as the velocity (speeding
          up) and in the opposite direction (slowing down). Be sure to
          include both directions of velocity. Emphasize that a negative

acceleration does not necessarily imply slowing down.

C. Use graphs of x(t) and v(t) to point out that an object may have zero velocity and non-zero acceleration.

IV. Motion in one dimension with constant acceleration.

A. Derive the kinematic equations for x(t) and v(t). If students know about integration, use methods of the integral calculus. In any event, show that v(t) is the derivative of x(t) and that a is the derivative of v(t).

B. Discuss kinematic problems in terms of a set of simultaneous equations to be solved. Examples: use equations for x(t) and v(t) to algebraically eliminate the time; to algebraically eliminate the acceleration. The equations of constant acceleration motion are listed in Table 2. Some instructors teach students to use the table, as is done in Sample Problem 7. Others ask students to always start with Eqs. 10 and 14, then use algebra to obtain the equations needed for a particular problem.

C. Sketch graphs of v(t) and x(t) for various initial conditions but the same acceleration. Draw a set of graphs for positive and negative acceleration.

V. Free fall.

A. Give the values for g in SI and British units. Point out that the acceleration due to gravity is directed toward the center of the earth but that for local problems the earth is essentially flat and the acceleration may be taken in the same direction at different points. Explain that a = +g if down is taken to be the positive direction and a = -g if up is the positive direction. Do examples using both choices.

B. Explain that all objects have the same acceleration due to gravity. Different objects may have different accelerations because air influences their motions differently. This can be demonstrated by placing a coin and a wad of cotton in a glass cylinder about 1 m long. Turn the cylinder over and notice the coin reaches the bottom first. Now use a vacuum pump to partially evacuate the cylinder and repeat the experiment. Repeat again with as much air as possible pumped out.

C. Point out that free fall problems are special cases of constant acceleration kinematics and the methods described earlier can be used.

D. Drop a small ball through two photogates, one at the top to turn on a timer and one further down to turn it off. Repeat for various distances and plot the position of the ball as a function of time. Explain that the curve is parabolic and indicates a constant acceleration.

SUGGESTIONS

1. To obtain some qualitative understanding of velocity and acceleration, ask the students to discuss questions 7, 8, 10, 11, and 12.

2. To make more use of the calculus assign problems 22, 23, 26, 27, 28, and 29.

3. Discuss differences between average and instantaneous velocity and acceleration. See problems 23, 26, and 27.

4. To emphasize the interpretation of graphs assign problems 18, 19, 20, and 21.

5. This is a good place to point out the HINTS sections of the text. Chapter 2 contains 10 of them and many others are scattered throughout the text. They contain excellent suggestions for students to follow in solving problems.

6. Computer projects

    Use a spreadsheet or your own computer program to demonstrate the limiting processes used to define velocity and acceleration. Given the functional form of $x(t)$, have the computer calculate and display the coordinate for some time t_0 and a succession of later times, closer and closer to t_0. For each interval have it calculate and display the average velocity.

    Have students use the root finding capability of Eureka or their own computer programs to solve kinematic problems. Nearly every one of them can be set up as a problem which involves finding the root of either the coordinate or velocity as a function of time, followed perhaps by substitution of the root in another kinematic equation. Problems need not be limited to those involving constant acceleration. Air resistance, for example, can be taken into account. See the calculator supplement for sample problems. The same program can be used to solve rotational kinematic problems in Chapter 11.

7. The Microcomputer Based Lab Project *Motion* (HRM Software, 175

Tompkins Avenue, Pleasantville, NY 10570-9973) is useful for helping students interpret kinematic graphs. As a student moves back and forth in front of a sonar ranging device, his position, velocity, or acceleration is plotted on the monitor of an Apple II computer. A graph can be designed by the instructor and the student asked to duplicate it by moving. Several sonic rangers are reviewed in TPT January 1988.

8. Demonstrations

Uniform velocity and acceleration, velocity as a limiting process: Freier and Anderson Mb10 - 13, 15, 18, 21, 22; Hilton M2, M3, M4, M5.

9. Films

Straight-Line Kinematics, 16 mm, b/w, 34 min. Modern Learning Aids, Division of Ward's Natural Science Establishment, P.O. Box 312, Rochester, NY 14601. Although this film is over 20 years old, the generation of distance, speed, and acceleration graphs using special test car equipment is an excellent teaching device.

Velocity from Position; Position from Velocity I; Position from Velocity II, S8, b/w, 3 min each. American Association of Physics Teachers, 5110 Roanoke Place, College Park, MD 20740. Average and instantaneous velocities are developed through the motion of an automobile. The latter two films discuss the often neglected concept of area under the curve.

Acceleration Due to Lunar Gravity, S8, color, 4 min. American Association of Physics Teachers (see address above). The "Guinea and Feather" experiment is demonstrated on the moon by astronauts, who drop a hammer and a feather in the vacuum of the moon.

10. Computer program

Motion, Cross Educational Software, P.O. Box 1536, Ruston, LA 71270. Apple II+. Generates graphs of coordinate, velocity, and acceleration for one dimensional motion. Presents problems in one dimensional dynamics (translational and rotational). Students can be assigned tutorial sections. Some sections can be used to illustrate lectures. Reviewed TPT September 1983.

11. Laboratory

MEOS Experiment 7-5: Analysis of Rectilinear Motion. Students measure the position as a function of time for various objects rolling down an incline, then use the data to plot speeds and accelerations as functions of time. No knowledge of rotational motion is required. This experiment emphasizes the definitions of velocity and acceleration as differences over a time interval.

MEOS Experiment 8-1: <u>Motion in One Dimension</u> (omit part dealing with conservation of energy). Essentially the same experiment except pucks sliding on a nearly frictionless surface are used. This experiment may be be done with dry ice pucks or on an air table or air track.

BE Experiment 7: <u>Uniformly Accelerated Motion</u>. The same technique as the MEOS experiments but a variety of setups are described: the standard free fall apparatus, the free fall apparatus with an Atwood attachment, an inclined plane, an inclined air track, and a horizontal air track with a pulley attachment.

Chapter 3   VECTORS

BASIC TOPICS

I.   Definition.
   A.   Explain that vectors have magnitude and direction, and that they obey certain rules of addition.
   B.   Example of a vector: displacement. Give the definition and physical interpretation of the sum of two displacements. Demonstrate vector addition by walking along two sides of the room. Point out that the distance traveled is not the magnitude of the displacement. Go back to the original position and point out that the displacement is now zero. Discuss question 1.
   C.   Compare vectors with scalars.
   D.   Insist that students clearly identify vectors by using vector notation.

II. Vector addition by the graphical method.
   A.   Draw two vectors tail to head, draw the resultant, and point out the direction of the resultant.
   B.   Define the negative of a vector and define vector subtraction as $\underline{A} - \underline{B} = \underline{A} + (-\underline{B})$. Graphically show that if $\underline{A} + \underline{B} = \underline{C}$ then $\underline{A} = \underline{C} - \underline{B}$.
   C.   Show that vector addition is both commutative and associative.

III. Vector addition by the analytic method.

A. Derive the expression for the components
of a vector, given its magnitude and the
angles it makes with the coordinate
axes. Point out that the components
depend on the choice of coordinate
system and compare the behavior of
vector components with the behavior of
a scalar. Find the components of a
vector using each of the coordinate
systems shown. Point out that it is
possible to orient the coordinate system
so that only one component of a given
vector is not zero. Overlays are useful
to show vector components.

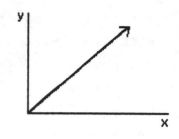

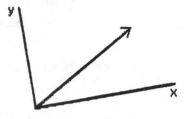

B. Define the unit vectors along the coordinate axes. Give the form
used to write a vector in terms of its components and the unit
vectors. Explain that unit vectors are unitless. Discuss question
8.

C. Vector addition. Give the
expressions for the components
of the resultant in terms of
the components of the addends.
Demonstrate the equivalence of
the graphical and analytic
methods of finding a vector sum.
See the diagram to the right.

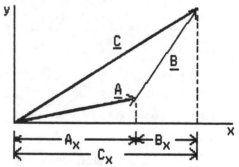

D. Show how to find the magnitude and angles with the coordinate
axes, given the components. Explain that calculators give only
one of the two possible values for the inverse tangent and show
how to determine which is correct for a given situation.

IV. Multiplication involving vectors.
A. Multiplication by a scalar. Give examples of both positive and
negative scalars multiplying a vector. Give the components of the
resulting vector as well as its magnitude and direction.
B. Scalar product of two vectors (may be postponed until Chapter 7).
Point out that $\underline{A} \cdot \underline{B}$ is the magnitude of $\underline{A}$ multiplied by the
component of $\underline{B}$ along an axis in the direction of $\underline{A}$, and vice
versa.

C. Vector product of two vectors (may be postponed until Chapter 12). Point out that $|\underline{A} \times \underline{B}|$ is the magnitude of $\underline{A}$ multiplied by the component of $\underline{B}$ along an axis perpendicular to $\underline{A}$, and vice versa.

D. Show the results of scalar and vector multiplication involving cartesian unit vectors. See problems 39 and 40. Give students the useful mnemonic for vector products: $\hat{i}$, $\hat{j}$, $\hat{k}$ written in that order clockwise around a circle. One starts with the first named vector in the vector product and goes around the circle to the second named vector. If the direction of travel is clockwise the result is the third vector. If if it is counterclockwise the result is the negative of the third vector.

D. Problems 47 and 51 cover the equations for scalar and vector products in terms of components. These are useful for the discussion and problems of Chapters 7 and 12.

SUGGESTIONS

1. Ask students to use graphical representations of vectors to think about questions 2, 3, 4, 5, and 6. Assign problems such as 3, 4, 6, and 7.

2. Problems that stress the calculation of vector components include 9, 10, 12, and 19. Some good problems to test understanding of analytic vector addition are 24, 27, and 29.

3. Computer project
    Have students use Eureka or write their own computer programs to carry out conversions between polar and cartesian forms of vectors, vector addition, scalar and vector products.

4. Demonstrations
    Vector addition: Freier and Anderson Mb2, 3; Hilton M10a, b, c.

5. Computer programs
    Vector Addition II, Vernier Software, 2920 89th Street, Portland, OR 97225. Apple II. User supplies magnitude and direction of up to 19 vectors. Individual vectors and their resultant are drawn on the monitor screen and the magnitude and direction of the resultant are given numerically. Handy for lectures and can also be used to drill students. Reviewed TPT February 1986.
    College Physics Series, Vol. I: Vectors and Graphics, Cross Educational Software (see Chapter 2 notes for address). Apple II+.

Tutorials on the resolution of vectors into components, vector
addition, scalar product, vector product. A quiz follows each
tutorial. Reviewed TPT January 1983.

6. Laboratory

    BE Experiment 4: <u>Composition and Resolution of Forces - Force-Table
    Method</u>. Students mathematically determine a force which balances 2 or
    3 given forces, then check the calculation using a commercial force
    table. They need not know the definition of a force, only that the
    forces in the experiment are vectors along the strings used, with
    magnitudes proportional to the weights hung on the strings. The focus
    is on resolving vectors into components and finding the magnitude and
    direction of a vector, given its components.

Chapter 4   MOTION IN A PLANE

BASIC TOPICS

I.   Definitions.
     A. Draw a particle path, showing the position vector at several
         times and the change in the position vector for several
         intervals.
     B. Define velocity as $d\underline{r}/dt$. Write in both vector and component
         form. Point out that the velocity vector is tangent to the path.
         Define speed of the magnitude of the velocity.
     C. Define acceleration as $d\underline{v}/dt$. Write in both vector and component
         form. Point out that $\underline{a}$ is not zero if either the magnitude or
         direction of $\underline{v}$ changes with time.
     D. Show that the particle is speeding up only if $\underline{a} \cdot \underline{v}$ is positive. If
         $\underline{a} \cdot \underline{v}$ is negative the particle is slowing down and if $\underline{a} \cdot \underline{v} = 0$ its
         speed is not changing.

II. Two dimensional motion with constant acceleration.
     A. Write down the kinematic equations for $\underline{r}(t)$ and $\underline{v}(t)$. Stress that
         these form two sets of one dimensional equations, linked by the
         common variable t and are to be solved simultaneously. Note that
         a_x affects only v_x and not v_y and that a_y affects only v_y. Throw
         a ball vertically, then catch it. Repeat while walking with
         constant velocity across the room. Ask students to observe the

motion of the ball relative to the blackboard and to describe its motion relative to your hand.

B. Point out that the shape of the trajectory can be found by eliminating t from the equations for x(t) and y(t). Show it is a parabola.

III. Projectile motion.

A. Draw the trajectory of a projectile, show the direction of the initial velocity and derive its components in terms of the initial speed and firing angle.

B. Write down the kinematic equations for x(t), y(t), v_x(t), and v_y(t). Include both a_x and a_y but then specialize to a_x = 0 and a_y = -g for positive y up. Point out that the acceleration is the same at all points of the trajectory.

C. Work examples. Use punted footballs, hit baseballs, or thrown basketballs according to season.

    1. Find the time for the projectile to reach its highest point, then find the coordinates of the highest point.

    2. Find the time for the projectile to hit the ground, at the same level as the firing point. Then find the range and the velocity components just before landing.

    3. Show how to work problems for which the landing point is not at the same level as the firing point.

D. Point out that all projectiles follow some piece of the full parabolic trajectory. For example, A to D could be the trajectory of a ball thrown at an upward angle from a roof to the street; B to D could be the trajectory of a ball thrown horizontally; C to D could be the trajectory of a ball thrown downward.

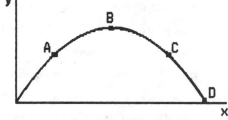

E. Demonstrate projectile motion by using a spring gun to fire a ball onto a surface at the firing height. Use various firing angles, including 45°, and point out that the maximum range occurs for a firing angle of 45°. Remark on the symmetry of the range as a function of firing angle. Mention that the maximum range occurs for a different angle when the ball is fired onto a surface at a different height.

F.   Use Fig. 13 to show the effect of air resistance.

IV.   Uniform circular motion.

A.   Draw the path and describe the motion, emphasizing that the speed remains constant. Remind students that the acceleration must be perpendicular to the velocity. By drawing the velocity vector at two times, argue that it must be directed inward.

B.   Derive $a = v^2/r$. As an alternative to the derivation given in the text, write the equations for the particle coordinates as functions of time, then differentiate twice.

C.   Example: calculate the speed of a satellite, given the orbit radius and the acceleration to due to gravity at the orbit. See Sample Problem 8.

V.   Relative motion.

A.   Relate the position of an object as given coordinate system A to the position as given in coordinate system B by $\underline{r}_A = \underline{r}_B + \underline{R}_{BA}$, where $\underline{R}_{BA}$ is the position of the origin of B relative to the origin of A. Differentiate to show that $\underline{v}_A = \underline{v}_B + \underline{V}_{BA}$ and $\underline{a}_A = \underline{a}_B + \underline{A}_{BA}$, where $\underline{V}_{BA}$ and $\underline{A}_{BA}$ are the velocity and acceleration, respectively, of B relative to A. Discuss examples of a ball thrown in accelerating and non-accelerating trains. See questions 17 and 23. The discussion may be carried out for motion in a plane or for one dimensional motion only.

B.   Remark that $\underline{a}_A = \underline{a}_B$ if the two systems are not accelerating with respect to each other. This is an important point for the discussion of inertial reference frames.

C.   Only notes A and B need to be covered, as an introduction to the idea of an inertial frame. If you do not intend to do more, they can be covered in connection with Chapter 5. If you intend to do more, work several problems dealing with airplanes flying in winds and boats sailing in moving water. See problems 69, 82, and 84.

SUPPLEMENTARY TOPIC

Relative motion at high speeds. Include this topic if the course contains an introduction to modern physics.

SUGGESTIONS

1. Use questions 3, 4, 5, 9, and 13 to generate discussions of ideal
    projectile motion. To include air resistance in the discussions, ask
    questions 11 and 12.

2. Ask question 14 in connection with centripetal acceleration.

3. Assign problems 6, 7, and 8 to have students think about the analysis
    of motion in two dimensions.

4. Assign problem 15 as the first problem in the study of projectile
    motion, then have the students go on to several of 16 through 20 and
    30 through 51.

5. Assign problems 56 and 63 in connection with uniform circular motion.

6. Computer project

    Have students use Eureka or their own root finding programs to solve
    projectile problems. It is instructive to have them plot the velocity
    components as functions of time for a projectile subject to air
    resistance. Consider initial velocities which are both greater and
    less than the terminal velocity. Also have them study the maximum
    height and range of projectiles with various coefficients of air
    resistance.

7. Demonstrations

    Projectile motion: Freier and Anderson Mb14, 16, 17, 19, 20, 23, 24,
    28; Hilton M13.

8. Films

    Accelerated Motion and Angle of Lean, 3/4" videocassette, color, 9
    min. North Texas State University, Department of Physics, Denton, TX
    76203. Reviewed AJP 49:383 (1981) and TPT 19:341 (1981). Runners on
    tracks, accelerating vehicles, and a merry-go-round illustrate the
    connection between acceleration and the lean of various bodies.
    Winner of the 1980 AAPT competition.

    Galilean Relativity: Ball Dropped from the Mast of a Ship, S8, color,
    approx. 3 min. Kalmia Company, Dept. P1, Concord, MA 01742. One of
    three titles from the Project Physics Galilean Relativity Series,
    this loop is best of the three and shows the motion of a falling all
    released from the mast of sailboat. Two different reference frames
    are used to view the motion.

    Velocity in Circular and Simple Harmonic Motion; Velocity and
    Acceleration in Circular Motion; The Velocity Vector; The
    Acceleration Vector, S8, b/w, approx. 4 min each. Kalmia Co. (address

given above). Four loops from the PSSC Vector Kinematics series.

9. Computer programs

Personal Problems, Addison-Wesley Publishing Company, Reading, MA 01867. Apple II. The program plots the trajectories of projectiles subjected to air resistance. It also plots the coordinates, velocity components, and acceleration components as functions of time and lists values for points along the trajectory. The user can specify the initial conditions and the coefficient of air resistance. Use this as an alternative to student programming. Reviewed TPT December 1985.

Physics Simulations I: Ballistic, Kinko's Service Corporation, 4141 State Street, Santa Barabra, CA 93110. Macintosh. Plots trajectories of projectiles, with or without a drag force proportional to velocity. Drag coefficient can be constant or depend exponentially on altitude. Excellent for illustrating lectures.

Motion. See Chapter 2 notes.

Newton's Laws, J&S Software, 140 Reed Avenue, Port Washington, NY 11050. Apple II. F = ma drill problems. Reviewed TPT February 1984.

Mechanics, EduTech, 634 Commonwealth Avenue, Newton Center, MA 02159. Apple IIe, II+. Demonstrates vertical fall with or without air resistance, hunter and monkey experiment, planetary motion.

10. Laboratory

MEOS Experiment 7-9: Ballistic Pendulum - Projectile Motion (first method only). Students find the initial velocity of a ball shot from a spring gun by measuring its range. Emphasizes the use of kinematic equations.

Also see Procedures A and B of BE Experiment 11: Inelastic Impact and the Velocity of a Projectile. In addition to using range data to find the initial velocity, students plot range as a function of firing angle.

Chapter 5   FORCES AND MOTION - I

BASIC TOPICS

I.   Inertia and Newton's first law.
      A.   State the law: if the net force on a body vanishes, its
           acceleration is zero. Stress that it is the resultant force on

the body that counts.

B. Give an eraser a shove across a table and note that it stops. Point out that the acceleration does not vanish while it is in motion. Push the eraser at constant velocity and explain that the force of your hand cancels the force of friction. There is now zero acceleration. Use an air track to show what happens when friction is eliminated. Now the force of the hand is not required for constant velocity motion. Point out that the upward force of the track on the cart cancels the downward force of gravity and there is zero acceleration in the vertical direction.

C. Define an inertial frame. Point out that the acceleration depends on the frame used to measure it and that the first law can be true for only a select set of frames. Cover the essential parts of sections 4-8 and 4-10 on relative motion, if they were not covered earlier. See notes VA and B for Chapter 4.

II. Newton's second law.

A. Explain that the environment influences the motion of an object and that force measures the extent of the interaction. The result of the interaction is an <u>acceleration</u>. Place a cart at rest on the air track. Push it to start it moving and note that it continues at constant velocity. Once started, push it to increase its speed, then push it to decrease its speed. In each case note the direction of the force and the direction of the acceleration.

B. Define force in terms of the acceleration imparted to the standard 1 kg mass. Explain how this definition can be used to calibrate a spring, for example. Point out that force is a vector, in the same direction as the acceleration. If two or more forces act on the standard mass, its acceleration is the same as when a force equal to the resultant acts.
Units: newton, pound.

C. Have three students pull on a rope, knotted together as shown. Ask one to increase his or her pull and ask the others to report what they had to do to remain stationary.

D. Define mass in terms of the ratio of the acceleration imparted to the standard mass and to the unknown mass, with the same force

acting. Attach identical springs to two identical carts, one
empty and the other containing a lead brick. Pull with the same
force (same elongation of the springs) and observe the difference
in acceleration. Units: kg, slug.

E.   State the second law. Stress that the force that appears is the
net or resultant force. Explain that the law holds in inertial
frames. Point out that this is an experimentally established law
and does not follow as an identity from the definitions of force
and mass.

F.   Discuss examples: calculate the force required to stop an object
in a given time, given the mass and initial velocity; calculate
the force required to keep an object in uniform circular motion,
given its speed and the radius of its orbit.

III. Newton's third law.

A.   State the law. Stress that the two forces in question act on
different bodies and each helps to determine the acceleration of
the body on which it acts.

B.   Examples: Hold a book stationary in your hand, identify
action-reaction pairs (hand-book, book-earth). Now allow your
hand and the book to accelerate downward with an acceleration
less than g and again identify action-reaction pairs.

IV. Applications of Newton's laws.

A.   Go over the steps used to solve a one body problem: identify the
body and all forces acting on it, draw a free body diagram,
choose a coordinate system, and finally apply the second law in
component form.

B.   Set up the situation described in Sample Problem 7 using an
inclined air track. Do the calculation, then cut the string.

C.   Extend the ideas to the case of more than one body. Point out
that the steps are repeated for each body separately and that the
coordinate system need not be the same for all bodies. Show how
rods, strings, and pulleys relate the motions of bodies in
various cases. Explain the approximations of light, inextensible
strings, massless rods, and massless pulleys. Show how to invoke
the third law when necessary. Do several examples, carefully
giving each step. The Atwood machine (Sample Problem 9) and
weighing an object in an elevator (Sample Problem 10) are good

examples. If possible, give a demonstration of each example.

SUGGESTIONS

1. Discuss questions 1 and 2 in connection with problems 1 and 2.
2. Use questions 5, 24, and 34 to help students think about the influence of forces on bodies. Assign a few problems from the group 3 through 10.
3. Discuss questions 35 and 37 in connection with problem 41.
4. Use questions 8, 11, and 15 and problems 17 and 18 to discuss mass and weight.
5. Use questions 13 and 14 and problems 12, 13, 14, and 36 to discuss Newton's third law.
6. Assign a few applications problems, such as 26, 31, 46, 58, 63, 64, or 69.
7. Demonstrations
    Inertia: Freier and Anderson Mc1 - 5, Me1; Hilton M6.
    $\underline{F} = \underline{ma}$: Freier and Anderson Md2, M11; Hilton M7.
    Action-reaction pairs: Freier and Anderson Md1, 3, 4; Hilton M8.
    Mass and weight: Freier and Anderson Mf1, 2.
    Tension in a string: Freier and Anderson M11.
8. Films
    Frames of reference, 16 mm, b/w, 28 min. Modern Learning Aids (see Chapter 2 notes for address). This classic film will never become dated. Frames of reference are illustrated in clever and imaginative ways by Hume and Ivey. Probably the best of the PSSC film series.
    Zero-G, 16 mm, color, 14 min. National Audiovisual Center, General Services Administration, Washington, DC 20409. (Free loan from NASA regional film library.) The skylab is a "natural" laboratory for "zero g" experiments.
    Inertial Forces - Translational Acceleration, S8, color, 8 min. Kalmia Company (see Chapter 4 notes for address). An excellent film showing how apparent weight changes as an elevator moves up and down.
9. Computer program
    Physics: Elementary Mechanics, Control Data Company, 3610 West 77th Street, Bloomington, MN 55435. Apple II, IBM PC. Drill problems in collisions, gravitation, satellite motion, rotational dynamics, harmonic motion. Statements of problems are not complete and students must ask the computer for additional data. Helps students think about

what information is required to solve mechanics problems. Reviewed TPT May 1986.

10. Laboratory

MEOS Experiment 8-2: <u>Concept of Mass: Newton's Second Law of Motion</u>. Students measure the accelerations of two pucks that interact via a spring on a nearly frictionless surface and compare the ratio to the ratio of their masses. This experiment may be done with dry ice pucks or on an air table or air track.

Chapter 6   FORCES AND MOTION - II

<u>BASIC TOPICS</u>

I.   Frictional forces.

A. Place a large massive wooden block on the lecture table. Attach a spring scale, large enough to be read easily. If necessary tape sandpaper to the table under the block. Pull weakly on the scale and note that the reading is not zero although the block does not move. Pull harder and note that the reading increases but the block still does not move. Remark that there must be a force of friction opposing the pull and that the force of friction increases as the pull increases. Now increase your pull until the block moves and note the reading just before it starts to move. Pull the block at constant speed and note the reading. Have the students repeat the experiment, in a qualitative manner, using books resting on their chair arms. To show that the phenomenon depends on the nature of the surface, the demonstration can be repeated after waxing the wooden block and table top.

B. Give a brief qualitative discussion about the source of frictional forces. Stress that the force of static friction has whatever magnitude and direction are required to hold the body at rest, up to a certain limit in magnitude.

C. Explain that the force of kinetic friction and the upper limit for the force of static friction are, to a good approximation, proportional to the normal force. Define the coefficients of friction. Explain that the coefficient of static friction is often used for bodies which are rolling without slipping. Also explain that if one surface is stationary the force of kinetic friction is directed opposite to the velocity of the second body and that the direction of the force of static friction is determined by the condition that the resultant force on the body be zero.

D. Go over problem solving strategy. For kinetic friction use $F_k = \mu_k N$ as a force law, substitute into the second law along with other forces and solve for the unknowns. In the static case, solve for the force of friction needed to hold to body at rest, then compare with $\mu_s N$ to see if the body is indeed at rest.

E. Examples:

1. Find the angle of an inclined plane for which sliding starts; find the angle at which the body slides at constant speed. These examples can be analyzed in association with a demonstration and the students can use the data to find the coefficients of friction.

2. To insure that the students do not think of the normal force as equal to the weight, hold an eraser against the blackboard by pushing on it horizontally. Analyze this situation.

3. Find the acceleration of a body being pulled across a floor by means of a force which is not horizontal. This demonstrates the dependence of the normal force and the force of friction on the externally applied force.

II. The drag force and terminal speed.

A. Make or buy a small toy parachute. Drop two weights side by side and note they reach the floor at the same time. Attach the parachute to one and repeat. Explain that the force of the air reduces the acceleration.

B. Explain that viscous forces are functions of the speed. Eq. 16 holds for many. Explain the meaning of terminal speed and derive $v_t = (2mg/\rho A)^{1/2}$.

III. Uniform circular motion.

A. Point out that for uniform circular motion to occur there must be a radially inward force of constant magnitude and something in the environment of the body supplies the force. Whirl a mass tied to a string around your head and explain that the string supplies the force. Set up a loop-the-loop with a ball or toy cart on a track and explain that the combination of the normal force of the track and the force of gravity supplies the centripetal force. Have students identify the source of the force in examples and problems as they are discussed.

B. Point out that $F = mv^2/r$ is just $F = ma$ with the expression for centripetal acceleration substituted for a.

C. Discuss problem solving strategy. After identifying the forces, find the radial component of the resultant and equate it to mv^2/r.

D. Examples: find the speed and period of a conical pendulum. Find the speed with which a car can round an unbanked curve, given the coefficient of static friction. Analyze the loop-the-loop and point out that the ball leaves the track when the normal force vanishes. Show that the critical speed at the top is given by $v^2/r = g$.

IV. The forces of nature.

This section can be used to emphasize that force laws describe the mutual interaction of bodies in terms of their inherent properties and positions.

SUGGESTIONS

1. To start students thinking about the physical basis of frictional forces, have them discuss question 1.

2. The ways in which the coefficients of friction are used is emphasized in problems 11 and 13. To help students understand the role played by the normal force, assign question 4 in connection with problems 3 and 6. Also assign problem 5. Assign problem 17 in connection with the measurement of coefficients of friction with an inclined plane. Problems 27, 28, and 31 provide interesting applications.

3. Emphasize that terminal speed is a limit which can be reached from either above or below. Discuss questions 12 and 14. To apply Eqs. 16 and 17, discuss question 11 and problem 48. Also consider assigning problem 46.

4. Discuss question 15 in connection with problems 51 and 52. Discuss question 25 in connection with problems 61 and 62.

5. Computer project

Write a computer program to integrate Newton's second law for time and velocity dependent forces, then have students investigate the motion of an object subjected to a force that is proportional to v or v^2. The program can be used later to study forced harmonic motion and resonance. A simple Euler method is adequate and quite instructive. See the calculator supplement for details.

6. Demonstrations

Friction: Freier and Anderson Mk; Hilton M11.

Inclined plane: Freier and Anderson Mj2.

Centripetal acceleration: Freier and Anderson Mb29, 31, Mml, 2, 4 – 8, Ms5; Hilton M16.

7. Film

Principles of Lubrication, 16mm, color, 23 min. International Film Bureau, 332 S. Michigan Ave., Chicago, IL 60604. Discusses the laws of friction and presents a variety of examples and models. Reviewed TPT 20:325 (1982).

8. Computer program

Mechanics. See notes for Chapter 4.

9. Laboratory

MEOS Experiment 7-6: Coefficient of Friction – The Inclined Plane. Students determine the coefficients of static and sliding friction for 3 blocks on an inclined plane. They devise their own experimental procedures.

MEOS Experiment 7-7: Radial Acceleration (Problem I only). The centripetal force and the speed of a ball on a string, executing uniform, circular motion, is measured for various orbit radii. Essentially a verification of $F = mv^2/r$.

MEOS Experiment 7-8: Investigation of Uniform Circular Motion, or BE Experiment 13: Centripetal Force. Students measure the force acting on a body undergoing uniform circular motion, with the centripetal force provided by a spring.

MEOS Experiment 8-3: Centripetal Force. Students measure the speed of a puck undergoing uniform circular motion on a nearly frictionless surface. The data is used to calculate the centripetal force.

Chapter 7   WORK AND ENERGY

BASIC TOPICS

I.   Work done by a constant force.

   A.   Write down W = $\underline{F} \cdot \underline{d}$. Explain that this is the work done on the
        object by the constant force $\underline{F}$ as the object undergoes a
        displacement $\underline{d}$. Explain that work can be calculated for each
        individual force and that the total work done on the object is
        the work done by the resultant force. Point out that work is a
        scalar quantity. Also point out that work is zero for a force
        perpendicular to the displacement and that, in general, only the
        component of $\underline{F}$ tangent to the path contributes to the work. The
        force does no work if the displacement is zero. Work can be
        positive or negative, depending on the relative orientation of $\underline{F}$
        and $\underline{d}$. For a constant force the work depends only on the
        displacement, not on details of the path. Units: joule, erg,
        ft·lb, eV.

   B.   Calculate the work done by the force of gravity as a mass falls a
        distance h and as it rises a distance h. Calculate the work done
        by a non-horizontal force used to pull a box across a floor. Also
        calculate the work done by the normal force, the force of
        gravity, and the force of friction. Consider both an accelerating
        box and one moving with constant velocity. Work problem 5.

II. Work done by a variable force.

   A.   For motion in one dimension, discuss the integral form for work
        as the limit of a sum over infinitesimal path segments. Explain
        that the sum can be carried out by a computer even if the
        integral cannot be evaluated analytically.

   B.   Examples: derive expressions for the work done by an ideal spring
        and a force of the form k/x^2.

   C.   For motion in more than one dimension, write down the integral
        form for the work and explain its interpretation as the limit of
        a sum over infinitesimal path segments. Explain that this is the
        general definition of work. Calculate the work done by each force
        as a pendulum is pulled to a height h by a horizontal applied
        force $\underline{F}$. This example can be done both as a path integral and by
        using the equation for work done by a constant force.

D. Calculate the work done by a frictional force which is constant in magnitude and tangent to the path of a particle. Point out that it is negative and that its magnitude is the magnitude of the force multiplied by the path length. This is not a constant force since its direction changes as the direction of motion changes.

III. The work-energy theorem.
   A. Define kinetic energy for a particle. Point out that kinetic energy is a scalar and depends on the speed but not on the direction of the velocity.
   B. Prove the theorem for motion in one dimension. If the students are mathematically sophisticated, extend the theorem to the general case. Stress that it is the total work (done by the resultant force) that enters the theorem.
   C. Point out that only the component of a force parallel or antiparallel to the velocity changes the speed. Other components change the direction of motion. Positive total work results in an increase in kinetic energy and speed, negative total work results in a decrease.
   D. Demonstrate the theorem by considering a ball thrown into the air. At first neglect air resistance and point out that during the upward part of the motion the force of gravity does negative work and the kinetic energy decreases. Use energy considerations to compute the maximum height of the ball. Note that $v^2 = v_0^2 +$ 2as (which was derived in the study of kinematics) follows from the work-energy theorem if the force is constant. As the ball falls, the force of gravity does positive work and the kinetic energy increases. Show that the ball returns with its initial speed. Then include air resistance and argue that the ball returns with less than its original speed. See question 15.

IV. Power.
   A. Define power. Units: watt, horsepower.
   B. Show that $P = \underline{F} \cdot \underline{v}$.

<u>SUGGESTIONS</u>

1. Go over Sample Problem 3, then ask the students to discuss question

3.

2. For practice in distinguishing between work done by one of several forces acting and the total work, ask students to discuss questions 4, 9, and 13, then have them work problem 1 or 2.

3. Assign problems 15 and 38 on springs.

4. Assign problems on the work-energy theorem, such as 21, 22, 31, 32, and 34.

5. To have students use the calculus, assign problem 30.

6. Computer project

    Have students use Eureka or write a computer program to numerically evaluate the integral for work, then use the program to calculate the work done by various forces, given as functions of position. Include a non-conservative force and use to the program to show the work done on a round trip does not vanish. See the calculator supplement for details.

7. Demonstrations

    Work: Freier and Anderson Mvl.

8. Computer program

    Work and Energy, J&S Software (see Chapter 4 notes for address). Apple II. Drill problems on computation of work, work-energy theorem, and conservation of energy.

9. Laboratory

    MEOS Experiment 7-16: Elongation of an Elastomer. Students measure the elongation of an elastomer for a succession of applied forces and use a polar planimeter to calculate the work done by the force. The experiment may also be done in connection with chapter 13.

    BE Experiment 10: Mechanical Advantage and Efficiency of Simple Machines. This experiment can be used to broaden the course to include these topics. A lever, an inclined plane, a pulley system, and a wheel and axle are studied. In each case the force output is measured for a given force input and the work input is compared to the work output.

Chapter 8   THE CONSERVATION OF ENERGY

BASIC TOPICS

I.   Conservative and non-conservative forces

A.   Definitions. Explain that a force is conservative if either of the following hold:
1.   the work done by the force on every round trip is zero.
2.   the work done by the force is independent of the path.
B.   Show that the two definitions are equivalent.
C.   Discuss as examples the force of gravity and the force of an ideal spring. For either or both of these show that the work done depends only on the end points and not on the path between, then argue that the work vanishes for a round trip. Point out that on some parts of the path the force does positive work and the kinetic energy increases while on other parts it does negative work and the kinetic energy decreases.
D.   Use a force of friction with constant magnitude as an example of a non-conservative force. Show that the work done by the force cannot vanish on a round trip since it is negative for each segment. Show that the work is not independent of the path by considering two paths of different length between the same two end points.
E.   Use a cart on a linear air track to demonstrate these ideas. Attach two springs as shown and explain the equivalence with Fig. 3 of the text. Point out that the kinetic energy returns to roughly the same value and that the springs do zero work during a round trip. Reduce or eliminate air flow to show the influence of a non-conservative force. If this is done rapidly and skillfully, you can cause the cart to stop at the opposite end from which it started.

II.   Potential energy.
A.   Give the definition of potential energy in terms of work for motion in one dimension. See Eq. 7 and emphasize that W is the work done __by__ the force responsible for the potential energy.
B.   Discuss the following properties:
1.   The zero is arbitrary. Only potential energy differences have physical meaning.
2.   The potential energy is a scalar function of position.
3.   The force is given by $F = -dU/dx$ in one dimension.
4.   Unit: joule.
C.   Derive expressions for the potential energy functions for the force of gravity (uniform gravitational field) and the force of

an ideal spring. Use the force of friction as an example to explain why a potential energy function cannot be defined for non-conservative forces.

D. Use the work-energy theorem to show that ΔU is the work that must be applied by an external agent to increase the potential energy by ΔU without changing the kinetic energy. Show that ΔU is recovered as kinetic energy when the external agent is removed. Example: raising an object in a gravitational field.

III. Conservation of energy.

A. Explain that if all the forces acting on a body are conservative then $K + U =$ constant. This follows from the work-energy theorem with the work of the conservative forces represented by the change in potential energy. The negative sign in Eq. 7 is essential to obtain this result. Emphasize that U is the sum of the individual potential energies if more than one conservative force acts.

B. Conversion of kinetic to potential energy and vice versa. Drop a superball on a rigid table top and point out when the energy is totally potential and when it is totally kinetic. The question of elasticity can be glossed over by saying that to a good approximation the ball rebounds with unchanged speed. Also discuss the energy in a spring-mass system. Return to the cart on the air track and discuss its motion in terms of $K + U =$ constant. To avoid later confusion in the students' minds, start the motion with neither K nor U equal to zero. In this discussion, emphasize that the energy remains in the system but changes its form during the motion. The agent of the change is the force acting.

C. Potential energy curves. Use the spring-mass curve, then a more general one, and show how to calculate the kinetic energy and speed from the coordinate and total energy. Show the turning points on the curves and discuss their physical significance. Use $F = -dU/dx$ to argue that the particle turns around at a turning point. For an object on a frictionless roller coaster track find the speed at various points and identify the turning points.

IV. Potential energy in 2 and 3 dimensions.

A. Define potential energy as a line integral and explain that it is the limit of a sum over infinitesimal path segments.

B. Example: simple pendulum. Since the gravitational potential energy depends on height, in the absence of non-conservative forces the pendulum has the same swing on either side of the equilibrium point and always returns to the same turning points. Demonstrate with a pendulum hung near a blackboard and mark the end points of the swing on the board. For a more adventurous demonstration, suspend a bowling ball pendulum from the ceiling and release the ball from rest in contact with your nose. Stand very still while it completes its swing and returns to your nose.

V. Non-conservative forces.

A. Derive $\Delta E = W_f$. Stress that ΔE can be positive or negative.

B. Examples: an external agent, an explosion (positive ΔE), friction (negative ΔE).

C. Go back to the air track experiment and again slowly reduce the air flow. Discuss the change in mechanical energy due to friction.

SUPPLEMENTARY TOPIC

To introduce topics from modern physics, discuss the equivalence of mass and energy and the quantization of energy.

SUGGESTIONS

1. Have the class account for the energy in various situations. See questions 1, 2, 3, 4, 6, 7, 9, and 10.

2. Discuss applications of the conservation of energy principle. See questions 14 and 21 as well as problems 3, 5, 7, 9, 17, 18, and 23.

3. Draw several potential energy curves and have the class analyze the particle motion for various values of the total energy. This can provide particularly useful feedback as to how well the students have mastered the idea of energy conservation.

4. Demonstrations
   Conservation of energy: Freier and Anderson Mnl - 3, 6; Hilton Ml4a, b, e.
   Non-conservative forces: Freier and Anderson Mwl.

5. Laboratory

BE Experiment 9: <u>Work, Energy, and Friction</u>. A string is attached to a car on an incline and passes over a pulley at the top of the incline. Weights on the free end of the string are adjusted so the car rolls down the incline at constant speed. The work done by gravity on the weights and on the car is calculated and used to find the work done by friction. The coefficient of friction is computed. The experiment is repeated for the car rolling up the incline and for various angles of incline. It is also repeated with the car sliding on its top and the coefficients of static and kinetic friction are found.

Chapter 9   SYSTEMS OF PARTICLES

<u>BASIC TOPICS</u>

I.   Center of mass.
   A.   Define the center of mass by giving its coordinates in terms of the coordinates of the individual particles in the system. Optional: extend the definition to include a continuous mass distribution. Note that if the object has a point, line, or plane of symmetry, the center of mass must be at that point, on that line, or in that plane. Point out that no particle need be at the center of mass. Example: a doughnut. Explain that the general motion of a rigid body may be described by giving the motion of the center of mass and the motion of the object around the center of mass. See Fig. 1.
   B.   Derive expressions for the velocity and acceleration of the center of mass in terms of the velocities and accelerations of the particles in the system.
   C.   Derive $\Sigma \underline{F}_{ext} = M\underline{a}_{cm}$. As an example, consider a two particle system with external forces acting on both particles and each particle interacting with the other. Invoke Newton's third law to show that the internal forces cancel when all forces are summed.
   D.   Spin an eraser as you toss it. Point out that, if the influence of air can be neglected, the center of mass follows the parabolic trajectory of a projectile although the motion of other points is more complicated.

II. Momentum.

  A. Define momentum for a single particle. Optional: give the
      relativistic definition.

  B. Show that Newton's second law can be written $\underline{F}$ = d$\underline{p}$/dt for a
      particle. Emphasize that the mass of the particle is constant and
      that this form of the law does <u>not</u> imply that a new term $\underline{v}$dm/dt
      has been added.

  C. State that the total momentum of a system of particles is the
      <u>vector</u> sum of the individual momenta and show that $\underline{P}$ = M$\underline{v}_{cm}$.

  D. Show that Newton's second law for the center of mass can be
      written $\Sigma\underline{F}_{ext}$ = d$\underline{P}$/dt, where $\underline{P}$ is the total momentum of the
      system. Stress that this equation is valid only if the mass of
      the system is constant.

III. Conservation of linear momentum.

  A. Point out that $\underline{P}$ = constant if $\Sigma\underline{F}_{ext}$ = 0. Stress that one
      examines the external forces to see if momentum is conserved in
      any particular situation. Point out that one component of $\underline{P}$ may
      be conserved when others are not. Put two carts, connected by a
      spring, on a horizontal air track and set them in oscillation by
      pulling them apart and releasing them from rest. Explain that the
      center of mass does not accelerate. Push one cart and explain
      that the center of mass is now accelerating. Use the conservation
      of momentum principle to derive an expression for the velocity of
      one cart in terms of the velocity of the other.

  B. Consider a projectile which splits in two and find the velocity
      of one part, given the velocity of the other. Point out that
      mechanical energy is conserved for the cart-spring system but is
      not for the fragmenting projectile. The exploding projectile idea
      can be demonstrated with an air track and two carts, one more
      massive than the other. Attach a brass tube to one cart and a
      tapered rubber stopper to the other. Arrange so that the tube is
      horizontal and the stopper fits in its end. The tube has a small
      hole in its side, through which a firecracker fuse fits. Start
      the carts at rest and light a firecracker in the tube. The carts
      rapidly separate, strike the ends of the track, come back
      together again, and stop. Arrange the initial placement so the
      carts strike the ends of the track simultaneously. Explain that $\underline{P}$
      = 0 throughout the motion.

SUPPLEMENTARY TOPICS

1.   Variable mass systems. To demonstrate, screw several hook eyes into a toy CO_2 propelled rocket, run a line through the eyes and string the line across the lecture hall. Start the rocket from rest and have the students observe its acceleration as it crosses the hall.

2.   Internal work and energy. This topic will come up naturally if the acceleration of a car is discussed. Mention that forces of the road on the tires do zero work if the tires do not slip.

SUGGESTIONS

1.   Use questions 1, 2, 4, 5, and 6 along with problems 3, 4, and 7 to generate discussion about the position of the center of mass.

2   Question 14 is a good test of understanding of the motion of the center of mass. Discuss it as an introduction to the problems. Assign problem 18.

3.   To stress that acceleration of the center of mass requires an external force, assign problems 13, 15, and 17.

4.   Assign conservation of momentum problems, such as 32, 36, 41, and 45.

5.   Assign problems such as 37 and 40, which are concerned with situations in which kinetic energy is not conserved.

6.   Demonstrations
      Center of mass, center of gravity: Freier and Anderson Mp7, 12, 13; Hilton M18a, b.
      Motion of center of mass: Freier and Anderson Mp1, 2, 16 - 19.
      Conservation of momentum: Freier and Anderson Mg4, 5, Mi2; Hilton M15.
      Rockets: Freier and Anderson Mh

7.   Films
      Center of Mass, S8, color, 3 min. Kalmia Company (see Chapter 4 notes for address). This film is part of the Mechanics on an Air Track series. Reviewed TPT 16:334 (1976).
      Human Moments, S8, color, 4 min. American Association of Physics Teachers (see Chapter 2 notes for address). This film is part of the NASA-Skylab film series. Reviewed AJP 44:1021 (1976).

8.   Laboratory
      MEOS Experiment 7-9: Ballistic Pendulum - Projectile Motion (do both

methods if the first was done in conjunction with Chapter 4, otherwise do the second method only). A ball is shot into a trapping mechanism at the end of a pendulum. The initial speed of the ball is found by applying conservation of momentum to the collision and conservation of energy to the subsequent swing of the pendulum. Also see BE Experiment 11: <u>Inelastic Impact and the Velocity of a Projectile</u>.

MEOS Experiment 8-6: <u>Center of Mass Motion</u>. Two pucks are connected by a rubber band or spring and move toward each other on a nearly frictionless surface. A spark timer is used to record their positions as functions of time. Students calculate and study the position of the center of mass as a function of time. They also find the center of mass velocity. Can be performed with dry ice pucks or on an air table or air track.

MEOS Experiment 8-7: <u>Linear Momentum</u>. Essentially the same as 8-6 but data is analyzed to give the individual momenta and total momentum as functions of time. Kinetic energy is also analyzed.

Chapter 10   COLLISIONS

BASIC TOPICS

I.   Properties of Collisions

    A.   Explain that for the collisions considered two bodies interact with each other over a short period of time and that the times before and after the collision are well defined. The force of interaction is great enough that external forces can be ignored during the interaction time. Explain that the identities of the bodies which exit the interaction may be different from those which enter: decays of fundamental particles and nuclei can be included in discussions of collisions.

    B.   Set up a collision between two carts on an air track. Point out the interaction period, the periods before and after the interaction.

II.   Impulse.

    A.   Define the impulse of a force as the time integral of the force. Note that it is a vector. Clearly distinguish between impulse (integral over time) and work (integral over path). Draw a force vs. time graph for the force of one body on the other during a collision and point out the impulse is the area under the curve.

B. Define the time average force and show that the impulse is the product $\underline{F}_{ave}\Delta t$, where Δt is the time of interaction.

C. Use Newton's second law to show that the impulse on a body equals the change in its momentum. Refer to the air track collision and point out that it is the impulse of one body on the other which changes the momentum of the second body.

D. Use the third law to show that two bodies in a collision exert equal and opposite impulses on each other and show that total momentum is conserved. Again stress that external forces are neglected during the collision.

III. Collisions in one dimension.

A. Two body elastic collisions.

1. Derive expressions for the final velocities in terms of the masses and initial velocities.

2. Specialize the general result to the case of equal masses and one body initially at rest. Demonstrate this collision on the air track using carts with spring bumpers. Point out that the carts exchange velocities.

3. Specialize the general result to the case of a light body, initially at rest, struck by a heavy body. Demonstrate this collision on the air track. Point out that the velocity of the heavy body is reduced only slightly and that the light body shoots off at high speed. Relate to a bowling ball hitting a pin.

4. Specialize the general result to the case of a heavy body, initially at rest, struck by a light body. Demonstrate this collision on the air track. Point out the low speed acquired by the heavy body and the rebound of the light body. Relate to a ball rebounding from a wall. A nearly elastic collision can be obtained with a superball.

B. Two body completely inelastic collisions.

1. Derive an expression for the velocity of the bodies after the collision in terms of their masses and initial velocities.

2. Demonstrate the collision on a air track, using carts with velcro bumpers. Point out that the kinetic energy of the

bodies is not conserved and calculate the energy loss. Note that $\frac{1}{2}(m_1+m_2)v_{cm}^2$ is retained and explain that this energy is sometimes called the kinetic energy of the center of mass.

C. Point out that while the greatest energy loss occurs when the interaction is completely inelastic, there are many other inelastic collisions in which less than the maximum energy loss occurs. Note that it is possible to have a collision in which kinetic energy increases (an explosive impact, for example).

IV. Optional: two body collisions in two dimensions.

A. Write down the equations for the conservation of momentum, in component form, for a collision with one body initially at rest. Mention that these can be solved for two unknowns and that the outcome of the interaction is not determined by the initial momentum.

B. Consider a completely inelastic collision for which the two bodies do not move along the same line initially. Calculate the final velocity.

C. Consider an elastic collision for which one body is at rest initially and the initial velocity of the second is given. Assume the final direction of motion of one is known and calculate the final speeds and the final direction of motion of the other.

SUPPLEMENTARY TOPIC

Reactions and decay processes. This topic provides background for Chapter 48 on nuclear physics and should be covered, now or later, if that chapter is included in the course.

SUGGESTIONS

1. Assign and discuss some problems dealing with inelastic collisions for which the mechanism of the kinetic energy loss is given explicitly. See problems 48 and 49. Also discuss question 12. Problem 44 may also be assigned with this group.

2 After discussing elastic collisions, ask question 7.

3. Demonstrate the ballistic pendulum and show how it can be used to measure the speed of a bullet. See problems 39 and 41. Problem 38 can also be assigned with this group.

4. Assign some problems such as 56 and 57, which deal with 3 dimensional collisions.

5. Computer project

    Have students use Eureka or write a program to graph the total final kinetic energy as a function of the final velocity of one object in a two body, one dimensional collision, given the initial velocities and masses of the two objects. Ask them to run the program for specific initial conditions and identify elastic, completely inelastic, and explosive collisions on their graphs.

6. Demonstrations

    Freier and Anderson Mg1 - 3, Mi1, 3, 4, Mw3, 4; Hilton M15d, e, f, M19j.

7. Films

    Two Dimensional Collisions, I and II, S8, color, 3 min. each. Kalmia Company (see Chapter 4 notes for address). Part of the Project Physics film series. Reviewed TPT 14:56 (1976).

    Collisions, S8, color, 4 min. American Association of Physics Teachers (see Chapter 2 notes for address). Another film from the NASA-Skylab Film Series. Reviewed TPT 14:56 (1976).

8. Computer program

    Collisions on an Air Track, Cambridge Development Laboratory, 1696 Massachusetts Avenue, Cambridge, NA 02138. Apple II. Chiefly tutorial but includes segments suitable for lecture illustrations. User specified collisions are simulated and in each case a numerical analysis of dynamic quantities (velocity, momentum, kinetic energy) is given. Reviewed TPT September 1985.

9. Interactive videodisk

    Physics and Automobile Collisions by Dean Zollman (John Wiley, 1984). The disk shows collisions of cars with fixed barriers and two car collisions (head-on, at 90°, and at 60°). One sequence shows the influence of bumper design, others show the influence of air bags and shoulder straps on mannikens. All are slow motion films of manufacturers' tests and many show grids and clocks. Students can stop the action to take measurements, then make calculations of momentum and energy transfers. For most exercises a commercial player is satisfactory; for a few a computer controlled player is required.

10. Laboratory

    MEOS Experiment 7-10: Impulse and Momentum. Students use a microprocessor to measure the force as a function of time as a toy truck hits a force transducer. They numerically integrate the force to find the impulse, then compare the result with the change in momentum, found by measuring the velocity before and after the collision.

BE Experiment 8: <u>Impulse and Momentum</u>. In part A a mass is hung on a string that passes over a pulley and is attached to an air track glider. The glider accelerates from rest for a known time and a spark timer is used to find its velocity at the end of the time. The impulse is calculated and compared with the momentum. In part B a glider is launched by a stretched rubber band and a spark record of its position as a function of time is made while it is in contact with the rubber band. A static technique is used to measure the force of the rubber band for each of the recorded glider positions and the impulse is approximated. The result is again compared with the final momentum of the glider.

MEOS Experiment 7-11: <u>Scattering</u> (for advanced groups). The deflection of pellets from a stationary disk are used to investigate the scattering angle as a function of impact parameter and to find the radius of the disk.

MEOS Experiment 8-5: <u>One-Dimensional Collision</u>. A puck moving on a nearly frictionless surface collides with a stationary puck. A spark timer is used to record the positions of the pucks as functions of time. Students calculate the velocities, momenta, and energies before and after the collision. May be performed with dry ice pucks or on an air table or track.

MEOS Experiment 8-8: <u>Two-Dimensional Collisions</u>. Same as MEOS 8-5 but the pucks are allowed to scatter out of the original line of motion. Students must measure angles and calculate components of the momenta. The experiment may be performed with dry ice pucks or on an air table.

BE Experiment 12: <u>Elastic Collision - Momentum and Energy Relations in Two Dimensions</u>. A ball rolls down an incline on a table top and strikes a target ball initially at rest at the edge of the table. The landing points of the balls on the floor are used to find their velocities just after the collision. The experiment is run without a target ball to find the velocity of the incident ball just before the collision. Data is used to check for conservation of momentum and energy. Both head-on and grazing collisions are investigated. A second experiment, similar to MEOS 8-8, is also described.

Chapter 11   ROTATIONAL MOTION

BASIC TOPICS

I.   Rotation about a fixed axis.
     A.   Spin an irregular object on a fixed axis. A bicycle wheel or
          spinning platform with the object attached can be used. Draw a
          rough diagram, looking along the axis. Explain that each point in
          the body has a circular orbit and that, for any selected point,
          the radius of the orbit is the perpendicular distance from the
          point to the axis. Contrast to a body that is simultaneously
          rotating and translating. See Fig. 2.
     B.   Define angular displacement (in radians and revolutions), angular
          speed (in rad/s and rev/s) and angular acceleration (in rad/s^2
          and rev/s^2). Treat both average and instantaneous quantities but
          emphasize that the instantaneous quantities are most important
          for us.
     C.   Note that θ is positive for an angle in one direction and
          negative for an angle in the other, then interpret the signs of ω
          and α. Give examples of spinning objects for which ω and α have
          the same sign and for which they have opposite signs.
     D.   Point out the analogy to linear motion. Use the first three
          entries of Table 3.

II. Vector properties of rotation (optional).
     A.   This topic is not covered in the text but it is a nice
          application of the vector cross product. Point out that ω and α
          can be thought of as the components of vectors $\underline{\omega}$ and $\underline{\alpha}$,
          respectively. For fixed axis rotation, the vectors lie along the
          axis, with the direction of $\underline{\omega}$ determined by a right hand rule: if
          the fingers curl in the direction of rotation, then the thumb
          points in the direction of $\underline{\omega}$. If $d\omega/dt > 0$ then $\underline{\alpha}$ is in the same
          direction; if $d\omega/dt < 0$ then it is in the opposite direction.
     B.   For the special case $\underline{\omega} = \omega\hat{\underline{z}}$, show that regardless of the choice
          of origin along the z axis, $\underline{v} = \underline{\omega} \times \underline{r}$. Differentiate this to show
          that $\underline{a} = \underline{\alpha} \times \underline{r} + \underline{\omega} \times \underline{v}$. Interpret the first term as the tangential
          acceleration and the second as the centripetal acceleration. Note

that this approach correctly gives the direction and magnitude of the velocity and acceleration of a point on the rotating body.

III. Rotation with constant angular acceleration.

    A. Emphasize that the discussion here is restricted to rotation about a fixed axis but that the same equations can be used when the axis is in linear translation. This type motion will be discussed in the next chapter.

    B. Write down the kinematic equations for $\theta(t)$ and $\omega(t)$. Make a comparison with the analogous equations for linear motion (see Table 1).

    C. Point out that the problems of rotational kinematics are similar to those for one dimensional linear kinematics and that the same strategies are used for their solution.

    D. Go over examples. Calculate the time and number of revolutions for an object to go from some initial angular speed to some final angular speed, given the angular acceleration. Calculate the time to go a given number of revolutions and the final angular speed, again given the angular acceleration.

IV. Linear speed and acceleration of a point rotating about a fixed axis.

    A. Write down $s = r\theta$ for the arc length. Explain that it is a rearrangement of the defining equation for the radian and that θ must be in radians for it to be be valid.

    B. Wrap a string on a large spool that is free to rotate about a fixed axis. Mark the spool so the angle of rotation can be measured. Slowly pull out the string and explain that the length of string pulled out is equal to the arc length through which a point on the rim moves. Compare the string length to $r\theta$ for $\theta = \pi/2$, π, $3\pi/2$, and 2π. Show that $s = r\theta$ reduces to the familiar result for $\theta = 2\pi$.

    C. Differentiate $s = r\theta$ to obtain $v = r\omega$ and $a_t = r\alpha$. Emphasize that radian measure <u>must</u> be used. Point out that v gives the speed and a_t gives the acceleration of the string as it is pulled provided it does not slip on the spool. Point out that all points in a rotating rigid body have the same value of ω and and the same value of α but points that are different distances from the axis have different values of v and different values of a_t.

D. Point out that the velocity is tangent to the circular orbit but that the total acceleration is not. a_t gives the tangential component while $v^2/r = r\omega^2$ gives the radial component. The tangential component is not zero only when the point on the rim speeds up or slows down in its rotational motion while the radial component is not zero as long as the object is turning. For students who have forgotten, reference the derivation of $a_r = v^2/r$, given in section 4.7.

E. Explain how to find the magnitude and direction of the total acceleration in terms of ω, α, and r.

V. Kinetic energy of rotation.
   A. Show that $K = \frac{1}{2}mr^2\omega^2$ for a particle moving around a circle with angular velocity ω and $K = \frac{1}{2}I\omega^2$, where $I = \Sigma m_i r_i^2$, for a body rotating about a fixed axis. I is called the rotational inertia of the body.

   B. Point out that rotational inertia depends on the distribution of mass and on the position of the rotation axis. Point out that two bodies may have the same mass but quite different rotational inertias. Table 2 gives the rotational inertia for various objects and axes. Optional: show how to convert the sum for I to an integral.

   C. Prove the parallel axis theorem. The proof can be carried out using a sum for I rather than an integral. Explain its usefulness for finding the rotational inertia when the axis is not through the center of mass. Emphasize that the actual axis and the axis through the center of mass must be parallel for the theorem to be valid.

VI. Torque
   A. Define torque for a force acting on a single particle. Here we consider forces which lie in planes perpendicular to the axis of rotation and take $\tau = rF\sin\phi$ where $\underline{r}$ is a vector from the axis to the point of application of the force and ϕ is the angle between $\underline{r}$ and $\underline{F}$ when they are drawn tail to tail. Note that $\underline{r}$ is perpendicular to the axis. The definition will be generalized in the next chapter. Explain that $\tau = rf_t = r_\perp F$, where F_t is the tangential component of $\underline{F}$ and $r_\perp$ is the moment arm.

    B.   Explain that the torque vanishes if $\underline{F}$ is along the same line as $\underline{r}$ and that only that part of $\underline{F}$ which is perpendicular to $\underline{r}$ produces a torque. This is a mechanism for picking out the part of the force which produces angular acceleration, as opposed to the centripetal part. Also explain that the same force can produce a larger torque if it is applied at a point farther from the axis.

    C.   Use a wrench tightening a bolt as an example. The force is applied perpendicular to the wrench arm and long arms are used to obtain large torques.

VII. Newton's second law for rotation.

    A.   Use a single particle on a circular orbit to introduce the topic. Start with $F_r = ma_r$ and show that $\tau = I\alpha$, where $I = mr^2$. Explain that this equation also holds for extended bodies, although I is then the sum given in A.

    B.   Remark that problems are solved similarly to second law problems. Tell students to identify torques, draw free a body diagram, choose the direction of positive rotation, and substitute the total torque into $\tau = I\alpha$.

    C.   Wrap a string around a cylinder, free to rotate on fixed horizontal axis. Attach the free end of the string to a mass and allow the mass to fall from rest. Note that its acceleration is less than g, perhaps by dropping a free mass beside it. Go over Sample Problem 10.

VIII. Work-energy theorem for rotation.

    A.   Use $dW = \underline{F} \cdot d\underline{s}$ to show that the work done by a torque is given by $W = \int \tau d\theta$ and that power delivered is $P = \tau\omega$.

    B.   Use $\tau d\theta = I\alpha d\theta = \frac{1}{2}I d(\omega^2)$ to show $W = \frac{1}{2}I\omega_f^2 - \frac{1}{2}I\omega_i^2$.

    C.   For the situation of Sample Problem 10 use conservation of energy to find the angular velocity of the cylinder after the mass has fallen a distance h. Use rotational kinematics and the value for the angular acceleration found in the text to check the answer.

SUGGESTIONS

1.   Use questions 1 and 6 and problems 1 and 8 to discuss radian measure.

2.   Use techniques of the calculus to derive the kinematic equations. That is, integrate α = constant twice with respect to time. Assign

some problems from the group 3, 5, 6, and 7.

3. To discuss the relationship between angular and linear variables, use questions 11 and 13. Assign problems 32, 33, and 36.

4. Use questions 14, 15, 16, and 18 to guide students through a qualitative discussion of rotational inertia. Assign problems 49 and 50. Use problem 56 to discuss the radius of gyration, if desired.

5. Use problems 59 and 60 to discuss the calculation of torque. To test for understanding of the sign associated with torque assign problem 62.

6. To help students think about $\tau = I\alpha$, discuss question 19 and problem 71. To deal with a situation in which the dynamics of more than one object is important, demonstrate the Atwood machine and discuss problem 72. Assign problem 85 in connection with the rotational work-energy theorem.

7. Computer project

Ask students to use Eureka or their own root finding programs to solve rotational kinematic problems. See the calculator supplement for sample problems.

8. Demonstrations

Rotational variables: Hilton M16a.

Rotational dynamics: Freier and Anderson Ms7, Mt 5, 6, Mo5.

Rotational work and energy: Freier and Anderson Mv2, Mr5, Ms2.

9. Computer program

Motion. See notes for Chapter 2.

10. Laboratory

MEOS Experiment 7-14: Rotational Inertia. The rotational inertia of a disk is measured dynamically by applying a torque (a falling mass on a string wrapped around a flange on the disk). A microprocessor is used to measure the angular acceleration. Small masses are attached to the disk and their influence on the rotational inertia is studied. The acceleration of the mass can also be found by timing its fall through a measured distance. Then $a = r\alpha$ is used to find the angular acceleration of the disk. Also see BE Experiment 14: Moment of Inertia.

Chapter 12   ROLLING, TORQUE, AND ANGULAR MOMENTUM

BASIC TOPICS

I.   Rolling.

   A.   Remark that a rolling object can be considered to be rotating
        about an axis through the center of mass while the center of mass
        is moving. The text considers the special case for which the axis
        of rotation does not change direction. Point out that the
        rotational motion obeys $\tau = I\alpha$ and the translational motion of
        the center of mass obeys $\underline{F} = m\underline{a}$. Emphasize that one of the forces
        acting is the force of friction produced by the surface on which
        the object rolls.

   B.   Explain that the speed of a point at the top of a rolling object
        is $v_{cm} + \omega R$ and the speed of a point at the bottom is $v_{cm} - \omega R$.
        Specialize to the case of rolling without slipping. Point out
        that the point in contact with the ground has zero velocity, so
        $v_{cm} = \omega R$.

   C.   Use Fig. 5 to explain that a wheel rolling without slipping can
        be viewed as rotating about an axis through the point of contact
        with the ground. Use this and the parallel axis theorem to show
        that the kinetic energy is $\frac{1}{2}Mv_{cm}^2 + \frac{1}{2}I_{cm}\omega^2$.

   D.   Consider objects rolling down an inclined plane and show how to
        calculate the speed at the bottom using energy considerations. If
        time permits, carry out an analysis using the equations of motion
        and show how to find the frictional force which prevents
        slipping.

   E.   Roll a sphere, a hoop, and a cylinder, all with the same radius
        and mass, down an incline. Start the objects simultaneously and
        ask students to pick the winner. Go over Sample Problem 3.

F. Another interesting demonstration
    can be performed using the object
    shown, built from three concentric
    cylinders with the string wrapped
    around the smallest, like a yo-yo.
    Place it on the table and pull the
    string at various angles. The three
    different possibilities are shown.
    See question 10.

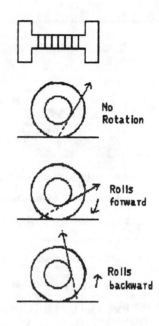

No Rotation

Rolls forward

Rolls backward

G. Consider a ball striking a bat.
    Show how to find the point at which
    the ball should be hit so that the
    instantaneous center of rotation is
    at the place where the bat is held.
    The striking point is called the
    center of percussion. When the ball is hit there the batter feels
    no sting.

II. Torque and angular momentum.

A. Define torque as $\underline{\tau} = \underline{r} \times \underline{F}$ and explain that this is the general
    definition. Review the vector product and give the right hand
    rule for finding the direction. Explain that $\underline{\tau} = 0$ if $\underline{r} = 0$, $\underline{F} =$
    0, or $\underline{r}$ is parallel (or antiparallel) to $\underline{F}$.

B. Define angular momentum for a single particle, using vector
    notation. Show that the definition reduces to the expected result
    for a particle moving in a circle. Also derive the relationship
    between angular momentum and angular speed. To show that a
    particle may have angular momentum even if it is not moving in a
    circle, calculate the angular momentum of a particle moving with
    constant velocity along a line not through the origin. Point out
    that both torque and angular momentum depend on the choice of
    origin.

C. Use Newton's second law to derive $\underline{\tau} = d\underline{\ell}/dt$ for a particle.
    Consider a particle moving in a circle, subjected to both
    centripetal and tangential forces. Show that $\underline{\tau} = d\underline{\ell}/dt$ reduces to
    $F_T = ma_T$, as expected.

D. Show that the torque about the origin exerted by gravity on a
    falling mass is mgd, where d is the perpendicular distance from
    the line of fall to the origin. Write down the velocity as a

function of time and show that the angular momentum is mgtd.
Remark that $\tau = d\ell/dt$ by inspection. See Sample Problem 6.

III. Systems of particles.
   A. Show that $\tau_{ext} = dL/dt$ for a system of particles for which
      internal torques cancel. Emphasize that τ_{ext} is the result of
      summing all torques on all particles in the system and that L is
      the sum of all individual angular momenta. Demonstrate in detail
      the cancellation of internal torques for two particles which
      interact via central forces. Point out that this equation is the
      starting point for investigations of the rotational motion of
      bodies.
   B. Make a connection to material of the last chapter by showing that
      the total angular momentum of a rigid body rotating about a fixed
      axis is given by $L = I\omega$. Point out this is really a vector along
      the axis. Assume all external forces are in planes perpendicular
      to the axis of rotation and show the net external torque is along
      the axis.

IV. Conservation of angular momentum.
   A. Point out that L = constant if $\tau_{ext} = 0$ and show that, for
      rotation about a fixed axis, $I\omega = I_0\omega_0$ if the net torque
      vanishes.
   B. As examples consider a mass dropped on the rim of a freely
      spinning platform, a person running tangent to the rim of a
      merry-go-round and jumping on, and a spinning skater whose
      rotational inertia is changed by dropping her hands.
   C. The third example can be demonstrated easily if you have a
      rotating platform which can hold a person. Have a student hold
      weights in each hand to increase the rotational inertia. Start
      him spinning with arms extended, then have him bring his arms in
      toward the sides of his body.

SUPPLEMENTARY TOPICS
1. Precession of a top.
2. Quantization of angular momentum.

SUGGESTIONS

1. To test for understanding of the relationships between variables for a rolling object, ask question 5 and assign problems 4 and 8.

2. In connection with the demonstration of objects rolling down an inclined plane, ask question 4 and assign problem 13.

3. Ask question 8 and assign problems involving rotational energy: 5, 6, and 9, for example.

4. Assign problems 17, 21, and 24 in connection with the calculation of torque and angular momentum. To emphasize that an object moving in a straight line may have angular momentum, assign problem 19 or 26.

5. Assign conservation of angular momentum problems: 41, 42, and 46, for example.

6. Demonstrations

   Rolling: Freier and Anderson Mb4, 7, 30, Mo3, Mp3, Mrl, 4, Msl, 3, 4, 6; Hilton Ml0d, Ml9c.

   Conservation of angular momentum: Freier and Anderson Mtl - 4, 7, 8, Mul; Hilton M8b, Ml9i.

   Gyroscopes: Freier and Anderson Mu2 - 18; Hilton Ml9a, b, f, g, h.

7. Films

   Angular Momentum: A Vector Quantity, 16 mm, b/w, 27 min. Modern Learning Aids (see Chapter 2 notes for address). Angular momenta are added vectorially using three spinning wheels.

   Conservation Laws in Zero-G, 16 mm, color, 18 min. National Audio-Visual Center (see Chapter 5 notes for address). This NASA film contains scenes both from earth-bound and the "zero g" skylab environment in which astronauts are shown spinning in space.

   Human Momenta; Moving Astronauts; Acrobatic Astronauts; Games Astronauts Play; Gyroscopes, S8, color, 4 min. each. American Association of Physics Teachers (see Chapter 2 notes for address). These loops are from the skylab series and take advantage of "weightlessness" conditions to demonstrate various rotational motions. Series reviewed AJP 44:1021 (1976).

8. Laboratory

   MEOS Experiment 7-12: Rotational and Translational Motion. Students measure the center of mass acceleration of various bodies rolling down an incline and calculate the center of mass velocities at the bottom. Results are compared to measured velocities. It is also instructive to use energy methods to find the final speeds.

   MEOS Experiment 7-13: Rotational Kinematics and Dynamics. Students find the velocity and acceleration of a ball rolling around a loop-the-loop and analyze the force acting on it.

MEOS Experiment 8-9: <u>Conservation of Angular Momentum</u>. Uses the Pasco rotational dynamics apparatus. A ball rolls down a ramp and becomes coupled to the rim of a disk that is free to rotate on a vertical axis. Students measure the velocity of the ball before impact and the angular velocity of the disk-ball system after impact, then check for conservation of angular momentum.

Chapter 13   EQUILIBRIUM AND ELASTICITY

<u>BASIC TOPICS</u>

I.   Conditions for equilibrium.

    A.   Write down the equilibrium conditions: $\Sigma \underline{F} = 0$, $\Sigma \underline{\tau} = 0$ (about any point). Remind students that only external forces and torques enter. Explain that these conditions mean that the acceleration of the center of mass and the angular acceleration about the center of mass both vanish. The body may be at rest or its center of mass may be moving with constant velocity or the body may be rotating with constant angular momentum. Point out that $\underline{P}$ = constant and $\underline{L}$ = constant form 6 equations which are to be solved for unknowns, usually the magnitudes of some of the forces or the angles made by some of the forces with fixed lines. Explain that we will be concerned chiefly with static equilibrium for which $\underline{P}$ = 0 and $\underline{L}$ = 0.

    B.   Show that, for a body in equilibrium, $\Sigma \underline{\tau} = 0$ about every point if $\Sigma \underline{\tau} = 0$ about any point.

    C.   Explain that the gravitational forces and torques, acting on individual particles of the body, can be replaced by a single force acting at a point called the center of gravity. If the gravitational field is uniform over the body the center of gravity coincides with the center of mass.

II.   Solution of problems.

    A.   Give the problem solving steps: isolate the body, identify the forces acting on it, draw a free body diagram, choose a reference frame for the resolution of the forces, choose a reference frame for the resolution of the torques, write down the equilibrium conditions in component form, and solve these simultaneously for the unknowns. Point out that the two reference frames may be different and that the reference frame for the resolution of torques can often be chosen so that one or more of the torques vanish.

B. Work examples. Consider an arm and hand holding a mass (Sample Problem 2) or a ladder leaning against a wall (Sample Problems 3 and 4).

C. Using 4 bricks, set up the situation depicted in Fig. 33. Point out how large the overhang can be, then assign problem 29 or discuss the solution in the lecture.

III. Elasticity.

A. Point out that we have been considering mainly rigid bodies until now. Real objects deform when external forces are applied. Explain that deformations are often important for determining the equilibrium configuration of a system.

B. Consider a rod of unstrained length L subjected to forces F applied uniformly at each end and define the stress as F/A, where A is the area of an end. Define strain as the fractional change in length $\Delta L/L$ caused by the stress. Explain that stress and strain are proportional if the stress is sufficiently small. Define Young's modulus E as the ratio of stress to strain and show that $\Delta L = FL/EA$. Explain that Young's modulus is a property of the object and point out Table 1.

C. Explain that if stress is small, the object returns to its original shape when the stress is removed and it is said to be elastic. Define yield strength and ultimate strength. Also explain that objects deform under shearing stresses.

D. Define pressure as the force per unit area acting at the surface of a volume of fluid. Define the bulk modulus B as the ratio of the pressure to the fractional change in volume of the fluid.

E. Calculate the fractional change in length for compressional forces acting on rods made of various materials. Use data from Table 1. Calculate the fractional change in volume for water subjected to atmospheric pressure.

SUGGESTIONS

1. Use questions 1, 2, 3, and 4 for a discussion of equilibrium conditions.

2. Use questions 11, 12, and 13 to help students gain understanding of the equilibrium conditions in specific situations. Assign problem 16 in connection with question 11.

3. Use questions 19 and 20 to discuss the incompleteness of the equilibrium conditions in some circumstances.

4. In discussions of elasticity use questions 21, 22, and 24.

5. Assign problem 2 to test for understanding of the conditions for equilibrium. Assign a few problems, such as 7, 8, and 10, for which only the total force is important. Assign others, such as 13, 15, 16, and 18, for which torque is also important. Assign problem 43 if the students need a challenge.

6. The fundamentals of elasticity are covered in problems 46 and 47. Shear is covered in problem 49.

7. Demonstrations

    Freier and Anderson Mo1, 2, 4, 6 - 9, Mp4 - 6, 9, 11, 14, 15, Mq1, 2.

8. Computer program

    Statics, Cross Educational Software (see Chapter 2 notes for address). Apple II+. Tutorial on solving equilibrium problems, examples, problems for students to solve. Reviewed TPT February 1983.

9. Laboratory

    BE Experiment 5: Balanced Torques and Center of Gravity. A non-uniform rod is pivoted on a fulcrum. A single weight is hung from one end and the pivot point moved until equilibrium is obtained. The data is used to find the center of gravity and mass of the rod. Additional weights are hung and equilibrium is again attained. The data is used to check that the net force and net torque vanish.

    BE Experiment 6: Equilibrium of a Crane. Students study a model crane, a rod attached to a wall pivot at one end and held in place by a string from the other end to the wall. Weights are attached to the crane and the equilibrium conditions are used to calculate the tension in the rod and in the horizontal string. The latter is measured with a spring balance.

    MEOS Experiment 7-16: Elongation of an Elastomer (see Chapter 7 notes).

    MEOS Experiment 7-17: Investigation of the Elongation of an Elastomer with a Microcomputer. Same as MEOS 7-16 but a microprocessor is used to plot the elongation as a function of applied force. A polar planimeter is used to calculate the work done.

Chapter 14   OSCILLATIONS

BASIC TOPICS

I.   Oscillatory motion.
   A.   Set up an air track and a cart with two springs, one attached to
        each end. Mark the equilibrium point, then pull the cart aside
        and release it. Point out the regularity of the motion and show
        where the speed is the greatest and where it is the least. By
        reference to the cart define the terms periodic motion,
        equilibrium point, period, frequency, cycle, and amplitude.
   B.   Explain that $x(t) = x_m \cos(\omega t + \phi)$ describes the coordinate of the
        cart as a function of time if $x = 0$ is taken to be the
        equilibrium point, where the force of the springs on the cart
        vanishes. State that this type motion is called simple harmonic.
        Show where $x = 0$ is on the air track, then show what is meant by
        positive and negative x. Sketch a mass on the end of a single
        spring and explain that the mass also moves in simple harmonic
        motion if dissipative forces are negligible.
   C.   Discuss the equation.
      1.   Explain that x_m is the maximum excursion of the mass from the
           equilibrium point and that the spring is compressed by x_m at
           one point in a cycle. x_m is called the amplitude of the
           oscillation. Explain that the amplitude depends on initial
           conditions. Draw several $x(t)$ curves, identical except for
           amplitude. Illustrate with the air track apparatus.
      2.   Note that ω is called the angular frequency of the
           oscillation and is given in radians/s. Define the frequency
           by $\nu = \omega/2\pi$ and the period by $T = 1/\nu$. Show that $T = 2\pi/\omega$ is
           in fact the period by direct substitution into $x(t)$: that is,
           show $x(t) = x(t+T)$. Explain that the angular frequency does
           not depend on the initial conditions. For the cart on the
           track, use a timer to show that the period, and hence ω, is
           independent of initial conditions. Draw several $x(t)$ curves,
           for oscillations with different periods. Replace the original
           springs with stiffer springs and note the change in period.

3. Define the phase of the motion and explain that the phase constant ϕ is determined by initial conditions. Draw several $x(t)$ curves, identical except for ϕ, and point out the different conditions at $t = 0$. Illustrate the various initial conditions with the air track apparatus.

D. Derive expressions for the velocity and acceleration as functions of time. Show that the speed is a maximum at the equilibrium point and is zero when $x = x_m$. Also show that the acceleration is a maximum when $x = x_m$ and is zero at the equilibrium point.

E. Show that the initial conditions are given by $x_0 = x_m\cos\phi$ and $v_0 = -x_m\omega\sin\phi$. Solve for x_m and ϕ: $x_m^2 = x_0^2 + v_0^2/\omega^2$ and $\tan\phi = -v_0/\omega x_0$. Calculate x_m and ϕ for a few special cases: $x_0 = 0$ and v_0 positive, $x_0 = 0$ and v_0 negative, x_0 positive and $v_0 = 0$, x_0 negative and $v_0 = 0$.

II. The force law.

A. State that $F = -kx$ for an ideal spring and point out that the minus sign is necessary for the force to be a restoring force. Hang identical masses on springs with different spring constants, measure the elongations and calculate the spring constants. Remark that stiff springs have larger spring constants than weak springs. Emphasize that the spring constant is a property of the spring.

B. Start with Newton's second law and derive the differential equation for $x(t)$. Show that $x = x_m\cos(\omega t+\phi)$ satisfies the equation if $\omega = \sqrt{k/m}$ and explain that this is the most general solution for a given force constant and mass.

C. Show a vertical spring-mass system. Point out that the equilibrium point is determined by the mass, force of gravity, and the spring constant. Show, both analytically and with the apparatus, that the force of gravity does not influence the period, phase, or amplitude of the oscillation.

III. Energy considerations.

A. Derive expressions for the kinetic and potential energies as functions of time. Show that the total energy is constant by adding the two expressions. Remark that the energy changes from kinetic at the equilibrium point to potential at a turning point, then back again.

B. Show how to use the conservation of energy to find the amplitude given the initial position and velocity and to find the speed as a function of position, given the amplitude.

IV. Applications.

A. Demonstrate a torsional pendulum and discuss it analytically. Derive the differential equation for the angle as a function of time and obtain the period in terms of the the spring constant and the rotational inertia.

B. Demonstrate a simple pendulum and discuss it analytically in the small amplitude approximation. Derive the differential equation for the angle as function of time and show how to obtain the period from the equation.

C. Demonstrate a physical pendulum. Use Newton's second law to analyze its motion and obtain an expression for its angular frequency in the small amplitude approximation.

SUPPLEMENTARY TOPICS

1. Simple harmonic motion and uniform circular motion. Unless time constraints are great, this topic should be covered. Mount a bicycle wheel vertically and arrange for it to be driven slowly with uniform angular speed. Attach a tennis ball to 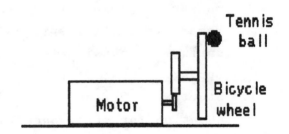 the rim and project the shadow of the ball on the wall. Note that the shadow moves up and down in simple harmonic motion. Point out that the period of the wheel and the period of the shadow are the same. It is possible to suspend a mass on a spring near the wall and adjust the wheel angular speed and the initial conditions so the mass and shadow move together for several cycles. A period of about 1 s works well. Analytically show that the projection of the position vector of a particle in uniform circular motion undergoes simple harmonic motion.

2. Damped harmonic motion, forced oscillations,
    and resonance. These are the basis for
    discussions of many interesting phenomena.
    If there is time, a brief discussion should
    be presented. Resonance can be demonstrated
    with three identical springs and two equal
    masses, as shown. Fasten the bottom spring
    to a heavy weight on the floor and drive the
    upper spring by hand (perhaps standing on a
    table). Obtain resonance at each of the
    normal modes (masses moving in the same and
    opposite directions). After showing the two
    resonances, drive the system at a low
    frequency to show a small response, then
    drive it at a high frequency to again show a
    small response. Repeat at a resonance
    frequency to show the larger response. To
    show pronounced damping effects, attach a
    large stiff piece of aluminum plate to each mass.

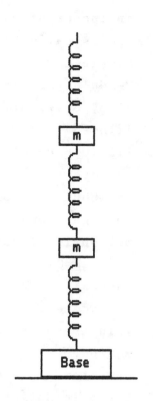

SUGGESTIONS

1. Assign questions 3, 5, 8, and 9 and problems 4 and 13 in support of
    the spring-mass demonstration and discussion. The maximum
    acceleration is covered in problems 6 and 13. Problem 9 deals with a
    vertical oscillator. Assign these for variety.

2. Assign problems 16, 17, 23, and 32 in support of the discussion of
    the mathematical form of $x(t)$ for simple harmonic motion.

3. Assign problem 42 in support of the discussion of energy. Also ask
    for the maximum speed of the mass.

4. If oscillators other than a spring-mass system are considered, assign
    problems 52 and 54 (angular simple harmonic motion), 57 (simple
    pendulum), and 64 and 66 (physical pendulum).

5. Computer project
    Have students use the numerical integration program developed in
    connection with Chapter 5 to investigate forced and damped harmonic
    motion. Have them plot the coordinate as a function of time to see
    transients. To study resonance, have them plot the amplitude as a
    function of forcing frequency. They can also investigate the

influence of damping on the amplitude and resonance width.

6. Demonstrations

    Simple harmonic motion: Freier and Anderson Mx1, 2, 3, 4, 7.

    Pendula: Freier and Anderson Mx6, 9, 10, 11, 12, My1, 2, 3, 8, Mz1, 2, 3, 6, 7, 9; Hilton M14d, f.

7. Films

    Simple Harmonic Motion, 16 mm, color, 17 min. Pennsylvania State University, Applied Research Laboratory, P.O. Box 30, State College, PA 16801. SHM of various phases, frequencies, and amplitudes is shown through computer animation. The relationship of SHM to circular motion is illustrated. Reviewed AJP 47:754 (1979)

    Coupled Oscillators - Equal Masses; Coupled Oscillators - Unequal Masses, S8, color, 3 min. each. Kalmia Company (see Chapter 4 notes for address). These loops are from the popular Miller series and illustrate the phenomenon of the combination of simple harmonic motion.

8. Computer programs

    Harmonic Motion Workshop, High Technology Software Products, P.O. Box 60406, 1611 NW 23rd Street, Oklahoma City, OK 73146. Apple II, II+, IIe. Simulation of simple harmonic motion. Displays velocity vector, acceleration vector, kinetic energy, potential energy. Damped and undamped. Useful for lectures. Reviewed TPT October 1983.

    Physics Simulations I: Oscillator, Kinko's Service Corporation (see Chapter 4 notes for address). Macintosh. Displays a mass in simple harmonic motion, damped or undamped. Plots position, potential energy, and kinetic energy as functions of time.

9. Laboratory

    MEOS Experiment 7-2: The Vibrating Spring. Students time a vertical vibrating spring with various masses attached, then use the data and a logarithmic plot to determine the relationship between the period and mass.

    BE Experiment 15: Elasticity and Vibratory Motion. The experiment is much the same as MEOS 7-2 in that a graph is used to determine the relationship between the mass on a spring and the period of oscillation. This measurement is preceded, however, by a static determination of the spring constant.

    MEOS Experiment 7-4: The Vibrating Ring. Students time the oscillations of various diameter rings, hung on a knife edge, then use the data and a logarithmic plot to determine the relationship between the period and ring diameter. A good example of a physical pendulum.

MEOS Experiment 7-15: <u>Investigation of Variable Acceleration</u>. A pendulum swings above a track and a spark timer is used to record its position as a function of time. Its velocity and acceleration are investigated.

MEOS Experiment 7-19: <u>Harmonic Motion Analyzer</u>. This apparatus allows students to vary the spring constant, mass, driving frequency, driving amplitude, and damping coefficient of a spring-mass system. They can measure the amplitude, period, and relative phase of the oscillating mass. A variety of experiments can be performed.

MEOS Experiment 8-4: <u>Linear Oscillator</u>. A spark timer is used to record the position of an oscillating mass on a spring, moving horizontally on a nearly frictionless surface. The period as a function of mass can be investigated and the conservation of energy can be checked.

MEOS Experiment 7-18: <u>Damped Driven Linear Oscillator</u>. The amplitude and relative phase of a driven damped spring-mass system are measured as functions of the driving frequency and used to plot a resonance curve.

MEOS Experiment 7-20: <u>Analysis of Resonance with a Driven Torsional Pendulum</u>. The driving frequency and driving amplitude of a driven damped torsional pendulum is varied and the frequency, amplitude, and relative phase are measured. Damping is electromagnetic and can be varied or turned off. A variety of experiments can be performed.

Chapter 15   GRAVITY

<u>BASIC TOPICS</u>

I.   Newton's law of gravity.
     A.   Write down the equation for the magnitude of the force of one point mass on another. Explain that the force is one of mutual attraction and lies along the line joining the masses. Explain that G is a universal constant determined by experiment. If you have a Cavendish balance, show it but do not take the time to demonstrate it. As a thought experiment dealing with the magnitude of G, consider a pair of 100 kg spheres falling from a

height of 100 m, initially separated by a bit more then their radii. As they fall, their mutual attraction pulls them only slightly closer together. Air resistance has more influence.

B. Explain that the same mathematical form holds for bodies with spherically symmetric mass distributions (this was tacitly assumed in A) and r is now the separation of their centers. Explain that the force on a point mass anywhere inside a uniform spherical shell is zero. (Optional: use integration to prove that this follows from Newton's law for point masses.) Use this to derive an expression for the force on a point mass inside a spherically symmetric mass distribution. See Sample Problem 3.

C. Point out the assumed equivalence of gravitational and inertial mass.

D. Calculate g for objects near the surface of the earth and justify the use of a constant acceleration due to gravity in previous chapters. Also show that the acceleration due to gravity is independent of the mass of the body.

E. Optional: Discuss factors which influence g and apparent weight.

II. Gravitational potential energy.

A. Use integration to show that the gravitational potential energy for two point masses is given by $U = -GMm/r$, if the zero is chosen at $r \to \infty$. Demonstrate that this result obeys $F = -dU/dr$.

B. Argue that the work needed to bring two masses to positions r apart is independent of the path. Divide an arbitrary path into segments, some along lines of gravitational force and others perpendicular to the gravitational force.

C. Consider a body initially at rest far from the earth and calculate its speed when it gets to the earth's surface. Calculate the escape velocity for the earth and for the moon.

D. Show how to calculate the gravitational potential energy of a collection of discrete masses. Warn the students about double counting the interactions - a term of the sum is associated with each <u>pair</u> of masses. Relate this energy to the binding energy of the system.

E. For a body held by gravitational force in circular orbit about another, much more massive body, show that the kinetic energy goes like $1/r$ and that the total mechanical energy is $-GMm/2r$. Explain that the energy is zero for infinite separation with the bodies at rest, that a negative energy indicates a bound system, and that a positive energy indicates an unbound system. Describe the orbits of recurring and non-recurring comets.

III. Planetary motion and Kepler's laws.

A. Consider a single planet in orbit about a massive sun. The center of mass for the system is essentially at the sun and it remains stationary.

B. Explain that the orbit is elliptical with the sun at one focus. This is so because the force is proportional to $1/r^2$ and the planet is bound.

C. Explain that the displacement vector from the sun to the planet sweeps out equal areas in equal time intervals. Sketch an orbit to illustrate. Show that the torque acting on the planet is zero because the force is along the displacement vector and that conservation of momentum leads to the equal area law. Note that the result is true for any central force.

D. For circular orbits show that the square of the period is proportional to the cube of the orbit radius and that the constant of proportionality is independent of the planet's mass. Explain that the result is also true for elliptical orbits if the radius is replaced by the semi-major axis.

E. Verify the result for planets in nearly circular orbits. The data can be found in Table 3.

SUPPLEMENTARY TOPICS

1. Detailed calculations of the gravitational force of a spherical distribution of mass on a point mass.
2. Maneuvering a satellite.

SUGGESTIONS

1. Review the distinction between weight and mass, then discuss questions 4, 5, 7, 10, 11, and 41.
2. Assign problem 2 to stress Newton's force law. Also assign problem 24 to show students where the value for g comes from.
3. Discuss problems 3, 9, and 12 in connection with calculations of the gravitational force of a spherically symmetric mass distribution on a

point mass.

4. Gravitational potential energy is covered in problems 35 and 42. These can be used as a model for electrostatic potential energy. Escape velocity and energy are covered in problems 36, 37, and 38.

5. Use questions 15, 17, 18, and 39 to discuss some interesting examples of planetary motion and tides. Assign problems 54, 55, 60, 61, and 63 (planetary orbits) and 29 (tides).

6. Films

Measurement of G, The Cavendish Experiment, S8, color, 3 min. Kalmia Company (see Chapter 4 notes for address). The oscillatory motion of a Cavendish balance is shown via time-lapse photography. Another loop from the Miller series.

Forces (excerpt), 16 mm or 3/4" videocassette, b/w, 8 min. Kalmia Company (see Chapter 4 notes for address). In this excerpt from the longer PSSC film, a Cavendish balance is constructed of recording tape, medicine bottles, and boxes of sand. Reviewed AJP 31:400 (1963).

7. Computer programs

Personal Problems, Addison-Wesley Publishing Company (see Chapter 4 notes for address). Apple II. Program on inverse square orbits plots open and closed orbits, given initial conditions. Numerical values of energy and angular momentum can be obtained for any point on orbit. An impulse can be applied to the object at any point in its motion. Excellent for lecture demonstrations.

Physics Simulations I: Kepler, Kinko's Service Corporation (see Chapter 4 notes for address). Macintosh. Plots orbits of one or two planets, with parameters set by user. Use to illustrate lectures or ask students to look at some interesting orbits.

Intermediate Physics Simulations: Three Bodies in 3-D; R.H. Good, Physics Department, California State University at Hayward, CA 94542. Apple II. Plots projections of orbits of either 2 or 3 interacting bodies on a user selcted plane. Use for lectures.

8. Laboratory

MEOS Experiment 7-21: Analysis of Gravitation. Students use the Leybold-Heraeus Cavendish torsional balance to determine G. Requires extremely careful work and a solid vibration free wall to mount the apparatus.

Chapter 16   FLUIDS

BASIC TOPICS

I.   Density and pressure.
    A.   Introduce the subject by giving a few examples of fluids,
         including both liquids and gases. Remark that fluids cannot
         support shear.
    B.   Define density as the mass per unit volume in a region of the
         fluid. Point out that the limit is a macroscopic limit: the
         limiting volume still contains many atoms. The density is a
         scalar and is a function of position in the fluid.
    C.   Explain that fluid in any selected volume exerts a force on the
         material across the boundary of the volume. The boundary may be a
         mathematical constuct and the material on the other side may be
         more of the same fluid. The boundary may also be a container
         wall. Explain that, for a small segment of surface area, the
         force exerted by the fluid is normal to the surface and is
         proportional to the area. The pressure is the force per unit area
         and $\underline{F} = p\underline{S}$, where the magnitude of $\underline{S}$ is the area and the
         direction of $\underline{S}$ is outward, normal to the surface. Units: N/m^2 =
         Pa, atmosphere, bar, torr.
    D.   Show that in equilibrium with y measured positive above some
         reference height $dp/dy = -\rho g$, where ρ is the fluid density. Then
         note that $p_2 - p_1 = -\int \rho g dy$, where the integral limits are y_1 and
         y_2. Finally, if the fluid is incompressible and homogeneous, then
         ρ is a constant. If $y_2 - y_1$ is sufficiently small that g is also
         constant, $p_2 - p_1 = -\rho g(y_2 - y_1)$. Point out that if p_0 is the surface
         pressure, then the pressure a distance h below the surface is p =
         $p_0 + \rho g h$. Note that the pressure is the same at all points at the
         same depth in the fluid.
    E.   Connect a length of rubber tubing to one arm of a U-tube
         partially filled with colored water. Blow into the tube, then
         suck on it. In each case note the change in water level. Insert
         the tube into a deep beaker of water. As the open end is lowered,
         the change in the level of the colored water will indicate the
         increase in pressure. Go over Sample Problem 3 to show the
         equilibrium positions of two unmixed liquids of different
         densities.

II. Pascal's and Archimedes' principles.

    A. State Pascal's principle. Start with $p = p_0 + \rho gh$, consider a change in p_0, and show $\Delta p = \Delta p_0$ if the fluid is incompressible. You can demonstrate the transmission of pressure with a soda bottle full of water, fitted with a tight rubber stopper. Wrap a towel around the neck of the bottle and hit the stopper sharply. With some practice you can blow the bottom out of the bottle cleanly.

    B. Apply the principle to a hydraulic jack. See problems 29 and 30. A hydraulic jack can be made from a hot water bottle, fitted with a narrow rubber tube. Put the bottle on the floor and fasten the tube to a tall ringstand so it is vertical. Place a thin wooden board on the bottle to distribute the weight and have a student stand on it. To change the pressure, use a plunger or rubber squeeze ball from an atomizer, or blow into the tube.

    C. State Archimedes' principle. Stress that the buoyant force acts as if through the center of gravity of the fluid before it is displaced. Also stress that the force is due to the surrounding fluid. Contrast the case of an immersed body surrounded by fluid with one placed on the bottom of the container. Consider a flat board floating on the surface of a liquid, compute the net upward force in terms of the difference in pressure and use $p = p_0 + \rho gh$ to show that this is the weight of the displaced liquid.

    D. Explain why some objects sink while others float.

    E. Fill a large mouthed plastic vessel with water precisely up to an overflow pipe. Immerse an object tied by string to a spring balance. Weigh the object while it is immersed and weigh the displaced water. Observe that the buoyant force is the same as the weight of the displaced water.

III. Fluid dynamics.

    A. Describe:

        1. Steady and non-steady flow.

        2. Rotational and irrotational flow.

        3. Compressible and incompressible flow.

        4. Viscous and nonviscous flow.

    B. Describe what is meant by a velocity field. Stress that it gives the velocity of the fluid at a specified point and time rather than the velocity of a particle as a function of time.

C. Describe streamlines for steady flow and point out that streamlines are tangent to the fluid velocity and that no two streamlines cross. Describe a _tube of flow_ as a bundle of streamlines. Sketch a tube of flow with streamlines far apart at one end and close together at the other.

D. Define volume flow rate and mass flow rate. Give the physical significance of Av and ρAv. State the equation of continuity: ρAv = constant along a streamline if there are no sources or sinks of fluid and if the flow is steady. Argue that if the equation were not true there would be a build up or depletion of fluid in some regions and the flow would not be steady. Discuss the special case of an incompressible fluid and explain that the fluid speed is greater where streamlines are close together than where they are far apart. Use the diagram of C as an example.

IV. Bernouli's equation.

A. Apply the work-energy theorem to a fluid flowing along a streamline to show that $p + \frac{1}{2}\rho v^2 + \rho gy$ = constant for steady, nonviscous, incompressible flow. Point out that this equation also gives the pressure variation in a static fluid (v = 0).

B. Discuss applications of Bernouli's equation. See section 11.

C. Hang two pith balls by strings of equal length, slightly separated. Blow gently between them. Discuss the reduction in pressure in the region between the balls. If you have a high pressure line available, shoot a high velocity air stream vertically upward and hang a smooth handled screw driver in the jet. About 80 psi is needed to lift it. Discuss applications to flying shapes.

SUPPLEMENTARY TOPICS

1. Measurement of pressure.
2. Flow of real fluids.

SUGGESTIONS

1. Use questions 4, 5, 6, 8, and 19 to discuss pressure.
2. Questions 9 through 17 and 20 through 31 deal with Archimedes'

principle. Pick several to illustrate applications of the principle.

3. Similarly, pick several questions from the group 33 through 45 and 47 through 51 to generate a qualitative discussion of Bernoulli's equation.

4. Use problems 2, 5, and 7 in connection with the definition of pressure, and problems 11 and 15 in connection with variations of pressure with depth. Pascal's principle is covered in problems 29 and 30.

5. Use problems 34 and 35 in connection with Archimedes' principle. Problem 36 provides a good test of understanding. Assign problem 47 to the better students.

6. The fundamental ideas of streamlines and the equation of continuity are covered in problems 52 and 53. Problems 57 and 58 are basic Bernouli's equation problems. Work and energy are covered in problems 61 and 62.

7. Demonstrations

Force and pressure: Freier and Anderson Fa, Fb, Fc, Fd, Fe, Ff, Fh; Hilton M20b, M20e, M22b, M22d, M22e, f.

Archimedes' principle: Freier and Anderson Fg; Hilton M20c; Hilton M22c.

Bernouli's principle: Freier and Anderson Fj, Fll.

8. Laboratory

MEOS Experiment 7-7: <u>Radial Acceleration</u> (Problem II only). Students measure the orbit radius of various samples floating on the surface of water in a spinning globe, analyze the forces on the samples. An application of the buoyancy force to rotational motion.

BE Experiment 16: <u>Buoyancy of Liquids and Specific Gravity</u>. Archimedes' principle is checked by weighing the water displaced by various cylinders. Buoyant forces are measured by weighing the cylinders in and out of water. The same cylinder is immersed in various liquids and the results used to find the specific gravities of the liquids.

Chapter 17   WAVES - I

BASIC TOPICS

I.   Qualitative introduction.
    A.   Explain that wave motion is the mechanism by which a disturbance
         created at one place travels to another. Use the example of a
         pulse on a taut string and point out that the displaced string
         causes neighboring portions of the string to be displaced. Stress
         that the individual particles have limited motion (perhaps
         perpendicular to the direction of wave travel), whereas the pulse
         travels the length of the string. Demonstrate by striking a taut
         string stretched across the room. Point out that energy is
         transported by the wave from one place to another. Ask the
         students to read the introductory section of the chapter for
         other examples of waves.
    B.   Explain the terms longitudinal and transverse. Demonstrate
         longitudinal waves with a slinky.

II. Sinusoidal traveling waves.
    A.   Write $y(x,t) = y_m \sin(kx - \omega t)$ and sketch y as a function of x for
         some arbitrary value of t. Below it sketch y for a time that is
         later by some fraction of a period. Explain that x identifies a
         point on the string and y gives the displacement of the string at
         that point, at time t. On the sketches point out the amplitude
         and wavelength. Show that at a given time the pattern on the
         string repeats in a distance equal to $2\pi/k$ and this is the
         wavelength. Remark that k is called the wave number of the wave.
    B.   Show that at a given place on the string the motion repeats in a
         time equal to $2\pi/\omega$. This is the period. State that the motion of
         the string at any point is simple harmonic and that ω is the
         angular frequency. Remind students that the frequency is $\omega/2\pi$.
    C.   Use the two sketches to explain that the disturbance on the
         string moves with constant velocity in the positive x direction.
         Remark that any given point on the string reaches its maximum
         displacement whenever a maximum on the wave passes that point.
         Since the time interval is one period a sinusoidal wave travels
         one wavelength in one period and $v = \lambda/T = \lambda\nu = \omega/k$. Show that
         $kx-\omega t$ = constant gives the position as a function of time of a

particular value of y and that that value moves with velocity $v = dx/dt = \omega/k$, in the positive x direction. Explain that $y(x,t) = y_m \sin(kx+\omega t)$ represents a wave traveling in the negative x direction.

D. Show that the string velocity is $u(x,t) = \partial y/\partial t = -\omega y_m \cos(kx-\omega t)$. Point out that x is held constant in taking the derivative since the string velocity is proportional to the difference in displacement of the <u>same</u> piece of string at two slightly different times. Remark that different parts of the string may have different velocities at the same time and the same part of the string may have a difference velocity at different times. Contrast this behavior with that of the wave velocity. Point out that for a transverse wave u is transverse.

E. Explain that the wave speed for an elastic medium depends on the inertia and elasticity of the medium. State that, for a taut string, $v = \sqrt{\tau/\mu}$, where τ is the tension in the string and μ is the linear mass density of the string. Show how to measure μ for a homogeneous, constant radius string. The expression for v may be derived as in the text or by showing that, for small displacements, the net force on a string segment is $\tau(\partial^2 y/\partial x^2)\Delta x$. Since the mass of the segment is $\mu\Delta x$, $\tau(\partial^2 y/\partial x^2) = \mu(\partial^2 y/\partial t^2)$. Show that $y = f(x \pm vt)$ satisfies this differential equation provided $v = \sqrt{\tau/\mu}$.

F. Point out that the frequency is usually determined by the source and that doubling the frequency for the same string with the same tension halves the wavelength. If a wave goes from one medium to another the speed and wavelength change but the frequency remains the same.

III. Energy considerations.

A. Point out that the energy in the wave is the sum of the kinetic energy of the moving string and the potential energy the string has because it is stretched in the region of the disturbance. Energy moves with the disturbance.

B. Multiply the transverse component of the string tension by the string velocity to show that the power transmitted past a point on the string is given by $P = -\tau(\partial y/\partial x)(\partial y/\partial t)$ for a transverse wave. Substitute the expression for $y(x,t)$ and find an equation for the power transmitted by a sinusoidal wave. Remark that the average over a cycle of $\cos^2(kx-\omega t)$ is $\frac{1}{2}$ and show that the average power is $P = \frac{1}{2}\mu v \omega^2 y_m^2$. Remark that it is proportional to the square of the amplitude. Relate the power transmitted to sound or light intensity.

IV. Superposition and interference.

A. Stress that displacements, not intensities, add. State that if y_1 and y_2 are waves which are simultaneously present, then $y = y_1 + y_2$ is the resultant wave. Using diagrams of two similar sinusoidal waves, show that the resultant amplitude can be twice the amplitude of one of them, can vanish, or can have any value between. Mention that the medium must be linear.

B. Start with the waves $y_1 = y_{1m}\sin(kx-\omega t+\phi)$ and $y_2 = y_{2m}\sin(kx-\omega t)$ and show that $y = y_m\sin(kx-\omega t+\alpha)$, where $y_{2m}^2 = y_{1m}^2 + y_{2m}^2 + 2y_{1m}y_{2m}\cos\phi$. Show that maximum constructive interference occurs when $\phi = 2n\pi$, where n is an integer and maximum destructive interference occurs when $\phi = (2n+1)\pi$, where n is again an integer. Remark that $y_m = 0$ if $y_{1m} = y_{2m}$ and $\phi = (2n+1)\pi$.

C. Explain that a phase difference can arise if waves start in phase but travel different distances to get to the same point. Show that $\phi = k\Delta x$ and find expressions for the path differences which result in maximum constructive and maximum destructive interference. In the first case Δx is a multiple of λ, while in the second it is an odd multiple of $\lambda/2$.

D. Interference can easily be demonstrated with a monaural amplifier, a signal generator, a microphone, an oscilloscope, and a pair of speakers. Fix the position of speaker S_1 and, with S_2 disconnected, show the wave form on the oscilloscope. Then connect S_2 and show the wave form as S_2 is moved. Because both speakers are driven by the same amplifier, the only phase difference is due to the path difference.

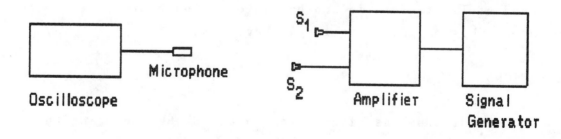

V.   Standing waves.

   A.   Use a vibrating tuning fork (driven, if possible) to set up a
        standing wave pattern on a string. Otherwise, draw the pattern.
        Point out nodes and antinodes. Explain that all parts of the
        string vibrate in phase and that the amplitude depends on
        position along the string. The disturbance does not travel. If
        possible, use a stroboscope to show the standing wave pattern.
        CAUTION: students with epilepsy should not watch this
        demonstration.

   B.   Explain that a standing wave can be constructed from two
        sinusoidal traveling waves of the same frequency and amplitude,
        traveling with the same speed in opposite directions. Use a
        trigonometric identity to show that $y_1 + y_2 = 2y_m \sin(kx)\cos(\omega t)$
        if the waves have the same phase constant. Find the coordinates
        of the nodes and show they are half a wavelength apart. Also find
        the coordinates of the antinodes and show they lie halfway
        between nodes.

   C.   Point out that standing waves can be created by a wave and its
        reflection from a boundary. Derive expressions for the standing
        wave frequencies for a string fixed at both ends. Place two
        speakers, driven by the same signal generator and amplifier, well
        apart on the lecture table, facing the class. Standing waves are
        created throughout the room. Have each student place a finger in
        one ear and move his head slowly from side to side in an attempt
        to find the nodes and antinodes. Use a frequency of about 1 kHz.

   D.   Consider a driven string and describe resonance. Explain that the
        amplitude becomes large when the driving frequency matches a
        standing wave frequency.

SUPPLEMENTARY TOPIC

The speed of light. This is another interesting topic that is important
for modern physics. Cover it if you have time.

SUGGESTIONS

1.   Include discussions of questions 5 and 6 when covering the
     mathematical description of traveling waves. Also assign some of

problems 6 through 8 and 11 through 13. Identification of the wave speed from the mathematical form of the wave is covered in problems 28 and 29. The wave and string speeds are compared in problem 29.

2. The fundamentals of interference are covered in problem 39. The amplitude of the combined wave is covered in problems 38 and 40.

3. Include questions 8 and 11 in a discussion of energy.

4. Discuss coherence and include question 12.

5. Assign problems 43 and 46 in connection with standing waves and problem 50 in connection with resonant frequencies.

6. Computer project

   Have students use Eureka, a spreadsheet, or their own computer programs to investigate energy in a string carrying a wave. The program should calculate the kinetic, potential, and total energies at a given point and time, given the string displacement as a function of position and time. Use the program to plot the energies as functions of time for a given position. Consider a pulse, a sinusoidal wave, and a standing wave. Demonstrate that energy passes the point in the first two cases but not in the third. For sinusoidal and standing waves, the program should also calculate averages over a cycle.

7. Demonstrations

   Traveling waves: Freier and Anderson Sa3, 4, 5, 6, 12, 13; Hilton S2a, c, d.

   Reflection: Freier and Anderson Sa7, 12, 14.

   Standing waves: Freier and Anderson Sa8, 9; Hilton S2b, g.

8. Films

   Standing Waves and the Principle of Superposition, 16 mm, color, 11 min. Encyclopaedia Britannica Educational Corporation, 425 N. Michigan Ave., Chicago, IL 60611. Standing waves by superposition, nodes in standing linear waves, Chladni plates, and soap films are the essential features of this film. Reviewed AJP 41:153 (1973).

   Standing Waves on a String; Standing Waves in a Gas; Vibrations of a Metal Plate; Vibrations of a Drum, S8, color, 3-4 min. each. Kalmia Company (see Chapter 4 notes for address). These loops from the Project Physics Series illustrate standing waves in a variety of media and in both one and two dimensions.

   Propagation of Waves, III: Interference, S8, color, 7 min. Walter de Gruyter, Inc., 200 Sawmill River Road, Hawthorne, NY 10532. Interference is illustrated by the ripple tank. Reviewed AJP 45:596

(1977).

Tacoma Narrows Bridge Collapse, S8, color, 4 min. Kalmia Company (see Chapter 4 notes for address). This is a classic. It shows the incredible amplitudes that can be built up in a macroscopic structure through resonance effects.

Dynamic Response of a Suspension Bridge, S8, color, 3.5 min. American Association of Physics Teachers (see Chapter 2 notes for address). To simulate the Tacoma Narrows Bridge collapse, a small suspension bridge is excited in a torsional mode.

9. Interactive videodisk

The Puzzle of the Tacoma Narrows Bridge Collapse by R.G. Fuller, D.A. Zollman, and T.C. Campbell (Wiley, 1982). This videodisk shows the film of the collapse, referenced above. It allows the students to select various demonstration experiments, which are then shown and used to investigate standing waves, resonance phenomena, and the effect of wind on the bridge.

10. Computer programs

Physics Disk 2: Waves, The 6502 Program Exchange, 2920 Moana, Reno, NV 89509. Apple II. Simulations useful for lectures include the reflection of a pulse at a fixed and at a free end of a string, superposition of two sine waves, standing waves, and beats. Reviewed TPT September 1986

Wave Addition II, Vernier Software (see Chapter 3 notes for address). Apple II. Simulation of the addition of two waves. In some segments the user chooses the parameters of the second wave. Useful as a lecture demonstration of beats and interference effects. Can also be used to demonstrate Fourier synthesis of sawtooth, square, and triangular waves. Reviewed TPT February 1986.

Animation Demonstration: Animated Waves, Conduit, The University of Iowa, Oakdale Campus, Iowa City, IA 52242. Apple II. Simulations which can be used to illustrate lectures on standing waves, traveling pulses, Doppler effect for sound, group velocity, and relativistic e-m waves. Reviewed TPT November 1986.

Intermediate Physics Simulations: Sum of Two Waves, R.H. Good (see Chapter 15 notes for address). Apple II. Shows two waves, one with user selected wavelength, and their sum. Use to demonstrate the addition of waves to create beats, standing waves. Can also be used to show group velocity. Illustrations for lectures.

11. Laboratory

MEOS Experiment 12-1: <u>Transverse Standing Waves</u> (Part A). Several
harmonics are generated in a string by varying the driving frequency.
Frequency ratios are computed and compared with theoretical values.
Values of the wave speed found using $\lambda\nu$ and using $\sqrt{\tau/\mu}$ are compared.
The experiment can be repeated for various tensions and various
linear mass densities.

BE Experiment 22: <u>A Study of Vibrating Strings</u>. A horizontal string
is attached to a driven tuning fork vibrator. It passes over a pulley
and weights are hung on the end. The weights are adjusted so standing
wave patterns are obtained and the wavelength of each is found from
the measured distance between nodes. Graphical analysis is used to
find the relationship between the wave velocity and the tension in
the string and to find the frequency. Several strings are used to
show the relationship between the wave velocity and the linear mass
density.

Chapter 18   WAVES - II

<u>BASIC TOPICS</u>

I.   Qualitative description of sound waves.
   A.   Explain that the disturbance that is propagated is a deviation
        from the ambient density and pressure of the material in which
        the wave exists. This comes about through the motion of
        particles. If Chapter 15 was not covered, you should digress to
        discuss density and pressure briefly. Point out that sound waves
        in solids can be longitudinal or transverse but sound waves in
        fluids are longitudinal: the particle move along the line of wave
        propagation. Use a slinky to show longitudinal waves and point
        out the direction of motion of the particles. State that sound
        can be propagated in all materials.
   B.   Draw a diagram, similar to Fig. 3, showing a compressional pulse.
        Point out regions of high, low, and ambient density. Also show
        the pulse at a later time.
   C.   Similarly, diagram a sinusoidal sound wave in one dimension and
        draw a rough graph of the density as a function of position for a
        given time. Give the rough frequency limits of audible sound and
        mention ultrasonic and infrasonic waves.

D. Discuss the idea that the wave velocity depends on an elastic property of the medium (bulk modulus) and on an inertia property (ambient density). Recall the definition of bulk modulus (or introduce it) and show by dimensional analysis that v is proportional to $\sqrt{B/\rho_0}$. Assert that the constant of proportionality is 1.

II. Mathematical description of one dimensional sound waves.

A. If desired, derive $v = \sqrt{B/\rho_0}$ as it is done in the text.

B. Write $s = s_m\cos(kx-\omega t)$ for the displacement of the material at x. Show how to calculate the pressure as a function of position and time. Relate the pressure amplitude to the displacement amplitude.

C. Define intensity as the power crossing a unit area perpendicular to the direction of propagation and show that the average intensity $\bar{I}$ is given by $\bar{I} = \frac{1}{2}v\omega^2 s_m^2 = p_m^2/(2v\rho_0)$.

D. Show a scale of the range of human hearing in terms of intensity. Introduce the idea of loudness, the bel, and the decibel. If you have a sound level meter, use an oscillator, amplifier, and speaker to demonstrate the change of a few db in sound level.

III. Standing longitudinal waves and sources of sound.

A. Use a stringed instrument or a simple taut string to demonstrate a source of sound. Point out that the wave pattern on the string is very nearly a standing wave, produced by combination of waves reflected from the ends. If the string is vibrating in a single standing wave then sound waves of the same frequency are produced in the surrounding medium. Demonstrate the same idea by striking a partially filled bottle, then blowing across its mouth. Also blow across the open end of a ball point pen case. If you have them, demonstrate Chladni plates.

B. Derive expressions for the natural frequencies and wavelengths of air pipes open at both ends and closed at one end. Stress that nodes occur at fixed ends of strings and closed ends of pipes and that antinodes occur near loose ends of strings and open ends of pipes.

C. Optional: Discuss the quality of sound for various instruments in terms of harmonic content. If possible, demonstrate the instruments.

D. Demonstrate voice patterns by connecting a microphone to an oscilloscope and keeping the set up running through part or all of the lecture. This is particularly instructive in connection with part C.

IV. Beats.

A. Demonstrate beats using two separate oscillators, amplifiers, and speakers, operating at nearly, but not exactly, the same frequency. Explain that this technique is used to tune an orchestra.

B. Write the expression for the sum of $s_1 = s_m \cos(k_1 x - \omega_1 t)$ and $s_2 = s_m \cos(k_2 x - \omega_2 t)$, where $\omega_1 \approx \omega_2$, but the two frequencies are not exactly equal. Note that the amplitude has an angular frequency of $\frac{1}{2}|\omega_1 - \omega_2|$ but the intensity has an angular frequency of $|\omega_1 - \omega_2|$. The latter is the beat angular frequency.

V. Doppler effect.

A. Explain that the frequency increases when the source is moving toward the listener, decreases when the source is moving away, and that similar effects occur when the listener is moving toward or away from the source. Use Figs. 15 and 16 to illustrate the physical basis of the phenomenon.

B. Derive expressions for the frequency when the source is moving and for the frequency when the listener is moving. Point out that the velocities are measured relative to the medium carrying the sound.

C. The effect can be demonstrated by placing an auto speaker and small audio oscillator (or sonalert type oscillator) on a rotating table. The sonalert can also be secured to a cable and swung in a circle. Show the effect of a passive reflector by moving a hand-held sonalert toward and away from the blackboard.

SUGGESTIONS

1. When discussing the qualitative aspects of sound waves, consider question 11.

2. In connection with the discussion of sound sources, consider questions 12, 13, 14, and 18.

3. The speed of sound is emphasized in problems 5, 7, and 8 while its

dependence on the bulk modulus and the density is covered in problem 9.

4. Interference of sound waves can be demonstrated by wiring two speakers to an audio oscillator and putting the apparatus on a slowly rotating platform. Students will hear the changes in intensity. Assign problems 19 and 44.

5. To help teach students the meaning of the bel and decibel, assign problem 25.

6. If you use a Knudt's tube in the lab, assign problem 47.

7. Tuning stringed instruments is covered in problems 50 and 51 while organ pipes are covered in problems 52 and 53.

8. Assign problems 61 and 62 in connection with beats.

9. Demonstrations

    Wavelength and speed of sound in air: Freier and Anderson Sa16, 17, 18, Sh1; Hilton S3e, f.

    Sound not transmitted in a vacuum: Freier and Anderson Sh2; Hilton S3a.

    Sources of sound, acoustical resonators: Freier and Anderson Sd3, Se, Sf, Sj6; Hilton S4, S7.

    Harmonics: Freier and Anderson Sj2 - 5

    Beats: Freier and Anderson Si4 - 6; Hilton S5.

    Doppler shift: Freier and Anderson Si1 - 3; Hilton S6.

10. Films

    Propagation of Waves II: Standing Waves and the Doppler Effect, S8, color, 4 min. Walter de Gruyter, Inc. (see Chapter 17 notes for address). A toy train moves the source in this ripple tank demonstration of the Doppler effect.

    Demonstrations in Acoustics, 3/4" videocassette, color, (various lengths). University of Maryland, Department of Physics and Astronomy, College Park, MD 20742. 29 different demonstrations in elementary acoustics are presented. A good source of demonstrations. Reviewed AJP 49:608 (1981). See the review for a complete list of the demonstrations.

11. Computer programs

    Animation Demonstration: Animated Waves. See Chapter 17 notes.

    The microcomputer Based Lab Project Sound, HRM Software (see Chapter 2 notes for address). Sound is picked up by a microphone and intensity is plotted as a function of time on the monitor screen. Use this as an alternative to an oscilloscope. It has the advantages that sound patterns can be stored on disk and recalled for later use and two patterns can be displayed simultaneously for comparison. Any portion of a pattern can be magnified for closer study.

12. Laboratory

MEOS Experiment 12-2: <u>Velocity of Sound in Air</u> and BE Experiment 23: <u>Velocity of Sound in Air - Resonance-Tube Method</u>. Resonance of an air column is obtained by holding a tuning fork of known frequency at the open end of a tube with one closed end. The length of the column is changed by adjusting the amount of water in the tube. The wavelength and speed of sound are found.

MEOS Experiment 12-3: <u>Velocity of Sound in Metals</u> and BE Experiment 24: <u>Velocity of Sound in a Metal - Kundt's-Tube Method</u>. A Kundt's tube is used to find the frequency of sound excited in a rod with its midpoint clamped and its ends free. Since the wavelength is known to be twice the rod length, $\lambda\nu$ can be used to find the speed of sound. In another experiment, a transducer and oscilloscope are used to time a sound pulse as it travels the length of a rod and returns.

MEOS Experiment 12-4: <u>Investigation of Longitudinal Waves</u>. The amplitude and phase of a sound wave are investigated as functions of distance from a speaker source. To do this, Lissajous figures are generated on an oscilloscope screen by the source signal and the signal picked up by a microphone. To eliminate noise, the speaker and microphone should be in a large sound proof enclosure with absorbing walls. Use MEOS Experiment 10-10 to familiarize students with the oscilloscope and Lissajous figures.

Chapter 19   TEMPERATURE

<u>BASIC TOPICS</u>

I.   The zeroth law of thermodynamics.
    A. Explain that two bodies are in thermal equilibrium if no changes take place in the macroscopic properties of either when they are placed in contact. For fluids the properties of interest include pressure and volume. Explain that if body A and body B are not in thermal equilibrium then changes do occur in one or more macroscopic properties when they are brought into contact.
    B. State the zeroth law: if body A and body B are each in thermal

equilibrium with body C, then A is in thermal equilibrium with B.

C. Explain that two bodies in thermal equilibrium are said to have the same temperature. The temperature of a body is measured by measuring some property of a thermometer in thermal equilibrium with it. Illustrate by reminding students that the length of the mercury column in an ordinary household thermometer is a measure of the temperature. Explain that the zeroth law guarantees that the same temperature will be obtained for two substances in thermal equilibrium with each other.

II. Temperature measurements.

A. Describe a constant volume gas thermometer. If available, demonstrate its use. The gas is placed in thermal contact with the substance whose temperature is to be measured and the pressure is adjusted so that the volume has some standard value (for that thermometer). After corrections are made, the temperature is taken to be proportional to the pressure: $T = ap$, where a is the constant of proportionality.

B. Describe the triple point of water and explain that water at the triple point is assigned the temperature $T = 273.16$ K. Solve for a and show that $T = 273.16(p/p_{tr})$.

C. Point out that thermometers using different gases give different values for the temperature when used as described. Explain the limit used to obtain the ideal gas temperature. See Fig. 6.

D. Explain that the absolute temperature or Kelvin scale is identical to the ideal gas scale for temperatures above 1 K and can be extended to lower temperatures, for which gas thermometers cannot be constructed.

E. Define the Celsius and Fahrenheit scales. Give the relationships between the degree sizes and the zero points. Give equations for conversion from one scale to another and give the temperature value for the ice and steam points in each system. Use Fig. 7. Explain that the Kelvin scale has an absolute zero and there can be no lower temperatures. $T = 0$ K has never be reached experimentally but it is possible, in principle, to come arbitrarily close.

III. Thermal expansion.

A. Describe linear expansion and define the coefficient of linear expansion: $\alpha = \Delta L / L \Delta T$. Point out Table 3. Obtain a bimetallic strip and use both a bunsen burner and liquid nitrogen (or dry ice) to show bending. Explain that these devices are often used in thermostats.

B. Discuss area and volume expansion. Carefully drill a $\frac{1}{2}$ inch hole in a piece of aluminum, roughly $1\frac{1}{4}$ inch thick. Obtain a 13 mm diameter steel ball bearing and place it in the hole. It will not pass through. Heat the plate on a bunsen burner and the ball passes through easily.

C. Demonstrate volume expansion of a gas using a flat bottomed flask, a bulbed tube, a two hole stopper, and some colored water. Partially evacuate the bulb so the colored water stands in the tube somewhat above the stopper. Place your hand on the bulb to warm the air inside and the water in the tube drops in response.

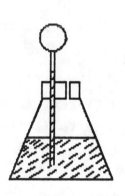

SUGGESTIONS

1. After discussing thermal equilibrium, ask questions 2 and 8.
2. After discussing gas thermometers, ask questions 4 and 13, assign problems 6 and 7.
3. Use questions 12, 14, 15, and 29 to discuss the general problem of defining and measuring temperature. Note that different properties are used to measure temperatures in different situations. To emphasize the great variety of temperature measuring techniques, assign problems 1, 2, 5, and 8. Problem 16 asks students to compare temperature readings on various scales.
4. Use question 18 and problem 21 in connection with the ball and hole demonstration, questions 19 and 20 and problem 44 in connection with the bimetallic strip demonstration.
5. Introduce Newton's law of cooling and assign problems 10 and 11.
6. Demonstrations
    Thermometers: Freier and Anderson Ha1 - 4; Hilton H1.
    Thermal expansion: Freier and Anderson Ha5 - 12; Hilton H2.
7. Laboratory

MEOS Experiment 9-3: <u>Linear Expansion</u> and BE Experiment 18: <u>Linear Coefficient of Expansion of Metals</u>. The length of a metal rod is measured at room temperature and at 100° (in a steam jacket), then the data is used to compute the coefficient of thermal expansion. The experiment can be repeated for several different metals and the results compared.

Chapter 20   HEAT AND THE FIRST LAW OF THERMODYNAMICS

<u>BASIC TOPICS</u>

I.   Heat.

A.   Explain that when thermal contact is made between two bodies at different temperatures, energy flows from the higher temperature body to the lower temperature body. The temperature of the hotter body decreases, the temperature of the cooler body increases, and the flow continues until the temperatures are the same. Energy also flows from warmer to cooler regions of the same body. State that heat is energy which is transferred because of a temperature difference. Distinguish between heat and internal energy. Emphasize that the idea of a body having heat content is not meaningful. Also emphasize that heat is not a new form of energy. It may be the kinetic energy of molecules or the energy in an electromagnetic wave. Examples: a bunsen burner flame, radiation across a vacuum. State that heat is usually measured in Joules but calories and British thermal units are also used. 1 kcal = 3.969 Btu = 4187 J.

B.   Explain that temperature changes can be brought about by mechanical work and emphasize that heat and work are alternate means of transferring energy. To demonstrate this, connect a brass tube, fitted with a rubber stopper, to a motor as shown. Make a wooden brake or clamp which fits tightly around the tube. Put a few

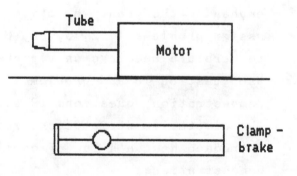

drops of water into the tube, start the motor, and exert pressure on the tube with the clamp. Soon the stopper will fly off. Note that mechanical work was done and steam was produced.

C. Define the heat capacity of a body as the amount of heat which must be absorbed to raise the temperature 1° without change of phase: C = Q/ΔT. Point out that it depends on the temperature and on the constraints imposed during the transfer. The heat capacity at constant volume is different from the heat capacity at constant pressure. Define the specific heat and the molar heat capacity. Point out Table 1.

D. Do a simple calorimetric calculation (see Sample Problem 3). Stress that the fundamental idea used is the conservation of energy.

E. Explain that energy must be transferred to or from a body when it changes phase (liquid to gas, etc.). The energy per unit mass is called the specific heat of transformation or specific latent heat. Point out Table 2.

II. Heat, work, and the first law of thermodynamics.

A. Describe a gas in a cylinder fitted with a piston. Explain that as the piston moves the gas volume changes and the gas does work W = $\int$p dV on the piston.

B. Draw a p-V diagram (Fig. 4) and mark an initial and final state with V_f > V_i. Explain that p and V are thermodynamic state variables and have definite, well defined values for a given thermodynamic state. They can be used to specify the state. Point out there are many paths from the initial to the final state. Define the term "quasi-static process" and explain that the various paths on the diagram represent quasi-static processes, for which the system is infinitesimally close to equilibrium states. Point out that for different paths p is a different function of V and different amounts of work are done by the gas. Also explain that different amounts of heat are transferred for different paths. Work and heat are not thermodynamic state variables.

C. Stress the sign convention for heat absorbed and work done as the system goes from one thermodynamic state to another. Q is positive if heat is absorbed by the system. W is positive if the system does work on its environment.

D. Explain that Q - W is independent of path. Define the internal energy by $\Delta U = Q - W$ and point out that ΔU is the same for any two selected states regardless of the path used to get from one to the other. State that ΔU is the change in mechanical energy (kinetic and potential energy) of all the particles which make up the system. Stress that the first law $\Delta U = Q - W$ is an expression of the conservation of energy.

III. Applications of the first law.

A. Adiabatic process. Explain that $Q = 0$ and $\Delta U = -W$. As an example, consider a gas in a thermally insulated cylinder and allow the volume to change by moving the piston. Explain that when the internal energy increases the temperature goes up for most materials. This can be achieved by compressing the gas. The opposite occurs when the piston is pulled out. Stress that no heat has been exchanged. Illustrate an adiabatic process on a pV diagram.

B. Constant volume process. Explain that $\Delta U = +Q$ since $W = 0$. Illustrate on a p-V diagram.

C. Isobaric process. Explain that $W = p(V_f - V_i)$. If the changes in temperature and volume are both known $\Delta U = mc_p \Delta T - p\Delta V$ can be used to find the change in internal energy, provided no phase change takes place. For a change in phase, show that $\Delta U = m\ell - p\Delta V$. Illustrate the two processes on a p-V diagram.

D. Describe adiabatic free expansion and note that $\Delta U = 0$. Explain that this process is not quasi-static and cannot be shown on a p-V diagram. The end points, however, are well defined thermodynamic states and are points on a p-V diagram.

E. Cyclical process. Explain that all state variables return to their original values and, in particular, $\Delta U = 0$. Thus $Q = W$. Illustrate on a p-V diagram.

IV. Transfer of heat.

A. Explain that steady state heat flow can be obtained if both ends of a slab are held at different temperatures. Define the thermal conductivity k of the material using $H = -kA \, dT/dx$ for a slab of uniform cross section A. Here H is the rate of heat flow. Emphasize that the negative sign appears because heat flows from hot to cold. Stress that H and T are constant in time in the steady state.

B. A demonstration which shows both thermal conductivity and heat capacity can be constructed from three rods of the same size, one made of aluminum, one made of iron, and one made of glass. Use red wax to attach small ball bearings at regular intervals

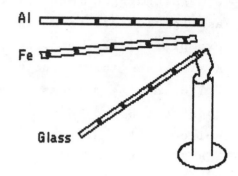

along each rod. Clamp the rods so that each has one end just over a bunsen burner. The rate at which the wax melts and the ball bearings drop off is mostly dictated by the thermal conductivity of the rods, but it is influenced a bit by the specific heats.

C. For a practical discussion, introduce the idea of R value and discuss home insulation. Consider a compound slab as in the text.

D. Qualitatively discuss radiation as a means of energy transfer. Place a heating element at the focal point of one spherical reflector and some matches, stuck in a cork, at the focal point of the another. Place the reflectors several meters apart and adjust the positions so that the heater is imaged at the matches. Use a 1 kW or so heater. The matches will ignite in about 1 minute.

E. Qualitatively discuss convection as means of energy transfer. Cut a circular groove in a wooden block and use this as the base for a glass tube. Light a candle and

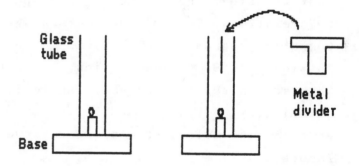

place it on the base, then cover it with the tube. The candle goes out. Cut a sheet metal divider with shoulders to sit on the tube edge. Relight the candle and insert the tube. Because of air circulation, the candle continues to burn. Remove the divider and the candle again goes out.

SUGGESTIONS

1. Understanding of many of the ideas of this chapter can be tested by asking students to state if certain processes can occur and to give examples if they can. Ask questions 2, 4, 27, 30, for example.

2. Assign questions 4 and 5 in conjunction with problem 27.

3. Following the discussion of heat exchange and heat capacity, ask questions 9 and 16.

4. Following the discussion of thermal conductivity, ask questions 22 and 24.

5. Use question 11 to stimulate discussion of convection and cooling.

6. Use questions 33, 37, and 38 as a review of energy transfer mechanisms.

7. Problems 17, 18, and 19 are good examples of calorimetry problems. Problems 5, 24, 27, 29 deal with changes of phase.

8. Problems 36, 38, 39, and 42 are all good tests of understanding of the first law.

9. Use problems 47, 48, and 56 in connection with heat transfer.

10. Assign problems such as 36, 38, 39, and 42, which involve the interpretation of p-V diagrams. Tell students to pay attention to signs.

11. Demonstrations.

    Heat capacity and calorimetry: Freier and Anderson Hb1, 2.

    Work and heat: Freier and Anderson He1 - 6.

    Heat transfer: Freier and Anderson Hc, Hd1 - 7, Hf; Hilton H3.

    p-V relations: Freier and Anderson Hg1 - 3; Hilton H5f.

12. Computer program

    Physics Vol. 6: Thermodynamics, Cross Educational Software (see Chapter 2 notes for address). Apple II. Tutorial programs on calorimetry, p-V, p-T, and V-T diagrams, thermodynamic cycles, heat engines, and molecular motion. Reviewed TPT April 1985.

13. Laboratory

    MEOS Experiment 9-1: Calorimetry - Specific Heat and Latent Heat of Fusion. Students use a calorimeter to find the specific heat of water and a metal sample. They also measure the latent heat of fusion of ice. Since the specific heat of the stirring rod and the calorimeter must be taken into account, this is a good exercise in experimental design.

    MEOS Experiment 9-2: Calorimetry - Mechanical Equivalent of Heat and BE Experiment 30: The Heating Effect of an Electric Current. A calorimeter is used to find the relationship between the energy dissipated by a resistive heating element and the temperature rise of the water in which it is immersed. Students must accept $P = i^2R$ for the power output of the heating element. With slight revision these experiments can also be used in conjuction with Chapter 28.

BE Experiment 19: <u>Specific Heat and Temperature of a Hot Body</u>. A calorimeter is used to obtain the specific heat of metal pellets. In a second part, a calorimeter and a metal sample with a known specific heat are used to find the temperature of a Bunsen burner flame.

BE Experiment 20: <u>Change of Phase – Heat of Fusion and Heat of Vaporization</u>. A calorimeter is used to measure the heat of fusion and heat of vaporization of water. If the lab period is long or writeups are done outside of lab, experiments 19 and 20 may be combined nicely.

MEOS Experiment 9-6: <u>Calorimetry Experiments</u> (with a microprocessor).

MEOS Experiment 9-4: <u>Thermal Conductivity</u>. The sample is sandwiched between a thermal reservoir and a copper block. The rate at which energy passes through the sample is found by measuring the rate at which the temperature of the copper increases. Temperature is monitored by means of a thermocouple.

MEOS Experiment 9-5: <u>Thermal Conductivity with Microprocessor</u>.

Chapter 21   THE KINETIC THEORY OF GASES

BASIC TOPICS

I.   Macroscopic description of an ideal gas.

   A.   Explain that kinetic theory treats the same type problems as thermodynamics but from a microscopic viewpoint. It uses averages over the motions of individual particles to find macroscopic properties. Here it is used to clarify the microscopic basis of pressure and temperature.

   B.   Define the mole. Define the Avogadro constant and give its value, 6.02×10^{23} mol^{-1}. Explain the relationships between the mass of a molecule, the mass of the sample, the molecular weight, the number of moles, the number of molecules, and the Avogadro constant. These often confuse students.

   C.   Write down the ideal gas equation of state in the form $pV = nRT$ and in the form $pV = NkT$. Give the values of R and k and state

that Boltzmann's constant is R/N_A. Explain that for real gases at
low density pV/T is nearly constant.

D. To show how the equation of state can be used in thermodynamic
calculations, go over Sample Problems 1 and 2. Derive the
equations needed from the first law and the equation of state.
Point out that the equation of state connects the three
thermodynamic variables p, V, and T and, for example, allows us
to draw isotherms on a p-V diagram.

II. Kinetic theory calculations of pressure, temperature, and specific
heat.

A. Go over the assumptions of kinetic theory for an ideal gas, as
given in Section 4. Consider a gas of molecules with only
translational degrees of freedom. Assume they are small and are
free except for collisions of negligible duration. Also assume
collisions with other molecules and with walls of the container
are elastic. At the walls the molecules are specularly reflected.

B. Discuss a gas in a cubic container and explain that the pressure
at the walls is due to the force of molecules as they bounce off.
By considering the change in momentum at the wall per unit time,
show that the pressure is given by $p = \frac{1}{3}\rho v_{rms}^2$. Define the rms
speed. Use Table 1 to give some numerical examples of v_{rms}^2 and
calculate the corresponding pressure.

C. Substitute $p = \frac{1}{3}\rho v_{rms}^2$ into the ideal gas equation of state and
show that $\frac{1}{2}Mv_{rms}^2 = (3/2)RT$, where M is the molecular weight. Show
this can be written $\frac{1}{2}mv_{rms}^2 = (3/2)kT$, where m is the mass of a
molecule. Remark that the left side is the mean kinetic energy of
a molecule and point out that the temperature is proportional to
the mean kinetic energy. See problems 52, 56, 57, and 82. For
many students the rms value of a quantity needs clarification.
Consider a system of 5 molecules and select numerical values for
their speeds. Make a numerical calculation of v_{rms}^2.

D. Explain that the internal energy of an ideal gas is the sum of
the kinetic energies of the molecules and write $U = \frac{1}{2}Nmv_{rms}^2 =$
$(3/2)NkT = (3/2)nRT$. For an ideal gas the internal energy is a
function of the temperature alone. This is an approximation for a
real gas.

E. Consider an ideal gas which undergoes a change in temperature at
constant volume and argue that $\Delta U = nC_v\Delta T$, where C_v is the molar

heat capacity at constant volume. Solve for C_v and show it is (3/2)R. Consider that same gas as it undergoes a change in temperature at constant pressure and show that $C_p = C_v + R$ and that this is (5/2)R. Compute the ratio of the specific heats. Show that for adiabatic processes pV^γ = constant. See Sample Problem 10.

III. Equipartition of energy.
  A. Point out that the values of C_v and C_p obtained above are close to experimental values for monatomic gases but are too low for gases of diatomic and polyatomic molecules. See Table 3. for these molecules there are more degrees of freedom and the energy may go into other motions instead of the translational motion of the molecules. Define the term degree of freedom and show how to count the number for monatomic, diatomic and polyatomic molecules.
  B. State the equipartition theorem: the energy is distributed equally among all degrees of freedom. Explain that this agrees with the previous result for monatomic gases: there are three degrees of freedom per molecule and each receives energy $\frac{1}{2}kT$.
  C. Discuss diatomic molecules and explain there are 2 new degrees of freedom, rotational in nature. Show that U = (5/2)nRT, C_v = (5/2)R, and C_p = (7/2)R. Compare with the values given in Table 3.
  D. Discuss polyatomic molecules. State that there are now 3 rotational degrees of freedom and show that U = 3nRT, C_v = 3R, and C_p = 4R. Compare with the values given in Table 3.
  E. Explain that vibrational degrees of freedom also contribute to the internal energy and to the specific heats. Point out using Fig. 12 that at low temperatures some degrees of freedom are frozen and do not absorb energy. This is a failure of the equipartition theorem and can be explained only when quantum mechanics is invoked.

SUPPLEMENTARY TOPICS

1. Mean free path.
2. Distribution of molecular speeds.

These topics add breadth to the coverage of kinetic theory. The distribution of molecular speeds is used in the discussion of thermonuclear fusion, in Chapter 49.

SUGGESTIONS

1. To start students thinking about the postulates of kinetic theory, ask questions 1, 3, 4, 10, 13, 31.

2. Use question 15 to discuss the difference between evaporation and boiling.

3. After discussing the relationship between the Maxwellian distribution and the temperature, ask questions 37 through 42.

4. After discussing the various specific heats, ask question 46. Assign problem 62 to emphasize that neither pressure not volume need be held constant when measuring the heat capacity.

5. For bright theoretically minded students, assign questions 32 and 37.

6. Assign a problem that deals with calculating the mass of an atom: see problems 1, 2, and 3.

7. Assign a problem, such as 7, that is a straightforward application of the ideal gas law. Then assign problems that show how the law is used to compute quantities when the gas changes state: 10 and 15, for example. You may also wish to assign problem 17, which deals with partial pressures. Problems 18 through 23 deal with real-life applications. If possible, assign one or two.

8. To stress the dependence of the molecular speed distribution on temperature, assign problem 28. The molecular basis of pressure is covered in problem 31, while problems 32 and 33 deal with molecular kinetic energies.

9. To help students understand the idea of rms speed, assign problem 50. Also consider problem 53.

10. Use problems 70 and 71 in connection with the adiabatic expansion of an ideal gas. Use problem 67 in connection with the equipartition of energy.

11. Demonstrations

    Avogadro's constant: Hilton H4a.

    Kinetic theory models: Freier and Anderson Hh1, 2, 4, 5.

    Brownian motion: Freier and Anderson Hh3.

    Mean free path and diffusion: Freier and Anderson Hh7, Hi.

12. Films

Pressure, Volume, and Boyle's Law, S8, color, 4 min. Kalmia Company (see Chapter 4 notes for address). Computer-animated particles in a box are used to illustrate the pressure-volume relationship. Reviewed Science Books & Films 121:169 (1976).

The Ideal Gas Law; Gravitational Distribution; The Maxwell-Boltzmann Distribution; Brownian Motion and Random Walk, Deviations from an Ideal Gas, S8 or 16 mm reels, color, 3-4 min. each. Kalmia Company (see Chapter 4 notes for address). Selected loops from the FITCH-MIT series dealing with various topics related to the kinetic theory of gases. Reviewed AJP 44:810 (1976).

13. Computer programs

Physics Vol. 6: Thermodynamics. See Chapter 20 notes.

Animation Demonstration: Animated Particles, Conduit (see Chapter 17 notes for address). Illustrations for kinetic theory lectures. The influence of gravitational and magnetic fields are also simulated. Reviewed TPT November 1986.

Intermediate Physics Simulations: Gas - with Collisions and Speed Distributions, R.H. Good (see Chapter 15 notes for address). Apple II. Large and small mass particles are represented on the screen. They are segregated to start and all particles of the same mass have the same velocity. Thermal equilibrium is achieved through collisions with the walls and with each other. Graphs of the velocity distributions are shown and students see them approach Maxwellian distributions. Excellent demonstration for showing the approach to thermal equilibrium.

Physics Simulations III: Gas, Kinko's Service Corporation (see Chapter 4 notes for address). An excellent simulation of gas molecules in a box. Use for illustration of lectures.

14. Laboratory

BE Experiment 17: Pressure and Volume Relations for a Gas. The volume of gas in a tube is adjusted by changing the amount of mercury in the tube and a U-tube manometer is used to measure pressure. A logarithmic plot is used to determine the relationship between pressure and volume.

MEOS Experiment 9-8: Kinetic Theory Model. The Fisher kinetic theory apparatus, consisting of a large piston-fitted tube of small plastic balls, is used to investigate relationships between pressure, temperature, and volume for a gas. A variable impeller at the base allows changes in the average kinetic energy of the balls; the piston

can be loaded to change the pressure. A variety of experiments can be
performed.

Chapter 22   THE SECOND LAW OF THERMODYNAMICS

BASIC TOPICS

I.   Engines and refrigerators.
     A.   Discuss heat engines and refrigerators in general, from the point
          of view of the first law only. Explain that an engine absorbs
          heat at a high temperature, rejects heat at a low temperature,
          and does work. Describe a refrigerator in similar terms. Define
          the efficiency of an engine and the coefficient of performance of
          a refrigerator.
     B.   Define a cycle as a process for which the system starts and ends
          in the same state. Point out that $\Delta U = 0$, $\Delta p = 0$, and $\Delta V = 0$ for
          a cycle. Explain that engines and refrigerators work in series of
          cycles.

II. The second law of thermodynamics.
     A.   Give both the Clausius and Kelvin-Planck statements. Remark that
          both statements imply the use of cycles, for otherwise the
          internal energy would be different for the initial and final
          states. The Clausius statement rules out the perfect refrigerator
          for which no work need be done; the Kelvin-Planck statement rules
          out the perfect heat engine, for which $Q_C = 0$. State that the two
          statements are equivalent. Optional: prove the equivalence.
     B.   Define a Carnot cycle. Assume the working substance is an ideal
          gas and draw the cycle on a p-V diagram. Stress that the working
          substance passes through a succession of equilibrium states. Show
          that the work done by the system during a cycle is $W = |Q_H| -$
          $|Q_C|$. Stress that $|Q_H|$ and $|Q_C|$ are magnitudes. For an engine Q_H
          is positive and Q_C is negative. Remark that the cycle can be run
          in reverse as a refrigerator.
     C.   Show that for a Carnot cycle $|Q_H|/T_H = |Q_C|/T_C$. Also show that
          the efficiency of a Carnot heat engine is given by $e = 1 -$
          $|Q_C|/|Q_H| = 1 - T_C/T_H$ and that the coefficient of performance of

a Carnot refrigerator is $K = T_C/(T_H-T_C)$. Remark that $e < 1$ unless $T_C = 0$. Explain that an engine with $e = 1$ has never been built. Point out Sample Problems 3 and 4.

D. Remind students of the difference between equilibrium and non-equilibrium states. For a gas in a non-equilibrium state, pressure and temperature may not have well defined values. For quasi-static processes the system is always arbitrarily close to equilibrium states throughout the process. Explain that these processes can be reversed by making the system pass through equilibrium states in the opposite direction. If the system passes through non-equilibrium states the process is irreversible. Give examples of reversible and irreversible compression of a gas. Remark that heat engines and refrigerators may be reversible or irreversible.

E. Use the second law to show that the efficiencies of all reversible heat engines operating between the same two temperatures are the same and that the efficiency of an irreversible engine is less than that of a reversible engine. In particular, show the converse allows the construction of an engine or refrigerator which violates the second law.

III. Entropy.

A. Remind students that Q_H and Q_C have opposite signs for a Carnot cycle so $Q_H/T_H + Q_C/T_C = 0$. Argue that any reversible cycle can be constructed from a series of alternating isotherms and adiabatic lines and, in the limit of infinitesimal differences between isotherms, $\int dQ/T = 0$. See Fig. 10.

B. Define the entropy difference between two infinitesimally close states as $dS = dQ/T$ and between any two states as $\Delta S = \int dQ/T$. Point out that the integral is independent of path and that S is therefore a thermodynamic state variable. Stress that the integral must be taken along a reversible path but that entropy differences are defined regardless of whether the actual process is reversible or irreversible. The end points must be equilibrium states, however.

C. Consider the adiabatic expansion of an ideal gas. Point out that the process is irreversible, $Q = 0$, and $\Delta U = 0$. Since the gas is ideal, $T_f = T_i$. Find the change in entropy by evaluating $\int dQ/T$

over a reversible isotherm through the initial and final states. Point out that the isothermal path does represent the actual process. Show that $\Delta S = nR\ln(V_f/V_i)$ and state this is positive.

D. Consider two rigid containers of ideal gas, at different temperatures, T_H and T_C. Place them in contact in an adiabatic enclosure. Show they reach equilibrium at temperature $T_m = (T_H + T_C)/2$. Then consider a reversible, constant volume process which connects the initial and final states and show that $\Delta S = C_v\ln(T_m^2/T_H T_C)$. Remark that this is positive.

E. Remark that for reversible processes the total entropy of the system and its environment does not change. This is because, for the combination of system and environment, the process is adiabatic and $dQ = 0$ for each segment of the reversible path. On the other hand, entropy increases for an adiabatic <u>irreversible</u> process.

F. State that the second law is equivalent to the statement that for processes which proceed from one equilibrium state to another the entropy of a closed system does not decrease. Point out that the previous two examples are consistent with this statement. Optional: show this statement is equivalent to the Clausius and Kelvin-Planck statements.

SUGGESTIONS

1. In order to make the ideas of thermodynamic equilibrium, reversible process, and irreversible process more concrete, ask a few of the questions in the group 6 through 12.

2. Consider practical engines and their efficiencies by approximating their operation by Carnot cycles. For a gasoline engine $T_H \approx 1000°$ F and $T_C \approx 400°$ F. Compare actual efficiencies with the carnot efficiency. Actual efficiencies can be obtained by considering the fuel energy available and the work actually obtained.

3. Consider practical refrigerators. Look in a catalog for typical values of the coefficient of performance and compare with the Carnot coefficient of performance.

4. Use questions 13 through 17 to discuss real and Carnot engines. Problems 4, 5, and 7 cover the fundamentals of cycles. Problem 10 covers work and heat transfer in a Carnot cycle. Problem 12 deals with the dependence of efficiency on the temperatures.

5. To start students thinking about entropy changes as they occur in common processes, ask a few of the questions in the group 22 through 32. Assign problems 34, 37, 46, and 50. To include entropy changes in calorimetry experiments, ask problems 43 and 44.

6. Ask bright students to draw a Carnot cycle on a T-S diagram. See question 25.

7. Demonstrations

    Engines: Freier and Anderson Hm5, Hn; Hilton H5a, b.

8. Film:

    The Reversibility of Time, S8, color, 4 min. Kalmia Company (see Chapter 4 notes for address). To illustrate that reverse motion is not always obvious, various scenes are run forward and backward. Part of the Project Physics Series.

Chapter 23   ELECTRIC CHARGE

BASIC TOPICS

I. Charge.

    A. Explain that there are two kinds of charge, called positive and negative, and that like charges repel each other, unlike charges attract each other. Give the SI unit (coulomb) and explain that it is defined in terms of current, to be discussed later. Optional: explain that current is the flow of charge and is measured in amperes. One coulomb of charge passes a cross section each second in a wire carrying a current of 1 A.

    B. Carry out the following sequence of demonstrations. These work best in dry weather.

        1. Suspend a pith ball by a string. Charge a rubber rod by rubbing it with fur, then hold the rod near the pith ball. The ball is attracted, touches the rod, then flies away after a short time. Use the rod to push the ball around without touching it. Explain that the rod and ball carry the same type charge. Hold the fur near the pith ball and explain that they are oppositely charged.

        2. Repeat using a second pith ball and a wooden rod charged by rubbing it on a plastic sheet (this replaces the traditional glass rod - silk combination and works much better). Place

the two pith balls near each other and explain they are
oppositely charged.

3. Suspend a charged rubber rod by a string. Use another charged
    rubber rod to push it around without touching it. Similarly,
    pull it with the charged wooden rod. Also show that only the
    rubbed end of the rubber rod is charged.

II. Conductors and insulators.

A. Explain the difference between a conductor and an insulator as
    far as the conduction of charge is concerned. Explain that excess
    charge on a conductor is free to move and generally does so when
    influenced by the electric force of other charges. Excess charge
    on a conductor is distributed so the net force on any of it is
    zero. Any excess charge on an insulator does not move far from
    the place where it is deposited. Remind students of the
    demonstration which showed that only the rubbed end of the rubber
    rod remains charged. Metals are conductors. The rubber rod is an
    insulator. Mention semiconductors and superconductors.

B. Use an electroscope to demonstrate the conducting properties of
    conductors. Charge the electroscope by contact with a charged
    rubber rod and explain why the leaves diverge. Discharge it by
    touching the top with your hand. Explain why the leaves converge.
    Recharge the electroscope with a charged wooden rod, then bring
    the charged rubber rod near the electroscope, but do not let it
    touch. Note the decrease in deflection and explain this by
    pointing out the attraction of the charge on the rod for the
    charge on the leaves. Throughout, emphasize the motion of the
    charge through the metal leaves and stem of the electroscope.

C. Demonstrate charging by induction. Bring a charged rubber rod
    near to but not touching an uncharged electroscope. Touch your
    finger to the electroscope, then remove it. Remove the rubber rod
    and note the deflection of the leaves. Bring the rubber rod near
    again and note the decrease in deflection. Observe that the
    electroscope and rod are oppositely charged. Confirm this with
    the wooden rod. Explain the process.

III. Coulomb's law.

A. Assert that experimental evidence convinces us that there are
    only two kinds of charge and that the force between a pair of

charges is along the line joining them, has magnitude proportional to the product of the magnitudes of the charges and is inversely proportional to the square of the distance between them. Further, the force is attractive for unlike charges and repulsive for like charges.

B. Write down Coulomb's law and include a unit vector along the line joining the charges. Give the SI value for ϵ_0 and for $1/4\pi\epsilon_0$. Stress that the law holds for point charges. Note in detail that the mathematical form of the law contains all the qualitative features discussed previously. If Chapter 15 was covered, point out the similarity with Newton's law of gravity and mention that, unlike charge, there is no negative mass.

C. Explain that a superposition law holds for electric forces and illustrate by finding the resultant force on a charge due to two other charges. Use the analogy with the law of gravity to show that the force of one spherical distribution of charge on another obeys the same law as two point charges and that the force on a charge inside a spherical charge distribution is zero.

IV. Quantization and conservation of charge.

A. State that all measured charge is an integer multiple of the charge on a proton: $q = ne$. Give the value of e: 1.60×10^{-19} C. Point out Table 1 and state that the charge on the proton is +e, the charge on the electron is −e, and the neutron is neutral. Go over Sample Problem 2.

B. State that charge is conserved in the sense that for a closed system the sum of all charge before an event or process is the same as the sum after the event or process. Stress that the charges in the sum must have appropriate signs. Example: rubbing a rubber rod with fur. The rod and fur are oppositely charged afterwards and the magnitude of the charge is the same on both. Also discuss the conservation of charge in the annihilation and creation of fundamental particles and note that the identity of the particles may change in an event but charge is still conserved. Examples: beta decay, electron-positron annihilation.

SUPPLEMENTARY TOPIC

The constants of physics. Include this section to add some breadth to the course.

SUGGESTIONS

1. Discuss questions 1, 2, 3, and 5, perhaps in connection with demonstrations or lab experiments.
2. Use questions 7 and 8 to see if students distinguish conductors from insulators.
3. Use questions 12 and 14 to test for qualitative understanding of the decrease in electric force with increasing separation of charges.
4. Question 11 is important for understanding the vanishing of the electrostatic field in a conductor, discussed later.
5. Problems 8, 9, 13 deal with the addition of electric forces in one dimension, problems 10 and 14 deal with the addition of electric forces in two dimensions. All cause students to think about the direction of the force.
6. Demonstrations

    Charging, electroscopes: Freier and Anderson Eal, 2, 11; Hilton Ela - f.

    Electric force: Freier and Anderson Ea5, 6, 8, 12, 15, 17, Eb3, 4, 9, 10, 12, Ec4 - 6.

    Induction: Freier and Anderson Eal2, 13, 14; Hilton Elg.

    Touch a grounded wire to several places within a small area of a wall. Rub a balloon with fur and place it in contact with that area. Ask students to explain why the balloon sticks.
7. Films

    The following film and loop series have material which is pertinent for Chapters 23 through 27. The films are listed in groups rather than divided among the various chapters. Appropriate films for a particular chapter can be selected by title.

    Electric Fields and Moving Media, 16 mm or 3/4" videocassette, color, 32 min. Educational Development Corporation, Distribution Center, 39 Chapel Street, Newton, MA 02160. Both static and time-varying electric fields and their effects on various media are illustrated. Natural phenomena and technological applications are shown. Reviewed SBF 11:104 (1975).

    Introduction to Electrostatics; Insulators and Conductors; Electrostatic Induction; The Electroscope; Charge Distribution: The Faraday Ice-Pail Experiment; Charge Distribution: Concentration and

Point Discharge; The Van de Graaff Generator; The Photoelectric
Effect; Capacitors and Dielectrics; Problems in Electrostatics, S8,
color, 4 min. each. Kalmia Company (see Chapter 4 notes for address).
These loops make up the Electrostatic Series developed by A.E.
Walters of Rutgers University. Large lecture hall apparatus is
utilized to illustrate the various phenomena. Student notes are
included with each film. Parts of this series are reviewed in TPT
12:507, 13:254, 13:371, and 14:58.

Capacitance of Capacitor Combinations – Parallel; Capacitance of
Capacitor Combinations – Series; Charge on the Outside of a
Conductor; Conductors, Insulators, and Capacitors; Coulomb's Law;
Discharging the Electroscope – Conduction and Ionization; Discharging
the Electroscope – the Photoelectric Effect; Electric Field and
Induced Charges; Electrostatic Attraction; Electrostatic Repulsion;
Increasing the Potential of a Capacitor; Polarity; Variation of
Charge with Curvature; A Working Model of a Van de Graaff Generator,
S8, color, 4 min. each. Encyclopaedia Britannica Educational
Corporation (see Chapter 17 for address). This popular Electrostatic
Series produced by Albert Baez illustrates a variety of electrostatic
phenomena and is a good source of demonstrations.

8. Computer program
Basic Concepts of Electricity, Series I: Basic Concepts, Merlan
Scientific Ltd., 247 Armstrong Avenue, Georgetown, Ontario L7G 4X6,
Canada. Apple II+, IIe. Introduction and drill on charging by
rubbing, current in simple circuits, electric potential difference.
Reviewed TPT November 1983.

9. Laboratory
MEOS Experiment 10-2: The Electrostatic Balance. A coulomb torsional
balance is used to find the functional relationship between the
electrostatic force of one small charged ball on another and the
separation of balls. An electrostatic generator is used to charge the
balls.

Chapter 24   THE ELECTRIC FIELD

BASIC TOPICS

I.   The electric field.
  A.   Use a fluid to introduce the idea of a field. The temperature of
       the fluid $T(x,y,z,t)$ is an example of a scalar field and the
       velocity $\underline{v}(x,y,z,t)$ is an example of a vector field. Point out
       that these functions give the temperature and velocity at the
       place and time specified by the dependent variables.
  B.   Explain that charges may be thought of as creating an electric
       field at all points in space and that the field exerts a force on
       another charge, if present. The important questions to be
       answered are: given the charge distribution what is the field?
       given the field what is the force on a charge?
  C.   Define the field as the force per unit charge on a positive test
       charge, in the limit of a vanishingly small test charge. Mention
       that the limiting process eliminates the influence of the test
       charge on the charge creating the field. SI units: N/C.
  D.   Use Coulomb's law to obtain the expression for the field of a
       point charge. Explain that the field of a collection of charges
       is the vector sum of the individual fields.

II. Field lines.
  A.   Explain that field lines are useful for visualizing the field.
       Draw field lines for a point charge and explain that, in general,
       the field at any point is tangent to the line through that point
       and that the magnitude of the field is proportional to the number
       of lines per unit area that cut a surface perpendicular to the
       lines.
  B.   By considering a sphere around a point charge and calculating the
       number of lines per unit area through the sphere, show that the
       $1/r^2$ law allows us to associate lines with a charge and to take
       the number of lines to be proportional to the charge. Explain
       that lines can be thought of as directed, that they originate at
       positive charge and terminate at negative charge.
  C.   Field lines can be illustrated by floating some long seeds in
       transformer oil in a shallow, flat bottomed dish. Place two metal
       plates in the dish and connect them to an electrostatic

generator. The seeds line
up along the field lines.
You can place the apparatus
on an overhead projector
and shadow project the seeds.

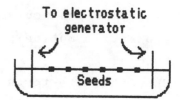

III. Calculation of the electric field.
A. Remind the students of the field of a point charge. Include the unit vector radially outward from the charge. Also remind them that the total field is the vector sum of the individual fields of the charge being considered.
B. Derive an expression for the field of an electric dipole by considering the field of two charges with equal magnitudes and opposite signs. Evaluate the expression in the limit of vanishingly small separation and finite dipole moment. Define the dipole moment and stress that it points from the negative toward the positive charge. Point out that the field is proportional to $1/r^3$ for points far from the dipole.
C. Consider a small set of discrete charges and calculate the electric field by evaluating the vector sum of the individual fields. Example: the field at the center of a square with various charges on its corners.
D. Show how to find the field of a continuous distribution of charge by deriving an expression for the field on the axis of a continuous ring of charge. Carefully explain how the integral is set up and how the vector nature of the field is taken into account by dealing with components.
E. Extend the calculation to find an expression for the field on the axis of a charged disk and for an infinite sheet of charge. This will be useful later when parallel plate capacitors are studied.

IV. Motion of a charge in an electric field.
A. Point out that the electric force on a charge is q$\underline{E}$ and explain that the electric field used is that due to all other charges (except q). Substitute the force into Newton's second law and remind the students that once the acceleration and initial conditions are known, kinematics can be used to find the subsequent motion of the charge.

B. Find the trajectory of a charge moving into a region of uniform field, perpendicular to its initial velocity. Compare to projectile motion problems studied in Chapter 4. See Sample Problem 8.

C. Show that the torque on a dipole in an electric field is $\underline{p} \times \underline{E}$ and that the potential energy of a dipole is $-\underline{p} \cdot \underline{E}$. To review oscillatory rotational motion calculate the angular frequency of small angle oscillations for a dipole with rotational inertia I in an electric field. Assume no other forces act.

SUGGESTIONS

1. Use question 3 to help explore the limit used to define the field in terms of the force on a test charge. Assign problem 2 to illustrate the measurement of an electric field.

2. Center a qualitative discussion of field lines on questions 5 through 8. Have students sketch field lines for various charge distributions. See problems 8, 9, 10.

3. Have students think about question 15 before working problem 20, 21, 22, or 23. These deal with the superposition of fields.

4. Use question 13 to test for understanding of the idea that a charge does not exert a force on itself.

5. Problem 31 is a good test of understanding of the derivation of the dipole field. Also assign problems 56 (torque on a dipole) and 57 (energy of a dipole).

6. To include the Milliken oil drop experiment, assign problems 47 and 48.

7. Computer projects
   Have students use Eureka or write programs to calculate the electric fields of discrete charge distributions. Have them use the programs to plot the magnitude of the field at various distances from a dipole, along lines that are perpendicular and parallel to the dipole moment.
   Have students write programs to trace field lines for discrete charge distributions. See the calculator supplement for details.

8. Demonstrations
   Lines of force: Freier and Anderson Eb1, Ec2 - 4.

9. Computer programs
   Physics Disk 3: Electric Fields and Potentials, The 6502 Program

Exchange (see Chapter 17 notes for address). Apple II. Generates field lines and equipotential surfaces for user supplied distribution of discrete charges. Diagrams can be stored for later display. Chiefly for lecture illustrations. Reviewed TPT September 1986.

Physics Simulations II: Coulomb, Kinko's Service Corporation (see Chapter 4 notes for address). Macintosh. User gives up to 15 charges and their positions, then the program displays electric field lines.

Laboratory Simulations in Atomic Physics, Norwalk High School Science Department, County Street, Norwalk, CT 06851. Apple II. Simulations of the deflection of an electron by an electric field, the Thompson e/m experiment, the Millikan oil drop experiment, and a mass spectrometer. Parameters are selected by the user. Excellent for illustrating lectures. Some parts can be used in connection with this chapter, some in connection with Chapter 30. Reviewed TPT March 1984.

Chapter 25   GAUSS' LAW

BASIC TOPICS

I. Electric flux.
    A. Start by discussing some of important concepts in a general way. Define a vector surface element. Define the flux of a vector field through a surface. Distinguish between open and closed surfaces and explain that for the latter the surface normal is taken to be outward. Interpret the surface integral for the flux as a sum over surface elements. If you covered Chapter 16 use the velocity field of a fluid as an example.
    B. Define electric flux. Point out that it is the normal component of the field which enters. Point out that the sign of the contribution of any surface element depends on the choice for the direction of d$\underline{A}$.
    C. Interpret the flux as a quantity that is proportional to the net number of field lines penetrating the surface. Stress that lines roughly in the same direction as the normal contribute positively to the sum, lines roughly in the opposite direction contribute negatively. Lines which pass completely through a closed surface contribute zero net flux.
    D. By considering surfaces with the same area but different

orientations, show that the net number of penetrating lines is proportional to the cosine of the angle between the lines and the normal to the surface. Conclude that $\underline{E} \cdot d\underline{A}$ is proportional to the number of lines through $d\underline{A}$.

E. As an example, calculate the flux through each side of a cube or cylinder in a uniform electric field.

II. Gauss' law.

A. Write down the law. Stress that the surface is closed and that the charge that appears in the law is the net charge enclosed. Illustrate by considering the surface of a sphere, with positive charge inside, with negative charge inside, with both positive and negative charge inside, and with charge outside. In each case draw representative field lines with the number of lines proportional to the net charge. Stress that the position of the charge inside is irrelevant for the flux through the surface. Also use Gauss' law to calculate the flux.

B. Use Gauss' law and symmetry arguments to obtain Coulomb's law.

C. Argue that the electrostatic field vanishes inside a conductor and use Gauss' law to show that there can be no net charge at interior points, under static conditions. Point out that exterior charge and charges on the surface separately produce fields in the interior but that the resultant field vanishes. Show that the electrostatic field inside a cavity surrounded by a conductor vanishes.

D. Demonstrate that any excess charge on a conductor resides on the exterior surface. Use a hollow metal sphere with a small hole cut in it. As an alternative, solder shut the top on an empty metal can and drill a small hole in it. This will not work as well because of the sharp edges. Charge a rubber rod by rubbing it with fur and touch it to the inside of the sphere. Be careful not to touch the edge of the hole. Repeat several times to build up charge. Now scrape at the interior with a metal transfer rod, again being careful not to touch the edge of the hole. Touch the transfer rod to an uncharged electroscope and note the lack of deflection. Scrape the exterior of the sphere with the transfer rod and touch the electroscope. Note the deflection.

E. Show how to calculate the charge on the inner and outer surfaces of neutral and charged conducting spherical shells when charge is placed in the cavities. See problem 45.

    F.   Show that the field just outside a charged conductor is given by $E = \sigma/\epsilon_0$, where σ is the surface charge density.

III. Applications of Gauss' law.

    A.   Derive expressions for the electric field at various points for a uniformly charged sphere and for a uniformly charged thick spherical shell. See problem 48. Remark that such distributions are possible if the sphere or shell is not conducting.

    B.   Derive an expression for the electric field at a point outside an infinite sheet with a uniform charge distribution.

    C.   Consider a point charge at the center of a neutral spherical conducting shell. Calculate the charges on the inner and outer surfaces. Repeat for for a charged shell. In each case find expressions for the electric field everywhere.

    D.   Note that Gauss' law can be used to find $\underline{E}$ only if there is adequate symmetry.

SUPPLEMENTARY TOPIC

Experimental verification of Gauss' law. Verification of the $1/r^2$ law.

SUGGESTIONS

1.   Use questions 5 and 7 through 10 to help students understand the flux integral and the charge which appear in Gauss' law. Problem 1 provides an example of flux from fluid mechanics. Use it if you covered Chapter 16. Examples from electrostatics are given in problems 3 and 4. These also demonstrate the vanishing of the total flux for a closed surface in a uniform field.

2.   Problems 5 and 6 illustrate the fundamental idea of Gauss' law. Problem 10 is also instructive.

3.   A more detailed discussion of the symmetry arguments used in connection with Gauss' law can be given with the aid of questions 13, 14, 21, 23, and 25.

4.   Use questions 16, 17, 18, and 20 and problems 17, 18, and 19 to discuss the electrostatic properties of conductors.

5.   Assign a variety of problems dealing with applications: 22, 23, 26 (cylinders of charge), 30 (planes of charge), 41, 43, 45 (spheres of

charge).

6. Computer project

    Have students use Eureka or their own computer programs to evaluate
    the flux integral in Gauss' law. Have them evaluate the flux through
    each face of a cube containing a point charge. Consider various
    positions of the charge within the cube to show that the flux through
    individual faces may change as the charge changes position but the
    total flux remains the same and obeys Gauss' law. Repeat for a point
    charge outside the cube.

7. Demonstrations

    Charges on conductors: Freier and Anderson Ea7, 18, 23, Eb7; Hilton
    E1h.

Chapter 26   ELECTRIC POTENTIAL

BASIC TOPICS

I.   Definition and properties.

     A. Define the potential difference between two points as the
         negative of the work per unit charge done by the electric field
         when a positive test charge moves from one point to the other. If
         you covered Chapter 15, use the similarity of Coulomb's law and
         Newton's law of gravity to argue that the electrostatic force is
         conservative, the work is independent of path, and the potential
         has meaning. If you did not cover Chapter 15, either derive or
         state these results. Stress the sign of the potential: the
         potential of the end point is higher than that of the initial
         point if the work is negative. The electric field points from
         regions of high potential toward regions of low potential and
         positive charge tends to be repelled from regions of high
         potential. The region near an isolated positive charge has a
         higher potential than regions far away. The opposite is true for
         a negative charge. Unit: volt. Define electron volt as a unit of
         energy.

     B. Show that the definition is equivalent to $V_B - V_A = -\int \underline{E} \cdot d\underline{s}$, where
         the integral is along a path from A to B. Point out that the
         potential is constant in regions of zero field. Note that N/C is
         the same as V/m and the latter is a more common unit for $\underline{E}$.

C. Point out that the potential is a scalar and that only potential differences are physically meaningful. One point can be chosen arbitrarily to have zero potential and the potential at other points is measured relative to the potential there. Often the potential is chosen to be zero where the field (or force) is zero. For a finite distribution of charge the potential is usually chosen to be zero at a point far away (infinity). Show a voltmeter and remark that the meter reads the potential difference between the leads.

D. As a first example, Consider a uniform electric field, like that outside a uniform plane distribution of charge, and show that potential is given by $-Ex + C$, where C is a constant. Since the distribution is infinite the point at infinity cannot be picked as the zero. As a more complicated example, consider one of the configurations discussed in the last chapter, a point charge at the center of a spherical conducting shell, say. Take the potential to be zero at infinity and compute its value at points outside the outer surface, inside the shell, and inside the inner surface.

E. Define the term equipotential surface. Show diagrams of equipotential surfaces for an isolated point charge and for the region between two uniformly charged plates. Point out that the field does zero work if a test charge is carried between two points on the same surface and note that this means that the force, and hence $\underline{E}$, is perpendicular to the equipotential surfaces. Note further that the work done by the field when a charge is carried from any point on one surface to any point on another is the product of the charge and the negative of the potential difference.

II. Calculation of the potential.
  A. Derive the expression for the potential due to a point charge, with the zero at infinity. Point out that the sign of the potential depends on the sign of the charge. Verify that the equipotential surfaces are spheres centered on the charge.

  B. Explain that the potential due to a distribution of charge is the sum of the potentials due to the individual charges. Derive an expression for the potential due to a dipole. Start with two charges at arbitrary separation and take the limit of zero separation but finite dipole moment. Write down an expression for the potential due to a small number of charges (at corners of a triangle or square, for example).

C. Remark that the calculation involves an integral over the charge distribution if the distribution is continuous. Derive expression for the potentials due to a charged ring and a disk.

III. Calculation of $\underline{E}$ from V.

A. Remind students that $\Delta V = -E\Delta x$ for a uniform field in the positive x direction. Note that E has the form $-\Delta V/\Delta x$ and $\underline{E}$ is directed from high to low potential. Use this result to reenforce the idea of an equipotential surface and the fact that $\underline{E}$ is perpendicular to equipotential surfaces.

B. Generalize the result to $E = -dV/ds$, where s is the distance along a normal to an equipotential surface. Then broaden this further to $\underline{E} = -\triangledown V$, expressed in component form. Verify that the prescription works for a point charge and for a dipole.

IV. Electric potential energy.

A. Remark that the potential energy of a charge Q in an electric field is QV, where V is the potential at the position of Q due to other charges. Both U and V are taken to be 0 when Q is far from the sources of the field.

B. Explain that the potential energy of a system of charges is the work an agent must do to assemble the system from infinite separation. This is the negative of the work done by the field. When charge Q is brought into position from infinity (where the potential is zero), the potential energy changes by QV, where V is the potential at the final position of Q due to charge already in place. Argue that the potential energy of a system of discrete charges is $\frac{1}{2}\Sigma QV$.

C. Calculate the potential energy of a simple system: charges at the corners of a triangle or square. First assume the charges are brought in from infinity one at a time and sum the potential energies, then use $\frac{1}{2}\Sigma QV$.

D. Remind students that potential energy can be converted to kinetic energy. Explain what happens if the charges used in the last example are released from their positions. Consider a proton fired directly at a heavy nucleus with charge Ze and find the distance of closest approach in terms of the initial speed.

IV. An insulated conductor.
   A. Recall that the electric field vanishes at points in the interior of a conductor. Argue that the surface must be an equipotential surface and that V at all points inside must have the same value as on the surface. State this is true whether or not the conductor is charged and whether or not an external field exists.
   B. Consider two spherical shells of different radii, connected by a very fine wire. Explain that $V_1 = V_2$ and show that $q_1/R_1 = q_2/R_2$, then show that the surface charge density varies inversely with the radius: $\sigma_1/\sigma_2 = R_2/R_1$. Recall that E is proportional to σ just outside a conductor and argue that σ and E are large near places of small radius of curvature and small near places of large radius of curvature. Use an electrostatic generator to show discharge from a sharp point and from a rounded (larger radius) ball. Discuss the function of lightning rods and explain their shape.

SUPPLEMENTARY TOPICS

The electrostatic generator. Explain how it works. This might be done in lab if they are used there. Spend some time explaining safety precautions.

SUGGESTIONS

1. Questions 1, 2, 7, 8, and 12 can be used to help students think about the arbitrariness of the zero of potential.
2. The relationship between the electric field and potential is explored in questions 13, 14, 15, and 18. Use them to test students' understanding. Problem 14 is a good test of understanding the scalar nature of the potential.
3. Assign some problems which ask for the potential of a given charge distribution: 8 (parallel plates), 9 (distance between equipotential surfaces), 13, 25, and 29 (spherical charge distribution), 28 (superposition of potentials due to point charges).
4. Use questions 16 and 18 to aid in a discussion of the field and potential of a conductor. Assign problems 63, 65, and 70.
5. Use problems 42, 43, and 53 in connection with the discussion of

electrostatic potential energy.

6. Computer project
    Have students use Eureka or their own root finding program to plot
    equipotential surfaces for a discrete charge distribution. It is
    instructive to consider two unequal charges (any combination of
    signs). See the calculator supplement for details.

7. Demonstrations
    Electrostatic generators: Freier and Anderson Ea22, Ec1; Hilton Eli,
    j.

8. Computer program
    Physics Disk 3: Electric Fields and Potentials. See Chapter 24 notes.

9. Laboratory
    MEOS Experiment 10-1: Electric Fields and BE Experiment 25: Mapping
    of Electric Fields. Students map equipotential lines on sheets of
    high resistance paper with metallic electrodes at two sides. In the
    MEOS experiment an audio oscillator generates the field and an
    oscilloscope or null detecting probe is used to find points of equal
    potential. If students are not familiar with oscilloscopes you might
    want to preface this experiment with Part A of MEOS Experiment 10-10.
    In the BE experiment the field is generated by a battery and a
    galvanometer is used as a probe.

Chapter 27   CAPACITANCE

BASIC TOPICS

I. Capacitance
    A. Describe a generalized capacitor. Draw a diagram showing two
        separated, insulated conductors. Assume they carry charge q and
        -q respectively, draw representative field lines, and point out
        that all field lines start on one conductor and terminate on the
        other. Explain that there is a potential difference between the
        conductors. Show a radio tuning capacitor and some commercial
        fixed capacitors.
    B. Define capacitance as $C = q/V$. Explain that V is proportional to
        q and that C is independent of q and V. C does depend on the
        shapes, relative positions, and orientations of the conductors
        and on the medium surrounding them. Point out that the positively

charged conductor is at the higher potential. Unit: 1 farad = 1 C/V. Mention that one usually encounters μF and pF capacitors. Capacitors on the order of 1 F have been developed for the electronics industry.

C. Remark that in circuit drawings a capacitor is denoted by $-||-$.

D. Show how to calculate capacitance in principle. Put charge q on one conductor, -q on the other, and calculate the electric field due to the charge, then the potential difference between the conductors. Except for highly symmetric situations, the charge is not uniformly distributed over the surfaces of the conductors and fairly sophisticated means must be used to calculate V. The text deals with symmetric situations for which Gauss' law can be used to calculate the electric field.

E. Examples: derive expressions for the capacitance of two parallel plates (neglect fringing) and two coaxial cylinders. Use Gauss' law to find the electric field, then evaluate the integral for the potential difference. Emphasize that the field is due to the charge on the plates.

F. Explain how the equivalent capacitance of a device can be measured. Consider a black box with two terminals. State that a potential difference V is applied the total charge q deposited is measured, from the first application of V to the time charge stops flowing. The capacitance is q/V.

G. If you want to demonstrate a parallel plate capacitor, one is available from Leybold. You can also make one using two $\approx$1 ft diameter circular plates of 1/8 inch aluminum sheet. Attach an aluminum disk to the center of each with a hole drilled for a support rod. Use an insulating rod on one and a metal rod on the other. By sliding the two conductors closer together, you can show the effect of changing d while holding q constant. An electroscope serves as a voltmeter.

H. Derive $1/C = 1/C_1 + 1/C_2$ for the effective capacitance of two capacitors in series and $C = C_1 + C_2$ for the effective capacitance of two capacitors in parallel. Emphasize that two capacitors in parallel have the same potential difference, two in series have the same charge. Explain the usefulness of these equations for circuit analysis.

II. Energy storage.

A. Derive the expression $W = \frac{1}{2}q^2/C$ for the work required to charge a capacitor. Explain that, as an increment of charge is transferred, work is done by an external agent (battery, for example) against the electric field of the charge already on the plates. Show that this expression is equivalent to $W = \frac{1}{2}CV^2$. Interpret the result as the potential energy stored in the charge system and explain that it can be recovered when the capacitor is discharged. Remark that if two capacitors are in parallel the larger one stores the greater energy. If two capacitors are in series the smaller one stores the greater energy.

B. Show that the energy density in a parallel plate capacitor is $\frac{1}{2}\epsilon_0 E^2$. State that this result is quite general and that its volume integral gives the work required to assemble charge to create the electric field E. Explain that the energy may be thought to reside in the field or it may be considered the potential energy of the charges. Calculate the energy stored in the electric field of a charged spherical conductor (see Sample Problem 7).

III. Dielectrics (optional).

A. Explain that when the region between the conductors of a capacitor is occupied by insulating material the capacitance increases by a factor $\kappa > 1$, called the dielectric constant of the material. Remark that $\kappa = 1$ for a vacuum.

B. Use a large commercial or home-made capacitor to show the effect of a dielectric. Charge the capacitor, then isolate it and insert a glass plate between the plates. The electroscope shows that V decreases and, since q is fixed, the capacitance increases.

C. Calculate the change in stored energy that occurs when a dielectric slab is inserted between the plates of an isolated parallel plate capacitor. (see Sample Problem 8). Also calculate the change in stored energy when the slab is inserted while the potential difference is maintained by a battery. Explain that the battery now does work in moving charge from one plate to the other.

D. Explain that dielectric material between the plates becomes polarized, with the positive charged ends of the dipoles attracted toward the negative conductor. This reduces the potential difference between the conductors for a given charge on them. The field of the dipoles opposes the external field, so the electric field is weaker between the plates than it would be if the material were not there.

E.   Show that if the polarization is uniform, the material behaves like neutral material with charge on its surfaces.

F.   Optional: Show how Gauss' law can be written in terms of $\kappa \underline{E}$ and the free charge. Show how to compute the polarization charge for a parallel plate capacitor with dielectric material between its plates.

SUGGESTIONS

1.   Use questions 1, 2, and 4 in the discussion about charging a capacitor.

2.   Use questions 3, 6, and 11 to emphasize the dependence of capacitance on geometry.

3.   To test understanding of induced polarization charge, ask question 12.

4.   The fundamental idea of capacitance is illustrated by problem 2. Assign problem 8 to have students compare spherical and plane capacitors. Problem 10 covers the dependence of the capacitance of a parallel plane capacitor on area and separation.

5.   Problem 25 is a good test of understanding of the derivation of the equations for parallel and series combinations of capacitors. Problems 22 and 23 cover equivalent capacitance, charge, and potential difference for series and parallel combinations. Also consider assigning some problems in which students must find the equivalent capacitance of various combinations. See problems 19 and 20, for example.

6.   Problems 40 covers most of the important points discussed in connection with energy storage. Also assign problem 42, which deals with the energy density around a point charge and problem 49, which deals with the energy needed to separate the plates of a parallel plate capacitor.

7.   Include questions 18 and 19 in the discussion of the influence of a dielectric on capacitance.

8.   Films
     Capacitor I: Voltage and Force, Capacitor II: Dipoles and Dielectrics, S8, color, 4 min. each. Walter de Gruyter, Inc. (see

Chapter 17 notes for address). Shows the relationship of the voltage on a capacitor to the plate separation and uses a beam balance to show the force-voltage relationship. Reviewed AJP 45:1014 (1977).

9. Demonstrations

    Charge storage: Freier and Anderson Eb8, Ed3, 7; Hilton E4b.

    Capacitance and voltage: Freier and Anderson Ed1; Hilton E4c, d.

    Energy storage: Freier and Anderson Ed8

    Dielectrics: Freier and Anderson Ed2, 4.

10. Laboratory

    MEOS Experiment 10-7 (Part B): <u>Measuring Capacitance with a Ballistic Galvanometer</u>. A ballistic galvanometer is used to measure the capacitance of individual capacitors and capacitors in series and parallel. Students must temporarily accept on faith that the deflection of the galvanometer is proportional to the total charge which passes through it.

    MEOS Experiment 11-2 (Part C): <u>Coulomb Balance Attachment (to the current balance)</u>. Students use gravitational force to balance the force of one capacitor plate on the other. The voltage and plate separation are used to find the charge on the plates, then ϵ_0 is calculated.

Chapter 28   CURRENT AND RESISTANCE

<u>BASIC TOPICS</u>

I. Current and current density.

    A. Explain that electric current is moving charge. Draw a diagram of a long straight wire with positive charge moving in it. Consider a cross section and state that the current is dq/dt if charge dq passes the cross section in time dt. Give the sign convention: both positive charge moving to the right and negative charge moving to the left constitute currents to the right. Early on it is good to use the words "conventional current" quite often. Later "conventional" can be dropped. Many high school courses now take the current to be in the direction of electron flow and it worthwhile making the effort to reduce confusion in students' minds. Unit: 1 ampere = 1 C/s.

    B. Explain that the current is the same for every cross section

under steady state conditions. Steady state means no charge is
building up or being depleted anywhere in the wire. Remark that
current is a scalar, but arrows are used to show the direction of
charge flow.

C. Explain that current is produced when charge is free to move in
an electric field. For most materials it is the negative
electrons which move and their motion is opposite to the
direction of the field. State that conventional current flow is
in the direction of the electric field.

D. Distinguish between the drift velocity and the velocities of
individual charges. Note that the drift velocity of electrons in
an ordinary wire is zero unless an electric field is turned on.

E. Explain that current density is a microscopic quantity that
describes current flow at a point. Use the same diagram but now
consider a small part of the cross section and state that $J = i/A$
as the area diminishes to a point. Derive $\underline{J} = ne\underline{v}_d$ and explain
that $i = \int \underline{J} \cdot d\underline{A}$ is the current through a finite surface. This
reduces to $J = iA$ for uniform current density.

F. Show how to calculate the drift speed from the free electron
concentration and current in the wire, assuming uniform current
density.

II. Resistance and resistivity.
A. Define resistance by $R = V/i$ and point out that R may depend on
V. Unit: 1 ohm = 1 V/A. Also define resistivity and conductivity.
Point out Table 1. Explain that the latter quantities are
characteristic of the material while resistance also depends on
the sample shape and the positions of the current leads. Make a
sketch similar to the one shown on the next page. Indicate
that $V_a - V_b = iR$ is
algebraically correct and
effectively defines the
resistance of that sample with
the leads connected at a and b.
Emphasize that the point at
which the current enters is iR
higher in potential than the
point at which it leaves.

B. Show that $R = \rho L/A$ for a conductor with uniform cross section A and length L, carrying a current that is uniformly distributed over the cross section.

C. Point out that for many samples the current is proportional to the potential difference and the resistance is independent of the voltage applied. These materials obey Ohm's law. Also point out that many important materials do not obey Ohm's law. Show Fig. 11. Use the circuit shown at the right and connect, in turn, samples of ohmic (carbon resistor) and non-ohmic material (solid state diode) across a-b. Use analog meters and vary the supply smoothly and fairly rapidly. For the ohmic material it will be apparent that i is proportional to V, while for the non-ohmic material it will be apparent that i is not proportional to V.

D. Remark that the resistivity of a sample depends on the temperature. Define the temperature coefficient of resistivity and point out the values given in Table 1.

E. Give a qualitative description of the mechanism which leads to Ohm's law behavior. Explain that collisions with atoms cause the drift velocity to be proportional to the applied field. Assume the electrons have zero velocity after each collision and that they accelerate for a time τ between collisions. Show that an electron goes the same distance on the average during the first five collisions as it does during the second five so the drift velocity is proportional to the field even though the electron accelerates between collisions. Now consider the quantitative aspects: derive the expression for the drift velocity in terms of $\underline{E}$ and the mean free time τ, then derive $\rho = m/ne^2\tau$. Point out that a long mean free time means a small resistivity because the electrons accelerate for a longer time between collisions and thus have a higher drift speed.

III. Energy considerations.

A. Point out that when current flows from the high to the low potential side of any device, energy is transferred from the current to the device at the rate P = iV. Draw the circuit shown

and note that $P = i(V_a - V_b)$
is algebraically correct
if P is the power supplied
to the device. Note that
if P is negative the device
is supplying energy at the
rate $-P$.

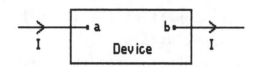

B. Give examples: Energy may be converted to mechanical energy (a motor), to chemical energy (a charging battery), to thermal energy (a resistor). Note also the converse. Mechanical energy (a generator), chemical energy (a discharging battery), and thermal energy (a thermocouple) may be converted to electrical energy.

C. Explain that in a resistor the electrons lose energy in collisions with atoms and this increases the thermal motion of the atoms. Show that the energy loss in a resistor is given by $P = i^2R = V^2/R$.

SUPPLEMENTARY TOPICS

1. Semiconductors
2. Superconductors

Both topics are important for modern physics and technology. Say a few words about them if you have time and encourage students to read on their own.

SUGGESTIONS

1. Use question 2 to expand the discussion of the current sign convention. Also assign problem 10.

2. Use questions 5, 9, 11, and 12 in the discussion of Ohm's law.

3. Definitions are covered in problems 2 (current), 5 (current density), and 6 (drift speed).

4. The dependence of resistance on length and cross section is emphasized in problems 27 and 28.

5. As part of the coverage of energy dissipation by a resistor, ask students to think about questions 21, 23, and 24 in connection with problems 47 and 48.

6. Demonstrations

    Model of resistance: Freier and Anderson Eg1.

    Thermal dissipation by resistors: Freier and Anderson Eh3.

Fuses: Freier and Anderson Eh5.

Ohm's law: Freier and Anderson Eg2, Eol; Hilton E2c.

Measurement of resistance, values of resistance: Freier and Anderson Eg3, 6; Hilton E3b.

Temperature dependence of resistance: Freier and Anderson Eg4, 5.

7. Laboratory

MEOS Experiment 10-3: <u>Electrical Resistance</u>. An ammeter and a voltmeter are used to find the resistance of a light bulb and wires of various dimensions, made of various materials. The dependence of resistance on length and cross section is investigated. Resistivities of the substances are calculated and compared.

BE Experiment 27: <u>Methods of Measuring Resistance</u>. Two voltmeter - ammeter methods and a Wheatstone bridge method are used to measure resistance and to check the equivalent resistance of series and parallel connections.

BE Experiment 29: <u>A Study of the Factors Affecting Resistance</u>. A Wheatstone bridge and a collection of wire resistors is used to investigate the dependence of resistance on length, cross section, temperature, and resistivity.

MEOS Experiment 10-8: <u>Temperature Coefficient of Resistors and Thermistors</u>. A Wheatstone bridge is used to measure the resistances of a resistor and thermistor in a water filled thermal reservoir. The temperature is changed by an immersion heater. Students see two different behaviors. A voltmeter-ammeter technique can replace the bridge if desired.

Also see MEOS Experiment 9-2 and BE Experiment 30, described in the Chapter 20 notes. These experiments can be revised to emphasize the power dissipated by a resistor. In several runs the students measure the power dissipated for different applied voltages.

Chapter 29   ELECTROMOTIVE FORCE AND CIRCUITS

BASIC TOPICS

I. Seats of emf.

A. Explain that a seat of emf moves positive charge from its negative to its positive terminal or negative charge in the opposite direction and maintains the potential difference between its terminals. Seats of emf are used to drive currents in circuits. Example: a battery <u>contains</u> a seat of emf, in series with an internal resistance. Note the symbol —|ı— used in circuit diagrams.

B. Explain that a direction is associated with an emf and that it is from the negative to the positive terminal, inside the seat. This is the direction current would flow if the seat acted alone in a completed circuit. Point out that when current flows in this direction the seat does positive work on the charge and define the emf as the work per unit positive charge: $\varepsilon = dW/dq$. Unit: volt. Point out that the rate at which energy is supplied by the seat is $i\varepsilon$. Also point out that the positive terminal of an ideal seat is ε higher in potential than the negative terminal.

II. Single loop circuits.

A. Consider a circuit containing a single seat of emf and a single resistor. Use energy considerations to derive the steady state circuit equation: equate the power supplied by the emf to the power loss in the resistor.

B. Derive the circuit equation by picking a point on the circuit, selecting the potential to be zero there, then traversing the circuit and writing down expressions for the potential at points between the elements until the zero potential point is reached again. Tell the students that if the current is not known a direction must be chosen for it and used to determine the sign of the potential difference across the resistor. When the circuit equation is solved for i, a negative result will be obtained if the wrong direction was chosen. As you carry out the derivation remind students that current enters a resistor at the high potential end and that the positive terminal of a seat of emf is at a higher potential than the negative terminal, regardless of the direction of the current.

C. Consider slightly more complicated single loop circuits. Include the internal resistance of the battery and solve for the current. Place two batteries in the circuit, one charging and the other discharging. Once the current is found, calculate the power gained or lost in each element.

D. For the circuits considered, show how to calculate the potential difference between two points on the circuit and point out that the answer is independent of the path used for the calculation. Explain the difference between the closed and open circuit potential difference across a battery.

III. Multiloop circuits.
A. Explain the junction theorem for steady state current flow. Explain that it follows from the conservation of charge and the fact that charge does not build up anywhere when the steady state is reached.
B. Using an example of a two loop circuit, go over the steps used to write down the loop and junction equations and to solve for the currents. Explain that if the current directions are unknown, an arbitrary choice must be made in order to write the equations and that if the wrong choice is made, the values obtained for the current will be negative. Warn students not to write duplicate junction equations. Point out that the total number of equations will be the same as the number of different currents and that each current must appear at least once in at least one loop equation.
C. Derive expressions for the equivalent resistance of two resistors in series and in parallel. Contrast with the expressions for the equivalent capacitance of two capacitors in series and in parallel.

IV. RC circuits.
A. Consider a series circuit consisting of an emf, a resistor, a capacitor, and a switch. Suppose the switch is closed at time $t = 0$ with the capacitor uncharged. Use the loop theorem and $i = dq/dt$ to show that $R(dq/dt) + (q/C) = \epsilon$. By direct substitution, show that $q = C\epsilon[1 - e^{-t/RC}]$ satisfies this equation and yields $q = 0$ for $t = 0$. Also find expressions for the potential differences across the capacitor and across the resistor. Show that $q = C\epsilon$ for times long compared to RC.
B. Explain that RC is called the time constant for the circuit and that it is indicative of the time required to charge the capacitor. If RC is large the capacitor takes a long time to charge. Show that $q/C\epsilon \approx 0.63$ when $t = RC$.
C. Derive the loop equation for a series circuit consisting of a capacitor and resistor. Suppose the capacitor has charge q_0 at

time t = 0 and show that $q = q_0 e^{-t/RC}$. Again find expressions for the potential differences across the capacitor and resistor. Point out that RC is now indicative of the time for discharge.

SUPPLEMENTARY TOPIC

Electrical measuring instruments. This material can be covered, as needed, in conjunction with the laboratory.

SUGGESTIONS

1.   Use questions 1, 4, 6, and 7 in the discussion of emfs and batteries. Problem 1 covers the fundamental idea of emf.

2.   Open and closed circuit terminal voltage of a battery are covered in problem 17.

3.   Use question 6 and problems 6 and 8 in the discussion of energy dissipation by a resistor. The dissipation of energy originally stored in a capacitor is investigated in problem 74. Assign it if you cover RC circuits.

4.   Assign some problems dealing with multiloop circuits. See problems 29, 30, 34, 42, 50.

5.   Demonstrations
     Seats of emf: Freier and Anderson Ee2, 3, 4; Hilton E3f.
     Measurement of emf: Freier and Anderson Eg7.
     Resistive circuits: Freier and Anderson Eh1, 2, 4, Eo2 - 8; Hilton E2b, E3a, c, d, g.

6.   Computer programs
     Circuit Lab, Mark Davids, 21825 O'Conner, St. Clair Shores, MI 48080. Atari 800 and 800XL, Apple II. One of four basic circuits can be selected. Light bulbs, switches, resistors, ammeters, and voltmeters are placed in the circuit by the user, who also selects values for the circuit elements. Ammeters and voltmeters then show correct values. Use as a drill or to illustrate circuits in lectures. Reviewed TPT April 1986.
     Basic Electricity, Programs for Learning, Inc., P.O. Box 954, New Milford, CT 06776. Apple II. Drill on circuits containing batteries and resistors. Reviewed TPT April 1984.

7.   Laboratory
     MEOS Experiment 10-7 (Part A): Measuring Current with a d'Arsonval

Galvanometer. Students determine the characteristics and sensitivity of a galvanometer. To expand this lab, ask the students to design an ammeter and a voltmeter with a full scale deflections prescribed by you. Students practice circuit analysis while trying to understand design considerations.

MEOS Experiment 10-9: The emf of a Solar Cell. Students study a slide wire potentiometer and use it to measure the emf of a solar cell. Another experiment which gives them practice in circuit analysis.

BE Experiment 28: Measurements of Potential Difference with a Potentiometer. Students study a slide wire potentiometer and use it to investigate the emf and terminal voltage of a battery and the workings of a voltage divider.

BE Experiment 26: A Study of Series and Parallel Electric Circuits. Students use ammeters and voltmeters to verify Kirchoff's laws and investigate energy balance for various circuits. They also experimentally determine equivalent resistances of resistors in series and parallel. This experiment can be extended somewhat by having them consider a network of resistors which cannot be reduced by applying the rules for series and parallel resistors.

BE Experiment 31: Circuits Containing More Than One Potential Source. Similar to Experiment BE 26 described above but circuits with more than one battery are considered. The two experiments can be done together, if desired.

MEOS Experiment 10-4: The R-C Circuit. Students connect an unknown resistor to a known capacitor, charged by a battery. The battery is disconnected and a voltmeter and timer are used to measure the time constant. The value of the resistance is calculated. In a second part an unknown capacitor is charged by means of a square wave generator and the decay is monitored on an oscilloscope. Again the time constant is measured, then it is used to calculate the capacitance. A third part explains how to use a microprocessor to collect data. Also see BE Experiment 32: A Study of Capacitance and Capacitor Transients.

Chapter 30   THE MAGNETIC FIELD

BASIC TOPICS

I.   Definition of the field and force on a moving charge.

   A.   Explain that moving charges create magnetic fields and that a
        magnetic field exerts a force on a moving charge. Both the field
        of a moving charge and the force exerted by a field depend on the
        velocity of the charge involved. The latter property
        distinguishes it from an electric field.

   B.   Define the magnetic field: the force on a moving test charge is
        $q_0 \underline{v} \times \underline{B}$ in the absence of an electric field. Review the rules for
        finding the magnitude and direction of a vector product. Point
        out that the force must be measured for at least two directions
        of $\underline{v}$ since the component of $\underline{B}$ along $\underline{v}$ cannot be found from the
        force. Units: 1 tesla = 1 N/A·m, 1 gauss = 10^{-4} T. Point out the
        magnitudes of the fields given in Table 1.

   C.   Explain that the magnetic force on any moving charge is $\underline{F}_B =$
        $q\underline{v} \times \underline{B}$. Point out that the force is perpendicular to both $\underline{v}$ and $\underline{B}$
        and is zero for $\underline{v}$ parallel or antiparallel to $\underline{B}$. Also point out
        that the direction of the force depends on the sign of q. Remark
        that the field can not do work on the charge and so cannot change
        its speed or kinetic energy. A magnetic field can be used to
        produce a centripetal force and can cause a charge to move in a
        circular orbit.

   D.   To show a magnetic force qualitatively, slightly defocus an
        oscilloscope so the central spot is reasonably large. Move a bar
        magnet at an angle to the face of the scope and note the movement
        of the beam.

   E.   Point out that the total force on a charge is $q(\underline{E} + \underline{v} \times \underline{B})$ when
        both an electric and a magnetic field are present.

II. Magnetic field lines.

   A.   Explain that field lines can be associated with a magnetic field.
        At any point the field is tangent to the line through that point
        and the number of lines per unit area that pierce a plane
        perpendicular to the field is proportional to the magnitude of
        the field.

   B.   To show field lines project Fig. 7 or place a sheet of clear plastic over a bar magnet and place iron filings on the sheet. Place the arrangement on an overhead projector. Explain that the filings line up along field lines.

C. Point out that magnetic field lines form closed loops; they continue into the interior of the magnet, for example. Contrast with electric field lines and remark that no magnetic charge has yet been found. Mention that magnetic field lines would start and stop at magnetic monopoles, if they exist. Remark that lines enter at the south pole of a magnet and exit at the north pole.

III. Charges in magnetic fields.

A. Derive $v = E/B$ for the speed of a charge passing through a velocity selector.

B. Outline the Thompson experiment and derive Eq. 12 for the mass to charge ratio.

C. Show how the Hall effect can be used to determine the sign and concentration of charge carriers in a conductor. Mention that these measurements are important for the semiconductor industry. Also mention that the Hall effect is used to measure magnetic fields. Show a Hall effect teslameter.

D. Consider a charge with velocity perpendicular to a constant magnetic field. Show that the orbit radius is given by $r = mv/qB$ and the period of the motion is $T = 2\pi m/qB$ (independently of v) for non-relativistic speeds. Remark that the orbit is a helix if the velocity of the charge has a component along the field. Mention that cyclotron motion is used in cyclotrons and synchrotrons.

IV. Force on a current loop.

A. Run a flexible non-magnetic wire near a strong permanent magnet. Observe that the wire does not move. Turn on a power supply so about 1 A flows in the wire and watch the wire move. Remark that magnetic fields exert forces on currents.

B. Consider a thin wire carrying current, with all charges moving with the drift velocity. Start with the force on a single charge and derive $d\underline{F} = id\underline{L} \times \underline{B}$ for an infinitesimal segment and $\underline{F} = i\underline{L} \times \underline{B}$ for a finite straight segment. Stress that $d\underline{L}$ and $\underline{L}$ are in the direction of the current.

C. Consider an arbitrarily

shaped segment of wire in
a uniform field. Show that
the force on the segment
between a and b is $\underline{F} = i\underline{L} \times \underline{B}$,
where $\underline{L}$ is the vector joining
the ends of the segment. This
expression is valid only if the field is uniform.

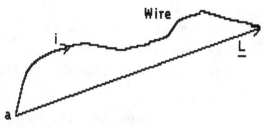

D. Point out that the force on a closed loop in a uniform field is
zero since $\underline{L} = 0$.

E. Calculate the force of a uniform field on a semicircular loop of
wire, in the plane perpendicular to $\underline{B}$. See Sample Problem 7. Do
this as in the text, then repeat using the result given in B
above.

V. Torque on a current loop.

A. Calculate the torque exerted by a uniform field on a rectangular
loop of wire arbitrarily oriented with two opposite sides
perpendicular to $\underline{B}$. See Fig. 26.

B. Define the magnetic dipole moment of a current loop (μ = NiA) and
give the right hand rule for determining its direction. For a
rectangular loop in a uniform field, show that $\underline{\tau} = \underline{\mu} \times \underline{B}$. State
that the result is generally valid for any loop in a uniform
field. Mention that other sources of magnetic fields, such as bar
magnets and the earth, have dipole moments.

C. Note that this is a restoring torque and that if the dipole is
free to rotate it will oscillate about the direction of the
field. If damping is present it will line up along the field
direction. Remark that this is the basis of magnetic compasses.

D. Explain how analog ammeters and voltmeters work. To demonstrate
the torque on a current carrying coil, remove the case from a
galvanometer and wire it to a battery and resistor so that it
fully deflects.

E. Show that the magnetic potential energy of a dipole is given by U
= $-\underline{\mu} \cdot \underline{B}$. Find the work required to turn a dipole through 90° and
180°, starting with it aligned along the field. Point out that U
is a minimum when $\underline{\mu}$ and $\underline{B}$ are parallel and is a maximum when they
are antiparallel.

SUGGESTIONS

1. Use questions 2, 5, and 6 to help in understanding the definition of the magnetic field.

2. Use questions 1 and 3 to discuss the magnetic force on a moving charge.

3. Use questions 13, 14, and 17 to include more detail in the discussion of the magnetic force on a current carrying wire.

4. Use question 9 to test for understanding of the motion of charges in a magnetic field.

5. The dependence of magnetic force on velocity and charge is emphasized in problems 2 and 3. Crossed electric and magnetic fields, used as a velocity filter, are explored in problems 10, 12, and 13.

6. Use problems 14, 15, and 16 to help students study the Hall effect.

7. Problems 23, 28, 29 are useful in the study of charges circulating in a magnetic field. Problems 31, 32, and 33 deal with a mass spectrometer.

8. Use problems 45 and 47 to stress the importance of the angle between the magnetic field and the current carrying wire on which it exerts a force. Problems 48 and 50 make use of kinematics and dynamics.

9. Use problems 53 and 58 to emphasize that the force of a uniform magnetic field on a loop is zero. Magnetic torques on loops are explored in problems 54, 56, and 61.

10. Use problem 62 in connection with magnetic dipoles.

11. Computer project

    Have students use numerical integration of Newton's second law to investigate the orbits of charges in magnetic and electric fields. See Chapter 15 of the calculator supplement for sample problems.

12. Films

    Because Chapters 30 and 31 are highly interrelated, the following films may be used profitably with either chapter.

    The Magnetic Field; The Field from a Steady Current; Field vs. Current; Uniform and Non-Uniform Fields, S8, color, 3 min. each. The Kalmia Company (see Chapter 4 notes for address). These loops are part of the Adler series produced at MIT.

    Magnetic Fields and Electric Currents, I, 16 mm, color, 14.5 min. BFA Educational Media, Division of Phoenix Films, 468 Park Avenue, New York, NY 10016. Various phenomena associated with magnetic fields are presented. Includes a wire carrying current and the magnetic properties of iron.

<u>Magnetic Fields and Electric Currents, II</u>, 16 mm, color, 12.5 min. BFA Educational Media (see address above). A companion film to the one listed above, it deals with the interaction of magnetic fields and illustrates the simple electric motor.

13. Demonstrations

Magnets and compasses: Hilton E6a, b, c, d.

Force on an electron beam: Freier and Anderson Ei18, Ep8, 11.

Forces and torques on wires: Freier and Anderson Ei7, 12, 13 – 15, 19, 20; Hilton E7a, b(1), c.

Meters: Freier and Anderson Ej1, 2.

Hall effect: Freier and Anderson Ei16.

14. Computer programs

<u>Charged Particle Workshop</u>, High Technology Software Products (see Chapter 14 notes for address). Apple II. Shows trajectories of charged particles in a uniform electric field, a uniform magnetic field, and crossed electric and magnetic fields. Velocity components can be displayed. Can be used to illustrate lectures.

<u>Laboratory Simulations in Atomic Physics</u>. See Chapter 24 notes.

15. Laboratory

BE Experiment 33: <u>A Study of Magnetic Fields</u>. A small magnetic compass is used to map field lines of various permanent magnets, a long straight current carrying wire, a single loop of current carrying wire, a solenoid, and the earth. Parts of this experiment might be performed profitably in connection with Chapter 31.

MEOS Experiment 11–3: <u>Determination of e/m</u>. See Chapter 31 notes.

MEOS Experiment 11–5: <u>The Hall Effect</u>. Students measure the Hall voltage and use it to calculate the drift speed and carrier concentration for a bismuth sample. The influence of the magnetic field on the Hall voltage is also investigated. Values of the magnetic field are given to them by the instructor.

132

Chapter 31   AMPERE'S LAW

BASIC TOPICS

I.   Magnetic field of a current.
    A.   Place a magnetic compass near a wire. Turn the current on and
         off, reverse the current. Note the deflection of the compass
         needle and remark that the current produces a magnetic field.
    B.   Write the Biot-Savart law for the field produced by an
         infinitesimal segment of a current carrying wire. Give the value
         for μ_0. Draw a diagram to show the direction of the current, the
         displacement vector from the segment to the field point, and the
         direction of the field. Explain that d$\underline{B}$ is in the direction of
         d$\underline{L}$x$\underline{r}$. Point out the angle between $\underline{r}$ and d$\underline{L}$. Mention that the
         integral for the field of a finite segment must be evaluated one
         component at a time. Point out that the angle between d$\underline{B}$ and a
         coordinate axis must be used to find the component of d$\underline{B}$.
    C.   Example: Show how to calculate the magnetic field of a straight
         finite wire segment. See the text, but use finite limits of
         integration. State that magnetic fields obey a superposition
         principle and point out that the result of the previous
         calculation can be used to find the field of a circuit composed
         of straight segments. Specialize the result to an infinite
         straight wire. Demonstrate the right hand rule for finding the
         direction of $\underline{B}$ due to a long straight wire.
    D.   Show how to find the force per unit length of one long straight
         wire on another. Treat currents in the same and opposite
         directions. Lay two long
         automobile starter cables
         on the table. Connect them
         in parallel to an auto
         battery, with a 0.5 Ω,
         500 W resistor and an
         "anti-theft" switch or
         starter relay in each
         circuit. Close one switch and note that the wires do not move.
         Close the other switch and note the motion. Show parallel and
         antiparallel situations. It is best to reconnect the wires or
         rearrange them rather than use a reversing switch.

E.   Give the definition of the ampere and remind students of the definition of the coulomb.

II. Ampere's law.
A.   Write the law in integral form. Explain that the integral is a line integral around a closed contour and interpret it as a sum over segments. Point out that it is the tangential component of <u>B</u> which enters. Explain that the current which enters is the net current through the contour. Two currents in opposite directions tend to cancel, for example.

B.   Explain the right hand rule which relates the direction of integration around the contour and the direction of positive current through the contour.

C.   Use Ampere's law to calculate the magnetic field <u>outside</u> a long straight wire. Either use without proof the circular nature of the field lines or give a symmetry argument to show that <u>B</u> at any point is tangent to a circle through the point and has constant magnitude around the circle. Point out that the integration contour is taken tangent to <u>B</u> in order to evaluate the integral in terms of the unknown magnitude of <u>B</u>.

D.   Use Ampere's law to calculate the field <u>inside</u> a long straight wire with a uniform current distribution. Note that the use of Ampere's law to find B has the same limitations as Gauss' law when used to find E: there must be sufficient symmetry.

E.   Use Ampere's law to calculate the field inside a solenoid. First argue that, for a long tightly wound solenoid, the field at interior points is along the axis and nearly uniform while the field at exterior points is nearly zero.

F.   Similarly, use Ampere's law to calculate the field inside a toroid.

III. Magnetic dipole field.
A.   Use the Biot-Savart law to derive an expression for the field of a circular current loop at a point on its axis. Stress the resolution of d<u>B</u> into components.

B.   Take the limit as the radius becomes much smaller than the distance to the field point and write the result in terms of the dipole moment. Explain that the result is generally true for loops of any shape as long as the field point is far from the

loop. Remind students that the dipole moment of a loop is
determined by its area and the current it carries.

SUGGESTIONS

1. Use questions 3 and 10 as part of the discussion of magnetic field
    lines.
2. Use questions 4, 15, 16, 17, and 20 to help students think about the
    interpretation of Ampere's law.
3. Use questions 8, 12, and 14 to discuss forces between current
    carrying wires.
4. The field of a long straight wire is considered in problem 3, while
    the superposition of fields due to two or more long straight wires is
    considered in problems 6, 27, 28, 29, and 33. Assign at least one of
    these. Combinations of long straight segments and circular segments
    are considered in problems 12 through 16 and finite straight segments
    are considered in problems 18 through 20. If you have not discussed
    it in class, have students work problem 17 before attempting the
    latter.
5. Problems 40 and 44 are good tests of fundamental understanding of
    Ampere's law. Also have students work problem 42, 46, or 50, which
    deal with practical calculations. Have better students attempt
    problem 48 or 51.
6. Solenoids and toroids are the subjects of problems 53 and 56. These
    configurations will be studied in the chapter on inductance. You
    might also include problems 51, 54, and 70, on current sheets and
    Helmholtz coils.
7. Fields of dipoles and torques on dipoles are considered in problems
    67 and 71. Assign these for their own value, but especially if you
    intend to cover Chapter 34.
8. Computer projects
    Have students use the Biot-Savart law and numerical integration to
    calculate the magnetic field due to a circular current loop at
    off-axis points. See Chapter 14 of the calculator supplement. They
    can use Eureka or their own programs.
    Use numerical integration to verify Ampere's law for several long
    straight wires passing through a square contour. Have them show the
    result of the integration is independent of the positions of the

wires, as long as they are inside the square. Also have them consider a wire outside. See Chapter 14 of the calculator supplement.

9.   Demonstrations

Magnetic fields of wires: Freier and Anderson Ei8 - 11; Hilton E7b, d, E9b, c.

Magnetic forces between wires: Freier and Anderson Eil - 6; Hilton E7e, f, g, E9a.

10.   Computer program

Physics Simulations II: Ampere, Kinko's Service Corporation (see Chapter 4 notes for address). Macintosh. Positions and currents of up to 9 coaxial loops are specified by the user, then the program displays magnetic field lines. Use to illustrate lectures.

11.   Laboratory

MEOS Experiment 11-1: The Earth's Magnetic Field. A tangent galvanometer is used to measure the earth's magnetic field. The dip angle is calculated.

MEOS Experiment 11-2: The Current Balance. The gravitational force on a current carrying wire is used to balance the magnetic force due to current in a second wire. The data can be used to find the value of μ_0 or to find the current in the wires. The second version essentially defines the ampere. Part B describes how a microprocessor can be used to collect and analyze the data.

BE Experiment 34: Measurement of the Earth's Magnetic Field. The oscillation period of a small permanent magnet suspended inside a solenoid is measured with the solenoid and the earth's field aligned. The reciprocal of the period squared is plotted as a function of the current in the solenoid and the slope, along with calculated values of the solenoid's field, is used to find the earth's field. If you are willing to postulate the expression for the field of a solenoid, this experiment can be performed in connection with Chapter 30.

MEOS Experiment 11-3: Determination of e/m. Students find the speed and orbit radius of an electron in the magnetic field of a pair of Helmholtz coils and use the data to calculate e/m. Information from Chapter 31 is used to compute the field, given the coil radius and current. If you are willing to postulate the field for the students, this experiment can be performed in connection with Chapter 30.

Chapter 32   FARADAY'S LAW OF INDUCTION

BASIC TOPICS

I.   The law of induction.

   A.   Connect a coil (50 to 100 turns) to a sensitive galvanometer and
        move a bar magnet in and out of the coil. Note that a current is
        induced only when the magnet is moving. Show all possibilities:
        the north pole entering and exiting the coil, the south pole
        entering and exiting the coil. In each case point out the
        direction of the induced current.

   B.   To show the current produced by changing the orientation of a
        loop, align the loop axis with the earth's magnetic field and
        rapidly rotate the loop once through 180°. Note the deflection of
        a galvanometer in series with the loop. Explain that this is the
        basic phenomenon of electric generators.

   C.   Connect a coil to a switchable DC power supply. Connect a
        sensitive galvanometer (digital, if possible) to the supply to
        show it is on. Place a second coil, connected to a voltmeter,
        near the first. Show that when the switch is opened or closed
        current is induced in the second coil, but that none is induced
        when the current in the first coil is steady.

   D.   Define the magnetic flux through a surface. Unit: 1 weber = 1
        $T \cdot m^2$. Point out that Φ_B measures the number of magnetic field
        lines that penetrate the surface. Remark the $\Phi_B = BA\cos\theta$ when $\underline{B}$
        is uniform over the surface and and makes the angle θ with its
        normal.

   E.   Give a qualitative statement of the law: an emf is generated
        around a closed contour when the magnetic flux through the
        contour changes. Stress that the law involves the flux through
        the surface bounded by the contour. Point out the surface and
        contour for each of the demonstrations done, then remark that the
        contour may be a conducting wire, a physical boundary of some
        material, or it may be a purely geometric construction. If the
        contour is conducting then current flows.

   F.   Give the equations for Faraday's law: $\epsilon = -d\Phi_B/dt$ for a single
        loop and $\epsilon = -Nd\Phi_B/dt$ for N tightly packed loops. Note that the
        emfs add for more than one loop.

II. Lenz's law.

    A.   Explain Lenz's law in terms of the magnetic field produced by the induced current if the contour is a conducting wire. Stress that the induced field must reenforce the external field in the interior of the loop if the flux is decreasing and must tend to cancel it if the flux is increasing. This gives the direction of the induced current, which is the same as the direction of the emf. Review the right hand rule for finding the direction of the field produced by a loop of current carrying wire. State that Lenz's law can be used even if the contour is not conducting. The current must then be imagined.

    B.   Optional: Give the right hand rule for finding the direction of positive emf. Point the thumb in the direction of $d\underline{A}$, then the fingers curl in the direction of positive emf. If Faraday's law gives a negative emf, then it is directed opposite to the fingers. Stress that the negative sign in the law is important if the equation, with the right hand rule, is to describe nature.

    C.   Consider a rectangular loop of wire placed perpendicular to a magnetic field. Assume a function B(t) and calculate the emf and current. Show how the directions of the emf and current are found. Point out that an <u>area</u> integral is evaluated to find Φ_B and a <u>time</u> derivative is evaluated to find ϵ. Some students confuse the variables and integrate with respect to time.

    D.   Consider a rectangular loop being pulled with constant velocity past the boundary of a uniform magnetic field. Calculate the emf and current. Show that the work done by the pulling agent equals the energy dissipated by the loop.

III. Induced electric fields.

    A   Explain that a changing magnetic field produces an electric field, which is responsible for the emf. The emf and electric field are related by $\epsilon = \oint \underline{E} \cdot d\underline{s}$, where the integral is around the contour. Remind students that this integral is the work per unit charge done by the field as a charge goes around the contour. Write Faraday's law as $\oint \underline{E} \cdot d\underline{s} = -\frac{d}{dt} \int \underline{B} \cdot d\underline{A}$. Note that $d\underline{s}$ and $d\underline{A}$ are related by a right hand rule: fingers along $d\underline{s}$ implies thumb along $d\underline{A}$. This is consistent with Lenz's law.

    B.   State that the induced electric field is like an electrostatic field in that it exerts a force on a charge but that it is unlike

an electrostatic field in that it is not conservative. For an
electrostatic field the integral defining the emf vanishes.

C. Consider a cylindrical region containing a uniform magnetic
field. Assume a time dependence for $\underline{B}$ and derive expressions for
the electric field inside the region and outside the region. See
Sample Problem 4. Point out that the lines of $\underline{E}$ form closed
circles.

SUPPLEMENTARY TOPIC

The betatron.

SUGGESTIONS

1. Answers to questions 10 through 16 and 18 through 22 depend on
understanding Lenz's law. Use several as examples and several to test
the students.

2. Questions 6 through 8 and 23 through 25 deal with some important
applications of Faraday's law. Use them in discussions or assign them
for students to think about. Also assign problems 30, 31, and 32,
which deal with AC generators. You might want to note that every
electric motor is also a generator. Discuss back emf and the
effective resistance of a motor.

3. If you do not plan to cover Chapter 36 but want to include some
practical application of Faraday's law, this is an appropriate time
to discuss transformers (see Section 6 of Chapter 36). Use questions
23 and 24 as an introduction to eddy currents, which lead to losses
in transformers.

4. Assign problems 1, 12, and 13 to give students some practice in
carrying out calculations of magnetic flux.

5. Assign problems 2, 4, and 5 to cover the emfs generated by various
time dependent magnetic fields. Addition of emfs is covered in
problem 19.

6. Motional emf is covered in problems 24 and 26. Problem 24 is a
particularly good test of understanding of Faraday's law. AC
generators are considered in problems 30, 31, and 32. If you use a
flip coil in the lab, assign problems 17 and 18.

7. Assign problems 43, 44, and 46 in connection with the discussion of
induced electric fields.

8. Demonstrations

As a supplementary demonstration, take
a large, long coil, insert a solid soft
iron rod with a foot or so sticking out,
and connect the coil via a switch to a
large DC power supply. Place a solid
aluminum ring around the iron rod. The
ring should fit closely but be free to
move. Close the switch and the ring will
jump up, then settle down. Repeat with a
ring that has a gap in it. Finally, use
an AC power supply. The effect can be
enhanced by cooling the ring with liquid
nitrogen.

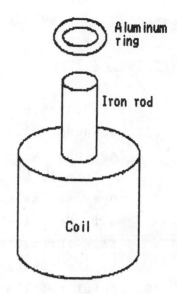

Generation of induced currents: Freier
and Anderson Ekl - 6; Hilton E8a.

Eddy currents: Freier and Anderson Eil - 6; Hilton E8d.

Generators: Freier and Anderson: Eq4 - 7, Erl; Hilton E8b, c.

9. Films

The Concept of a Changing Flux; Faraday's Law of Induction, S8,
color, 3 min. each. The Kalmia Company (see Chapter 4 notes for
address). Two more loops in the Adler-MIT electromagnetism series.
Lenz's Law; Large Inductance: Current Buildup, S8, color, 3 min.
each. American Association of Physics Teachers (see Chapter 2 notes
for address). Demonstrations of Lenz's law using different
conductors, an aluminum disk, and a hollow aluminum conductor.

10. Laboratory

BE Experiment 35:Electromagnetic Induction. Students measure the
magnitude and observe the direction of current induced by a changing
magnetic flux in a simple galvanometer circuit. Changing flux is
produced by moving permanent magnets, by moving current carrying
coils, and by changing current in a coil.

MEOS Experiment 11-4: The Magnetic Field of a Circular Coil. The emf
generated in a small search coil when a low frequency AC current
flows in a given circuit (a circular coil in this case) is used to
determine the magnetic field produced by the circuit. The field is
investigated as a function of position, specified in spherical
coordinates.

Chapter 33   INDUCTANCE

BASIC TOPICS

I.   Self-inductance.

    A. Connect a light bulb and choke coil in parallel across a switchable DC supply. Close the switch and note that the lamp is initially brighter than it is when the steady state is reached. Open the switch and note that the light brightens before going off. Remark that this behavior is due to the changing magnetic flux through the coil and that the flux is created by the current in the coil itself.

    B. Point out that when current flows in a loop it generates a magnetic field and the loop contains magnetic flux due to its own current. If the current changes so does the flux and an emf is generated around the loop. The total emf, due to all sources, determines the current. Remark that the self flux is proportional to the current and the induced emf is proportional to the rate of change of the current.

    C. Consider a tightly wound coil and define the inductance by $N\Phi_B = Li$, then show that the induced emf is $\epsilon = -Ldi/dt$. State that this expression is valid for any coil. Unit: 1 henry = 1 V·s/A.

    D. Inductors are denoted by $-\ell\ell\ell\ell\ell-$ in circuit diagrams. Point out that if the circuit element looks like $a-\ell\ell\ell\ell\ell-b$, then $V_a - V_b$ = Ldi/dt is algebraically correct. As an example use $i(t) = I\sin(\omega t)$. Note that i is positive when it is directed from a to b and negative when it is directed from b to a. Compute $V_a - V_b = LI\omega\cos(\omega t)$. Graph i and the potential difference as functions of time to show the phase relationship. Remark that a real inductor can be regarded as a pure inductance in series with a pure resistance.

    E. Show how to calculate the inductance of an ideal solenoid. Use the current to calculate the field, then the flux, and finally equate $N\Phi_B$ to Li and solve for L. Point out that L is independent of i but depends on geometric factors such as the cross sectional area, length, and density of turns.

    F. Optional: Show how to calculate the inductance of a toroid.

II. An LR circuit.

    A. Derive the loop equation for a single loop containing a seat of emf (battery), resistor, and inductor: $iR + Ldi/dt + \epsilon = 0$. Use the prototypes developed earlier:

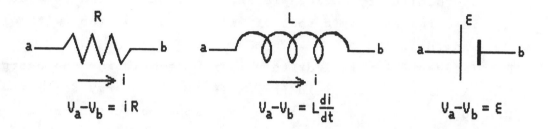

$$V_a - V_b = iR \qquad\qquad V_a - V_b = L\frac{di}{dt} \qquad\qquad V_a - V_b = \epsilon$$

    and remark that these are correct no matter whether the current is positive for negative or whether it is increasing or decreasing. Write down the solution for the current as a function of time for the case $i(0) = 0$: $i = (\epsilon/R)[1 - e^{-Rt/L}]$. Show that the expression satisfies the loop equation and meets the initial conditions. Show Fig. 7 and point out the asymptotic limit $i = \epsilon/R$ and the time constant $\tau = L/R$. Remark that if L/R is large the current approaches its limit more slowly.

    B. Explain the qualitative physics involved. When the battery is turned on and the current increases, the emf of the coil opposes the increase and the current approaches its steady state value more slowly than if there were no inductance. At long times the current is nearly constant so di/dt and the induced emf are small. The current is nearly the same as it would be in the absence of an inductor.

    C. Repeat the calculation for a circuit with an inductor and resistor, but no battery. Take the initial current to be i_0 and show that $i(t) = i_0 e^{-t/\tau}$. Graph the solution and point out that the emf of the coil opposes the decrease in current.

    D. Demonstrate the two circuits by connecting a resistor and coil, in series, to a square wave generator. Observe the current by placing oscilloscope leads across the resistor. Observe the voltage drop across the coil. Vary the time constant by varying the resistance.

III. Energy considerations.

    A. Consider a single loop circuit containing a seat of emf, a resistor, and an inductor. Assume the current is increasing.

Write down the loop equation, multiply it by i, and identify the power supplied by the seat of emf and the power lost in the resistor. Explain that the remaining term describes the power being stored by the inductor, in its magnetic field. Point out the similarity between $i\epsilon$ and $-iLdi/dt$ for the rate at which work is being done by a seat of emf and an inductor (with emf $-Ldi/dt$).

B. Integrate $P = iLdi/dt$ to obtain $U_B = \frac{1}{2}Li^2$ for the energy stored in the magnetic field (relative to the energy for $i = 0$).

C. Consider the energy stored in a long current carrying solenoid and show that the energy density is $u_B = \frac{1}{2}B^2/\mu_0$. Explain that this gives the energy density at a point in any magnetic field and that the energy required to establish a given magnetic field can be calculated by integrating the expression over the volume occupied by the field.

IV. Mutual induction.

A. Repeat the demonstration experiment discussed in note IC for Chapter 32. Explain it in terms of the concept of mutual induction. Point out that the flux through the second coil is proportional to the current in the first. Define the mutual inductance of the second coil with respect to the first by $M_{21} = N_2\Phi_{21}/i_1$. Show that $\epsilon_2 = -M_{21}di_1/dt$ is the emf induced in the second coil. Explain that $M_{12} = M_{21}$.

B. Example: Derive the mutual inductance of a small coil placed at the center of a solenoid or of a small tightly wound coil placed at the center of a larger coil. See Sample Problem 7.

C. Show that two inductors connected in series and well separated have an equivalent inductance of $L = L_1 + L_2$. Then show that if their fluxes are linked $L = L_1 + L_2 \pm 2M$, where the minus sign is used if the field lines have opposite directions. See problem 49. Also consider inductors in parallel.

SUGGESTIONS

1. Use questions 2, 5, 6, and 15 when discussing the calculation of inductance.

2. Use some of questions 7 through 14 when discussing the LR circuit.

3. After discussing energy flow in a simple LR circuit with increasing

current, ask question 10, then assign problems 33 and 36.

4. Assign problem 2 (coil), 3 (solenoid), or 8 (two parallel wires) as an example of a typical inductance calculation.

5. LR time constants are considered in problems 15, 16, and 18. Problems 25 and 26 are good tests of understanding of initial and steady state conditions in an LR circuit.

6. Problems 32 and 37 deal with energy storage and energy density in an inductor.

7. Good problems in connection with mutual inductance: 51 (coil and toroid), and 52 (two coaxial solenoids).

8. Demonstrations
   Self-inductance: Freier and Anderson Eq1 - 3; Hilton E12d.
   LR circuit: Freier and Anderson Eo11, En5 - 7; Hilton E12c.

Chapter 34   MAGNETISM AND MATTER

BASIC TOPICS

I. Magnetic dipoles in matter.

   A. Explain that current loops and bar magnets produce magnetic fields which, for points far away, are dipole fields. Review the expressions for the magnetic field of a dipole and for the dipole moment of a loop in terms of the current and area. Place a bar magnet under a piece of plastic sheet on an overhead projector. Sprinkle iron filings on the sheet and show the field pattern of the magnet. Remind students that field lines emerge from the north pole and enter at the south pole.

   B. Explain that electrons in atoms create magnetic fields by virtue of their orbital motions. Derive Eqs. 11 and 12, which give the relationship between orbital angular momentum and dipole moment.

   C. Explain that the electron and many other fundamental particles have intrinsic dipole moments, related to their intrinsic spin angular momentum. give the magnitude of the electron's intrinsic dipole moment: 9.27×10^{-24} J/T.

   D. Remark that it is chiefly the orbital and spin dipole moments of electrons that are responsible for the magnetic properties of materials.

II. Gauss' law for magnetism.

  A. Remark that no magnetic monopole has been observed yet but it is
      currently being sought. Write down Gauss' law for the magnetic
      field and state that magnetic field lines form closed contours so
      the flux through any closed surface vanishes. If monopoles were
      found to exist, the law would be modified to include them.
      Compare with Gauss' law for the electric field.

  B. To show that the ends of a magnet are not monopoles, magnetize a
      piece of hard iron wire. Use a compass to locate and mark the
      north and south poles. Break the wire into pieces and again use
      the compass to show that each piece has a north and a south pole.

III. Paramagnetism and diamagnetism.

  A. Explain that the dipole moment of an atom is found by summing the
      orbital and spin moments of all its electrons. A moment may be
      associated with the nucleus but it is extremely small. Some atoms
      have net moments while others do not.

  B. Give a qualitative discussion of paramagnetism. Explain that
      paramagnetic substances are composed of atoms with net dipole
      moments and, in the absence of an external field, the moments
      have random orientations, so that no net magnetic field is
      produced. An external field tends to align the moments and the
      the material produces its own field. Since the moments, on
      average, are aligned with the external field, the total field is
      stronger than the external field alone. Alignment is opposed by
      thermal agitation and both the net magnetic moment and magnetic
      field decrease as the temperature increases.

  C. Define magnetization and give the Curie law for small applied
      fields. Point out that $\underline{M}$ is proportional to $\underline{B}$ and inversely
      proportional to T. Describe saturation and explain that there is
      an upper limit to the magnetization. The limit occurs when all
      atomic dipoles are aligned. Use a teslameter or flip coil to
      measure the magnetic field just outside the end of a large, high
      current coil. Put a large quantity of manganese in the coil and
      again measure the field.

  D. Give a qualitative discussion of diamagnetism. Explain that an
      external field changes the electron orbits so there is a net
      dipole moment and that the induced moment is directed opposite to
      the field. This tends to make the total field weaker than the

external field alone. Bismuth is an example of a diamagnetic substance.

E. Explain that diamagnetic effects are present in all materials but are overshadowed by paramagnetic or ferromagnetic effects if the atoms have dipole moments.

IV. Ferromagnetism.

A. Explain that, for iron and other ferromagnetic substances (such as Co, Ni, Gd, and Dy), the atomic dipoles are aligned by an internal mechanism (exchange coupling) so the substance can produce a magnetic field spontaneously, in the absence of an external field. At temperatures above its Curie temperature, a ferromagnetic substance becomes paramagnetic. Gadolinium is ferromagnetic with a Curie temperature of about 20°C. Put a sample in a beaker of cold water (T < 20° C) and use a weak magnet to pick it up from the bottom of the beaker but not out of the water. Add warm water to the beaker and the sample will drop from the magnet.

B. Describe ferromagnetic domains and explain that the dipoles are aligned within any domain but are oriented differently in neighboring domains. The magnetic fields produced by the various domains nearly cancel for an unmagnetized sample. When the sample is placed in a magnetic field domains with dipoles aligned with the field grow in size while others shrink. The dipoles in a domain may also be reoriented as a unit.

C. Define hysteresis (see Fig. 15) and explain that the growth and shrinkage of domains are not reversible processes. Domain size is dependent not only on the external field but also on the magnetic history of the sample. When the external field is turned off, the material remains magnetized. Explain the difference between soft and hard iron in terms of hysteresis. Use a large, high current coil to magnetize a piece of hard iron and show that it remains magnetized when the current is turned off. Also magnetize a piece of soft iron and show it is magnetized only as long as the current remains on. When the current is turned off, very little permanent magnetization remains. Soft iron is used for transformer coils.

SUPPLEMENTARY TOPICS

The magnetism of the earth.

Nuclear magnetism.

SUGGESTIONS

1. Ask students to think about a permanent bar magnet which pierces the surface of a sphere and explain why the net magnetic flux through the surface is zero. Also ask them about the flux as a single charge crosses the surface.

2. Ask students to think about questions 1 and 2 in connection with induced magnetism.

3. In order to emphasize the different mechanisms for paramagnetism and diamagnetism, ask questions 12 and 13.

4. To test for understanding of spin and orbital magnetic moments, assign problems 1 and 2.

5. To test for understanding of Gauss' law for magnetism, assign problems 8 and 11.

6. Magnetization in a paramagnetic substance is covered in problems 19, 24, and 25. If Chapter 21 was included in the course also consider assigning problem 8. The attraction of a bar magnet for a paramagnetic substance is covered in problem 20

7. The repulsion of a diamagnetic substance by a bar magnet is covered in problem 26. This makes a nice companion to problem 20.

8. The Curie temperature of a ferromagnet is covered in problem 28 and the magnetization of a ferromagnet is covered in problem 31.

9. Demonstrations

   Field of a magnet: Freier and Anderson Er4

   Gauss' law: Freier and Anderson Er12

   Paramagnetism: Freier and Anderson Es3, 4

   Diamagnetism: Hilton E10b.

   Ferromagnetism: Freier and Anderson Es1, 2, 6 - 10; Hilton E10a, E10c, d.

   Levitation: Freier and Anderson Er10, 11

10. Films

    Monopoles and Dipoles, S8, color, 3 min. The Kalmia Company (see Chapter 4 notes for address). A film loop in the Adler-MIT electromagnetism series.

    Ferromagnetic Domain Wall Motion, S8, color, 4 min. The Kalmia

Company (see Chapter 4 notes for address). Domain boundary movements are illustrated in this film loop by Franklin Miller.

Magnetic Domains and Magnetization Processes in a Gadolinium Iron Garnet, S8, color, 4 min. American Association of Physics Teachers (see Chapter 2 notes for address). A transparent ferromagnetic material is subjected to an alternating external magnetic field.

11. Laboratory

MEOS Experiment 11-6: Magnetization and Hysteresis. Faraday's law is used to measure the magnetic field inside an iron toroid for various applied fields. A plot of the field as a function of the applied field shows hysteresis. A method for obtaining the hysteresis curve as an oscilloscope trace is also given.

Chapter 35   ELECTROMAGNETIC OSCILLATIONS

BASIC TOPICS

I.   LC oscillations.

A.   Draw a diagram of an LC circuit and assume the capacitor is charged. Explain that as charge flows, energy is transferred from the electric field of the capacitor to the magnetic field of the inductor and back again. When the capacitor has maximum charge, the current (dq/dt) vanishes, so no energy is stored in the inductor. When the current is a maximum the charge on the capacitor vanishes and no energy is stored in that element.

B.   Write down the loop equation, then convert it so the charge q on the capacitor is the dependent variable. If the direction of positive current is into the capacitor plate with positive charge q, then $i = dq/dt$. If it is out of that plate, then $i = -dq/dt$. Note that the form of the differential equation is the same as that of the displacement of a simple harmonic oscillator.

C.   Write down the solution: $q(t) = Q\cos(\omega t+\phi)$. Show by direct differentiation that this is a solution if $\omega^2 = 1/LC$. Show that ϕ is determined by the initial condition and treat the special case for which $q = Q$, $i = 0$ at $t = 0$.

D.   Once the solution is found, derive expressions for the current, the energy stored in the capacitor, and the energy stored in the inductor, all as functions of time. Sketch graphs of these quantities. Show that the total energy is constant. In preparation for the next chapter, it is also worthwhile deriving expressions for the potential differences across the capacitor and the inductor, then draw graphs of them as well. Mention that the charge on the capacitor is proportional to the potential difference across its plates and the time rate of change of the current is proportional to the voltage across the terminals of the inductor.

II. Damped and forced oscillations.

A. Write down the loop equation for a single LCR loop, then convert it so q is the dependent variable. State that $q(t) = Qe^{-Rt/2L}\cos(\omega' t + \phi)$ satisfies the differential equation. Here ω' is somewhat less than $1/\sqrt{LC}$. If time permits, the expression for ω' can be found by substituting the assumed solution into the differential equation.

B. Draw a graph of q(t) and point out that the envelope decreases exponentially. Each time the capacitor is maximally charged, the charge on the positive plate is less than the previous time. Explain that this does violate the conservation of charge principle since the total charge on the capacitor is always zero. Energy is dissipated in the resistor.

C. To show the oscillations, wire a resistor, inductor, and capacitor in series with a square wave generator and connect an oscilloscope across the capacitor. The scope shows a function proportional to the charge. Show the effect of varying C (use a variable capacitor), R (use a decade box), and L (insert an iron rod into the coil). If time permits, show that oscillations occur only if $1/LC > (R/2L)^2$.

D. Consider an LCR circuit with a sinusoidal oscillator and write down the loop equation. You may also want to write down the solution. In any event, show the solution by sketching graphs of the current amplitude as a function of the impressed frequency for several values of the resistance (see Fig. 6). Point out that the current amplitude is greatest when the impressed frequency matches the natural frequency of the circuit and that the peak becomes larger as the resistance is reduced.

E. Demonstrate resonance phenomena by wiring an LCR loop in series with a sinusoidal audio oscillator. Look at the current by putting the leads of an oscilloscope across the resistor. Use a decade box for the resistor and measure the current amplitude for various frequencies and for several resistance values. Be sure the amplitude of the oscillator output remains the same. Explain that similar circuits are used to tune radio and TV's.

F. Use a sweep generator to show the current amplitude. Set the oscilloscope sweep rate to accommodate that of the generator and put a small diode in series with the scope leads. Usually this will have enough capacitance that only the envelope will be displayed.

G. If you do not intend to cover the next chapter, briefly introduce the ideas of reactance, impedance, and Q value.

SUGGESTIONS

1. Questions 1, 3, 4, 5, and 6 can be used to help students think about the LC circuit discussions.

2. Discuss the initial conditions and the determination of the phase constant for an LC circuit, then ask questions 2 and 7.

3. Compare an oscillating LC circuit to an oscillating mass on a spring, then ask question 12. To fully answer the question you will need to consider a circuit with resistance and a mechanical oscillator subjected to a drag force.

4. Assign problems 1, 2, 4, 5, and 14 to test for understanding of the fundamentals of LC oscillations. Resonant frequency is covered in problems 8, 9, 10, and 23 and the relationship of the resonant frequency to the inductive and capacitive time constants is explored in problem 11. Problem 24 deals with the phase of the oscillations.

5. Demonstrations
   LCR series circuit: Freier and Anderson En12, E013; Hilton E13c, d, e. By making the resistance small you can demonstrate many of the ideas of this chapter.

Chapter 36   ALTERNATING CURRENTS

BASIC TOPICS

I.   Elements of circuit analysis.

   A. Consider a circuit consisting of a resistor and a seat of
       sinusoidally varying emf. State that the potential drop across
       the resistor is in phase with the current and the amplitudes are
       related by $I_R R = V_R$. Draw a phasor diagram: 2 arrows along the
       same line with length proportional to I_R and V_R respectively.
       Both make the angle ωt with the horizontal axis and rotate in the
       counterclockwise direction. Point out that the vertical
       projections represent $i_R(t)$ and $v_R(t)$ and these vary in
       proportion to $\sin(\omega t)$ as the arrows rotate in the
       counterclockwise direction.

   B. Consider a circuit consisting of a capacitor and a seat of
       sinusoidally varying emf. Start with $i_C = dq_C/dt = C dv_C/dt$, then
       show that v_C lags i_C by 90° and that the amplitudes are related
       by $I_C = V_C/X_C$, where $X_C = 1/\omega C$ is the capacitive reactance. Draw
       a phasor diagram to show the relationship. Mention that the unit
       of reactance is the ohm.

   C. Consider a circuit consisting of an inductor and a seat of
       sinusoidally varying emf. Start with $v_L = -L di_L/dt$, then show
       that v_L leads i_L by 90° and that the amplitudes are related by I_L
       $= V_L/X_L$, where $X_L = \omega L$ is the inductive reactance. Draw a phasor
       diagram to show the relationship.

   D. Wire a small resistor in series with a capacitor and a signal
       generator. Use a dual trace oscilloscope with one set of leads
       across the resistor and the other set across the capacitor.
       Remind students that the potential difference across the resistor
       is proportional to the current, so the scope shows the relative
       phase of i_C and v_C. Point out the difference in phase. Repeat
       with an inductor in place of the capacitor.

II. The LCR circuit.

   A. Draw the circuit, then construct a phasor diagram step by step
       (see Figs. 6a and 6b). First draw the current and resistor
       voltage phasors, in phase. Remind students that the current is
       the same in every element of the circuit so voltage phasors for

the other elements can be drawn using the phase relations between voltage and current developed earlier. Draw the capacitor voltage phasor lagging by 90° and the inductor voltage phasor leading by 90°. Make $V_C > V_L$, Their lengths are $X_C I$ and $X_L I$ respectively. Draw the projections of the phasors on the vertical axis and remark that the algebraic sum must be ϵ.

B. Draw the impressed emf phasor. Remark that its projection on the capacitor phasor must be (V_C-V_L). Make the analogy to a vector sum.

C. Use the phasor diagram to derive the expression for the current amplitude: $I = \epsilon/Z$, where $Z = [R^2 + (X_C-X_L)^2]^{1/2}$ is the impedance of the circuit. Show that the impedance is frequency dependent by substituting the expressions for the reactances. Also show that I is greatest for $X_C = X_L$ or $\omega^2 = 1/LC$.

D. Use the phasor diagram to derive the expression for the phase angle of i relative to the ϵ: $\tan\phi = (X_L-X_C)/R$. Point out that the phase angle vanishes at resonance and ϵ leads i if $X_L > X_C$, but ϵ lags i if $X_L < X_C$.

III. Power considerations.

A. Discuss average values over a cycle. Show that the average of $\sin^2(\omega t+\phi)$ is $\frac{1}{2}$ and define the rms value of a sinusoidal quantity. Point out that AC meters are usually calibrated in terms of rms values.

B. Derive the expression for the power input of the AC source: $P = i\epsilon = I\epsilon\sin(\omega t+\phi)\sin(\omega t)$. Show that $P_{av} = \epsilon_{rms}i_{rms}\cos\phi$. Do the same for the power dissipated in the resistor. In particular, show that its average value can be written $i_{rms}^2 R$ or $\epsilon_{rms}i_{rms}R/Z$. Use the phasor diagram to show that $R/Z = \cos\phi$ and then use this relationship to show that the average power input equals the average power dissipated in the resistor. Remark that the average power for the capacitor and inductor are zero.

C. Remark that $\cos\phi$ is called the power factor. If it is 1 the source delivers the greatest possible power for a fixed amplitude.

SUPPLEMENTARY TOPIC

The transformer. If this topic was not covered in connection with Chapter

32, you may wish to cover it now. A dual trace oscilloscope can be used to demonstrate transformer voltages.

SUGGESTIONS

1. When discussing solutions to the LCR loop equation, include questions 4 and 8.

2. Use questions 5 and 6 in the discussion of phasor diagrams.

3. Use questions 8, 10, 11, and 13 in the discussion of reactance and the relative phases of the potential and current.

4. More detail can be added to the discussion of the power factor by including questions 16, 17, 18, 20, and 21.

5. Assign problems 8 and 9 in connection with discussions of the phase and amplitude of separate inductive and capacitive circuits. Phase is also covered in problems 10, 11, 22, and 23.

6. Assign problem 20 to have students think about voltages around an LCR circuit.

7. Power in an LCR circuit is covered in problems 37 and 38 and the power factor in problem 40.

8. Demonstrations
   See LCR circuit demonstrations listed in Chapter 35 notes.
   Measurements of reactance and impedance: Freier and Anderson Eo9.
   Transformers: Freier and Anderson Ek7, Eml, 2, 4, 5, 7, 8, 10; Hilton Ell.

9. Films
   Electromagnetic Oscillator I: Free Oscillations; Electromagnetic Oscillator II: Forced Oscillations, S8, color, 3.3 min. (I) and 7.4 min. (II). Walter de Gruyter, Inc. (see Chapter 17 notes for address). I. Damped, critically damped, and over damped oscillations are illustrated in an LCR circuit. II. Using a double beam oscilloscope, the phase, voltage, and current of weakly damped resonant circuits are shown. Reviewed AJP 45:1014 (1977).

10. Laboratory
    MEOS Experiment 10-11: A.C. Series Circuits. Students use an oscilloscope and AC meters to investigate voltage amplitudes, phases, and power in CR and LCR circuits. Voltage amplitudes and phases are plotted as functions of the driving frequency to show resonance. Reactances and impedances are calculated from the data.
    BE Experiment 37: A Study of Alternating Current Circuits. An AC

voltmeter is used to investigate the voltages across circuit elements in R, CR, LR, and LCR circuits, all with 60 Hz sources. Reactances and impedances are computed. A section labelled optional describes the use of an oscilloscope. If possible, scopes should be used. This experiment is pedagogically similar to the text and can be used profitably to reenforce the ideas of the chapter. Warning: the lab book uses the word vector rather than phasor.

Chapter 37   MAXWELL'S EQUATIONS

<u>BASIC TOPICS</u>

I.   The Ampere-Maxwell law.

    A.   Use Table I to review the equations of electricity and magnetism discussed so far. Note the absence of any counterpart to Faraday's law, i.e. the creation of magnetic fields by changing electric flux. Tell students it should be there and you will now discuss its form.

    B.   Consider the charging of a parallel plate capacitor. Remind students that in Ampere's law d$\underline{s}$ and d$\underline{A}$ are related by a right hand rule and

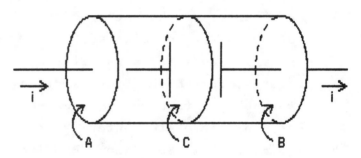

the surface integral is over any surface bounded by the closed contour. In the diagram, surfaces A, B, and C are all bounded by the contour which forms the left end of the figure. If we choose surface A or B then Ampere's law as we have taken it gives $\oint \underline{B} \cdot d\underline{s} = \mu_0 i$, but if we choose surface C, it gives $\oint \underline{B} \cdot d\underline{s} = 0$. Since the integral on the left side is exactly the same in all cases, something is wrong.

    C.   Note that the situation discussed and the lack of symmetry in the electromagnetic equations suggests that Ampere's law as used so far must be changed. Experiment confirms this conjecture. See question 11.

    D.   Explain that if the electric flux through an open surface changes

with time then there is a magnetic field and the magnetic field
has a tangential component at points on the boundary. Write down
the Ampere-Maxwell law: $\oint \underline{B} \cdot d\underline{s} = \mu_0 i + \mu_0 \epsilon_0 d\Phi_E/dt$, where Φ_E is the
electric flux through the surface. Compare to Faraday's law and
point out the interchange of $\underline{B}$ and $\underline{E}$, the change in sign, and the
appearance of the factor $\mu_0 \epsilon_0$.

E. Give the right hand rule that relates the normal to the surface
used to calculate Φ_E and the direction of integration around its
boundary. State that the surface may be a purely mathematical
construction and that the law holds for any surface.

II. Displacement current.
   A. Define the displacement current $i_d = \epsilon_0 d\Phi_E/dt$. Explain that it
does not represent the flow of charge and is not a true current,
but that it enters the Ampere-Maxwell law in the same way as a
true current. Discuss the direction of i_d. Consider a region in
which the electric field is uniform and is changing. Find the
direction for both an increasing and a decreasing field.
   B. Write down the Ampere-Maxwell law. Explain that in the examples
of Chapter 31 there are no changing electric fields so only true
currents were considered.
   C. Example: Consider a parallel plate capacitor with circular
plates, for which dE/dt is given. Derive expressions for $\underline{B}$ at
various points along the perpendicular bisector of the line
joining the plate centers. Consider points between the plates and
outside them. See Sample Problem 1.

III. Maxwell's equations.
   A. Write down the four equations in integral form and review the
physical processes that each describes.
   B. Carefully distinguish between the line and surface integrals that
appear in the equations and give the right hand rules that relate
the direction of integration for the contour integrals and the
normal to the surface for the surface integrals.
   C. Review typical problems: the electric field of a point charge,
the magnetic field of a uniform current in a long straight wire,
the magnetic field at points between the plates of a capacitor
with circular plates, the electric field accompanying a changing
uniform magnetic field with cylindrical symmetry.

D. State that these equations describe all electromagnetic phenomena
    to the atomic level and the natural generalizations of them
    provide valid descriptions of electromagnetic phenomena at the
    quantum level. They are consistent with modern relativity theory.

SUGGESTIONS

1. To test for understanding of the direction of the magnetic field
    induced by a changing electric field, assign questions 2 and 5. The
    directions of the displacement current, E, and B, are also covered in
    question 7. It is worthwhile asking this question before question 5.
2. Problems 13 and 15 help students think carefully about displacement
    current. The continuity of current and displacement current through a
    charging capacitor is covered in problem 9.
3. To have students compare the magnitudes of induced electric and
    magnetic fields, ask question 3 and assign problem 14.
4. To give students practice in associating Maxwell's equations with
    various electromagnetic phenomena, assign problem 16.

Chapter 38   ELECTROMAGNETIC WAVES

BASIC TOPICS

I. Qualitative features of electromagnetic waves.
    A. Explain that an electromagnetic wave is composed of electric and
        magnetic fields. The disturbance, analogous to the string shape
        that moves on a taut string, is made up of the fields themselves,
        moving through space or a material medium. Also explain that
        electromagnetic waves carry energy and momentum.
    B. State that the wave speed in empty space is given by $c = 1/\sqrt{\mu_0 \epsilon_0}$
        and is about 3.00×10^8 m/s. The existence of waves and the empty
        space speed are predicted by Maxwell's equations. Since the value
        of c is fixed, this fixes ϵ_0.
    C. Show the electromagnetic spectrum (Fig. 1) and point out the
        visible, ultraviolet, infrared, x-ray, microwave, and radio
        regions.
    D. State that an oscillating electric dipole creates a traveling
        electromagnetic wave and show a diagram of the fields (see Fig.

4). Point out that $\underline{E}$ and $\underline{B}$ are perpendicular to each other and to the direction of propagation and that they oscillate in phase with each other at any point.

II. Traveling sinusoidal waves.

A. Take $E = E_0\sin(kx-\omega t)$, along the y axis, and $B = B_0\sin(kx-\omega t)$, along the z axis. Remark that both fields travel in the positive x direction and that they are in phase.

B. Consider a rectangular area in the x,y plane, with width dx and length h (along y). Evaluate $\oint\underline{E}\cdot d\underline{s}$ and Φ_B, then show that Faraday's law yields $\partial E/\partial x = -\partial B/\partial t$. Substitute the expressions for E and B to show that $E = cB$, where $c = \omega/k$. Stress that the magnitudes of $\underline{E}$ and $\underline{B}$ are related. Remark that $\underline{E}$ is different at different points because $\underline{B}$ changes with time.

C. Consider a rectangular area in the x,z plane, with width dx and length h (along z). Evaluate $\oint\underline{B}\cdot d\underline{s}$ and Φ_E, then show that the Ampere-Maxwell law yields $-\partial B/\partial x = \mu_0\epsilon_0\partial E/\partial t$. Combine this with the result of part B to show that $c = 1/\sqrt{\mu_0\epsilon_0}$. Remark that $\underline{B}$ is different at different points because $\underline{E}$ changes with time.

III. Energy and momentum transport.

A. Define the Poynting vector $\underline{S} = (1/\mu_0)\underline{E}\times\underline{B}$ and explain that it is in the direction of propagation and that its magnitude gives the rate at which electromagnetic energy crosses a unit area perpendicular to the direction of propagation. Remark that for a plane wave $S = EB/\mu_0 = E^2/\mu_0 c$.

B. Consider the plane wave of section II, propagating in the positive x direction. Consider a volume of width dx and cross section A (in the y,z plane) and show that the electromagnetic energy in it is $dU = (EBA/\mu_0 c)dx$. This energy passes through the area A in time $dt = dx/c$ so the rate of energy flow per unit area is EB/μ_0, as previously postulated.

C. Explain that electromagnetic waves transport momentum and that S/c gives the momentum which crosses a unit area per unit time. The momentum is in the direction of $\underline{S}$. Also explain that if an object absorbs energy U then it receives momentum U/c. If the object reflects energy U then it receives momentum 2U/c.

D. As an example of radiation pressure, you may wish to consider solar pressure. S can be determined from the solar constant 1.4

kW/m^2 (valid just above the earth's atmosphere). Problems 27, 38, and 39 provide interesting examples.

IV. Polarization.

   A. Define the term plane polarization and, as an example, state that the wave considered in note IIA is plane polarized with the electric field at all points parallel to the y axis. Explain that the wave emitted by a dipole antenna is plane polarized and that a second antenna must be oriented along the electric field to pick up the maximum signal. Example: TV antennas are horizontal.

   B. Explain that many sources produce radiation that is a mixture of many independent waves with randomly distributed polarization directions. State they can be represented as two plane polarized waves with their polarization directions perpendicular to each other and with a randomly varying phase difference.

   C. Describe the action of a Polaroid sheet: it transmits light with $\underline{E}$ along a certain direction and absorbs light with $\underline{E}$ perpendicular to that direction. Show that for a plane polarized incident wave the transmitted intensity is proportional to $\cos^2\theta$, where θ is the angle between $\underline{E}$ and the polarizing direction for the sheet.

   D. Demonstrate the effect by placing two Polaroid sheets back to back and rotating one with respect to the other. Explain that if the incident light is unpolarized the first sheet cuts the intensity in half and renders the transmitted light polarized along the axis of the sheet. The second sheet is called an analyzer and passes all the light if its axis is parallel to that of the first but passes none if its axis is perpendicular to that of the first. If possible, pass out Polaroid sheets to the class.

SUPPLEMENTARY TOPICS

1. Measuring the speed of light.
2. The speed of light in special relativity.

SUGGESTIONS

1. In discussing the speed of light in terms of ϵ_0 and μ_0, remind students that these constants enter electrostatics and magnetostatics respectively and were first encountered in situations that had

nothing to do with wave propagation. Have them answer question 34.

2.   Remind students that, when the relationship between $\underline{E}$ and $\underline{B}$ for a plane wave was discussed, Faraday's law and the Ampere-Maxwell law were used. Then ask them to discuss question 15.

3.   The relationship between frequency, wavelength, and speed is explored in problems 3 and 4. These also give some examples of high and low frequency electromagnetic radiation.

4.   To stress the relationship between $\underline{E}$ and $\underline{B}$ in an electromagnetic wave, assign problem 12 or 18.

5.   To stress that electromagnetic waves need not be sinusoidal, assign problems 13 and 14. Remark that the wave associated with any function obeys the wave equation and travels with speed c in empty space.

6.   To emphasize the magnitude of the energy and momentum carried by an electromagnetic wave, discuss questions 17 and 21 and assign problems 15 and 41. Problems 22 deals with the average energy in terms of the magnetic field amplitude and problems 22 and 23 are numerical examples.

7.   Ask questions 23, 24, 28, and 29 in connection with the discussion of polarization. The situation described in question 28 can be demonstrated easily. Problems 53, 54, and 55 provide good tests of the ideas involved. You might also assign problems 56 and 57, dealing respectively with unpolarized and polarized incident light. Have students compare answers.

8.   Demonstrations
     Radiation: Freier and Anderson Ep4, 5.
     Speed of light: Freier and Anderson Oa4.
     Polarization: Freier and Anderson Oml, 2, 7 - 11, 14, 15, 17 - 19;
     Hilton O8b, c.

9.   Computer programs
     Physics Simulations II: Radiation, Kinko's Service Corporation (see Chapter 4 notes for address). Macintosh. Shows electric field lines of an accelerating charge in linear, circular, or oscillatory motion. User selects the velocity and can view either the near or far field.
     Intermediate Physics Simulations: Moving Charge, R.H. Good (see Chapter 15 notes for address). Apple II. Shows electric field lines of a moving charge with user selected velocity. The user can change the velocity and radiation fields are shown when the charge accelerates.
     Intermediate Physics Simulations: Radiating Dipole, R.H. Good (see

Chapter 15 notes for address). Apple II. Animated diagrams showing the electric field lines of a radiating electric dipole. Excellent illustration for a lecture on sources of radiation.

10. Laboratory

MEOS Experiment 13-7: <u>Polarization of Light</u>. Students investigate polarization by Polaroid sheets, reflection, refraction, and scattering. They also observe the rotation of the polarization direction by a sugar solution and view stressed objects through crossed polarizers. Essentially a collection of demonstrations done by students.

BE Experiment 46: <u>Polarized Light</u>. Essentially the same set of demonstrations but a photodetector is used to obtain quantitative data. The transmitted intensity for crossed polarizers is plotted as a function of the relative angle between them to verify the $\cos^2\theta$ law.

Chapter 39   GEOMETRICAL OPTICS

<u>BASIC TOPICS</u>

I.   Reflection and refraction
    A. Define a ray as a line which gives the direction of travel of a wave. It is perpendicular to the wave fronts (surfaces of constant phase).
    B. Draw a plane boundary between two media and show an incident, a reflected, and a refracted ray. Label the angles these rays make with the normal to the surface. Write down the laws of reflection and refraction, the latter in terms of the indices of refraction relative to the vacuum.
    C. Remark that the index of refraction is a property of the material and depends on the wavelength of the light. Point out Table 1, which gives the indices of refraction of various materials. Note that the index of refraction for a vacuum is 1 and is nearly 1 for air. Point out Fig. 2.
    D. Shine an intense, monochromatic, well collimated beam on a prism and point out the reflected and refracted beams. A laser works reasonably well but it is difficult for the class to see the beam. Use smoke or chalk dust to make it visible. To avoid the

mess, use an arc beam or the beam from a 35 mm projector, filtered by red glass. Make a $\frac{1}{2}$" hole in a 2" by 2" piece of aluminum and insert it in the film gate. Use white light from the projector and the prism to shown that light of different wavelengths is refracted through different angles.

E. Explain total internal reflection. Show that no wave is transmitted when the angle of incidence is greater than the critical angle and derive the expression for the critical angle in terms of the indices of refraction. Stress that the index for the medium of incidence must be greater than the index for the medium of the refracted light. Total internal reflection can be demonstrated with some pieces of solid plastic tubing having a diameter larger than that used for fiber optics. The beam inside is quite visible. If time permits, discuss fiber optics and some of its applications.

F. Discuss Brewster's law. Explain that unpolarized light incident on a boundary is partially or completely polarized on reflection. When the angle of incidence and the angle of refraction sum to 90° the reflected light is completely polarized, with $\underline{E}$ perpendicular to the plane of the incident and reflected rays. Show that $\tan\theta_p = n_2/n_1$, where medium 1 is the medium of the reflected ray. Aim the projector beam on a plane sheet of glass held by a ring stand. Darken the room and obtain a reflection spot on the ceiling. Adjust the angle of the beam and the normal to the glass sheet and use a sheet of Polaroid to show the change in polarization.

II. Mirrors.

A. Discuss point sources, which emit spherical waves in all directions. Explain that the rays extend radially outward from these sources.

B. Consider a point source in front of a plane mirror. On a diagram, trace several rays from the source that are reflected to an observer's eye, in front of the mirror. Point out that when the reflected rays are extended backward they intersect at the same point behind the mirror. The image is formed at that point. Explain that the image is virtual since no light is actually emitted from that point. Point out that the object and image both lie on a normal to the mirror and that they are the same distance from the mirror. Explain the sign in i = -o; i is negative for virtual images.

C. Consider a point source in front of a concave spherical mirror. Draw a diagram that shows the optic axis, the center of curvature, and the source on the axis. Show that paraxial rays form an image and the object and image distances are related by $1/o + 1/i = 2/r$. See Section 11. To emphasize the paraxial approximation, consider the case o = 2r and use a full hemispherical surface. The paraxial formula predicts all rays cross the axis at i = (2/3)r, but the ray which strikes the edge of the mirror crosses at the vertex.

D. Explain that the mirror equation is valid for convex mirrors and for any position of the object, even virtual objects for which incoming rays converge toward a point behind the mirror. Give the sign convention: o and i are positive for real objects and images (in front of the mirror) and are negative for virtual objects and images; r is positive for concave mirrors and negative for convex mirrors. Remark that a surface is concave or convex according to its shape as seen from a point on the incident ray.

E. Define the focal point as the image point when the incident light is parallel to the axis. By considering a source far away, show that f = r/2. Consider a concave mirror and show that for o > f the image is real, for o < f the image is virtual. Also show that for o = f parallel rays emerge after reflection.

F. Describe a geometric construction for finding the image of an extended source. Trace rays from an off axis point: one through the center of curvature, one through the focal point, and one parallel to the axis. Use both concave and convex mirrors as examples.

G. Define lateral magnification and show that m = -i/o. Explain the sign: m is positive for erect images and negative for inverted images.

H. Take the limit r → ∞ and show that the mirror equation makes sense for a plane mirror.

III. Spherical refracting surfaces.

A. Draw a convex spherical boundary between two media, use the law of refraction to trace a paraxial ray from a source on the axis, and show that $n_1/o + n_2/i = (n_2-n_1)/r$. See Section 11. You can

162

demonstrate the bending of the light using a laser and a round
bottom flask. Use a little smoke or chalk dust to make the beam
visible in air and just a pinch of powdered milk in the water to
make it visible inside the flask.

B. Point out that real images are on the opposite side of the
boundary from the incident light and virtual images are on the
same side. o and i are positive for real objects and images,
negative for virtual objects and images. r is positive for convex
surfaces, negative for concave. With this sign convention the
equation holds for concave or convex surfaces and for $n_2 > n_1$ or
$n_1 > n_2$.

C. Consider the limit $r \to \infty$, which yields $i = -on_2/n_1$. This is the
solution to the apparent depth problem. For water 4 inches deep,
a ball on the bottom appears to be at a depth of about 3 inches.
Use an aquarium filled with water and a golf ball to make a
hallway display.

IV. Thin lenses.

A. By considering two refracting surfaces close together, derive the
thin lens equation: $1/o + 1/i = (n-1)(1/r' - 1/r'')$. See Section
11. Stress that the equation holds for paraxial rays and for a
thin lens in vacuum (or, to a good approximation, in air). r' is
the radius of the first surface struck by the light and r'' is
the radius of the second. They are positive or negative according
to whether the surfaces are convex or concave when viewed from a
point on the incident ray. To may wish to generalize the equation
by retaining the indices of refraction. The result is $1/o + 1/i =
(n_2/n_1 - 1)(1/r' - 1/r'')$. This allows you to consider a thin
glass or air lens in water.

B. By considering $o \to \infty$ show that the focal length is given by $1/f =
(n-1)(1/r' - 1/r'')$ or more generally by $1/f = (n_2/n_1 - 1)(1/r' -
1/r'')$. Then show that $1/o + 1/i = 1/f$. Point out that there are
two focal points, the same distance from the lens but on opposite
sides. Rays from a point source at f on one side are parallel on
the other side. Incident parallel rays converge to f on the other
side.

C. Show how to locate the image of an extended object by tracing a
ray parallel to the axis, a ray through the lens center, and a
ray through the front focal point.

D.   Define magnification and show that m = -i/o. Explain that the sign tells whether the image is erect or inverted.

E.   Consider all possible situations: converging lens with o > f, o < f, and o = f; diverging lens with o > f, o < f, and o = f. In each case show whether the image is real or virtual, erect or inverted, and find its position relative to the focal point.

F.   Note that optical instruments are constructed from a combination of two or more lenses. Point out that to analyze them, one considers one lens at a time, with the image of the previous lens as the object of the lens being considered. This sometimes leads to virtual objects. Note that the overall magnification is given by m = $m_1 m_2 m_3$... and that the sign of m tells whether the image is erect or inverted. If the image lies on the opposite side of the system from the object and is outside the system then it is real, otherwise it is virtual.

SUPPLEMENTARY TOPIC

Optical instruments. This section may be studied in the laboratory. Ask the students to experiment with the image forming properties of positive and negative lenses, then construct one or more optical instruments. Display several instruments in the lab.

SUGGESTIONS

1.   Following the derivation of the mirror equation, ask questions 28, 29, and 30.

2.   Following the derivation of the lens equation, ask questions 36, 39, 39, 40, and 42.

3.   Simple examples of refraction are provided by problems 2, 7, 9, and 10. Assign one or two of them. The relationship between the index of refraction and the speed of light is covered in problem 3.

4.   The basics of total internal reflection are covered in problem 23. Problems 24, 25, and 26 form a nice set dealing with various aspects. Optical fibers are explored in problems 31 and 32.

5.   Use problems 35 and 36 to have students think about Brewster's angle.

6.   Problems 39, 40, and 43 deal with the image position for an object in front of a plane mirror. Problems 42, 50, and 52 deal with multiple mirrors. Assign at least one of each set.

7. Assign problem 55 to cover most of the properties of spherical mirrors. For a less involved assignment, consider problems 53 and 56.

8. Assign problem 59 to cover most of the properties of spherical refracting surfaces. To cover image location only, assign problem 60.

9. Assign problem 72 to cover most of the properties of thin lenses. Problems 64 and 66 make use of the lensmaker equation; problems 65 and 69 deal with magnification.

10. Computer projects

    Chapters 19 and 20 of the second volume of the calculator supplement contain programs for tracing rays as they are reflected or refracted as boundaries. The paraxial assumption need not be made and the fuzziness of an image can be studied. Assign some of the problems given in these chapters.

    Chapter 21 of the calculator supplement describes simple matrix optics and shows how to apply it to the study of telescopes, microscopes, and camera lenses. This material can be used for individual projects or in conjunction with the lab. Ask a bright camera buff to work out the exercises dealing with zoom lenses.

11. Demonstrations

    Refraction at a plane surface: Freier and Anderson Od1 - 7
    Total internal reflection: Freier and Anderson Oe1 - 7; Hilton O2c, d.
    Polarization by reflection: Freier and Anderson Om2
    Plane mirrors: Freier and Anderson Ob1 - 6, Ob8, 9; Hilton O1c, d.
    Spherical mirrors: Freier and Anderson Oc1 - 8, 10, 11; Hilton O1f.
    Prisms: Freier and Anderson Of1 - 4; Hilton O2a, b, O3a.
    Lenses: Freier and Anderson Og1 - 7, 9 - 13; Hilton O4a.
    Optical instruments: Hilton O5.

12. Computer program

    Optics and Light, Focus Media, Inc., 839 Stewart Avenue, Garden City, NY 11530. Apple II. Demonstration and tutorial on Snell's law and thin lenses. User selects the parameters, then the program draws a ray diagram. Reviewed TPT January 1985.

13. Laboratory

    MEOS Experiment 13-1: Laser Ray Tracing. A laser beam is used to investigate the laws of reflection and refraction and to observe total internal reflection and the formation of images by spherical mirrors. Measurements are used to calculate the index of refraction of several bodies, including liquids, and the focal length of mirrors. Tracing is done by arranging the apparatus so the laser beam grazes a piece of white paper on the lab table. Much the same set of activities are described in BE Experiment 38: Reflection and Refraction of Light, but pins are used as objects rather than a laser source and rays are traced by positioning other pins along them. The technique can be used if you do not have sufficient lasers for the class.

BE Experiment 39: The Focal Length of a Concave Mirror. Several methods are described, including a technique which involves finding the radius of curvature. Others involve finding the image when the object distance is extremely long, when it is somewhat greater than 2f, and when it is somewhat less than 2f. Then the mirror equation is used to solve for f.

MEOS Experiment 13-2: Lenses. A light source and screen on an optical bench are used to find the focal lengths and magnifications of both convex and concave lenses. Chromatic and spherical aberrations are also studied. Also see BE Experiment 40: Properties of Converging and Diverging Lenses, a compendium of techniques for finding focal lengths.

BE Experiment 41: Optical Instruments Employing Two Lenses. Students construct simple two lens telescopes and microscopes on optical benches, then investigate their magnifying powers. By trying various lens combinations they learn the purposes of the objective and eyepiece lenses.

MEOS Experiment 13-3: Prism Spectrometer. Helium lines are used to determine the index of refraction as a function of wavelength for a glass prism. A good example of dispersion and excellent practice in in carrying out a rather complicated derivation involving Snell's law. Also see BE Experiment 43: Index of Refraction with the Prism Spectrometer and BE Experiment 44: The Wavelength of Light. In the second of these experiments students use a prism spectrometer to determine the wavelength of lines from a sodium source.

Chapter 40   INTERFERENCE

BASIC TOPICS

I.   Huygens' principle
    A.   Shine monochromatic light through a double slit and project the
        pattern on the wall. Either use a laser or place a single slit
        between the source and the double slit. Use a diagram to explain
        the setup. Point out the appearance of light in the geometric
        shadow and the occurrence of dark and bright bands. You can make
        acceptable double slits by coating a microscope slide with lamp
        black or even black paint. Tape a pair of razor blades together
        and draw them across the slide. By inserting various thicknesses
        of paper or shim stock between the blades you can obtain various
        slit spacings.
    B.   Explain that Huygens principle will be used to understand the
        pattern (and other phenomena), then state the principle. Draw a
        plane wave front, construct spherical wave fronts of the same
        radius centered at several points along the plane wave front,
        then draw the plane tangent to these. Describe plane wave
        propagation in terms of Huygens' wavelets.
    C.   Use Huygens' principle to derive the law of refraction. Assume
        different wave speeds in the two media and show that the
        wavelengths are different. Consider wave fronts one wavelength
        apart and show that $\sin\theta_1/\sin\theta_2 = v_1/v_2$. Explain that $n = c/v$ and
        obtain the law of refraction.
    D.   Go back to the double slit pattern and explain that those parts
        of an incident wave front that are within the slit produce
        spherical wavelets which travel to the screen while wavelets from
        other parts are blocked. Some wavelets reach the geometric
        shadow. The spreading of the pattern beyond the shadow is called
        diffraction and will be studied in the next chapter. Wavelets
        from different slits arrive at the same point on the screen and
        interfere to produce the bands. This phenomena will be studied in
        this chapter.

II.   The double slit interference pattern.
    A.   Draw a diagram of a plane wave incident on a two slit system and
        draw a ray from each slit to a screen far away. Remark that the

waves are in phase at the slits but they travel different distances to get to the same point on the screen and may have different phases there. The electic fields sum to the total electric field. At some points the two fields cancel, at other points they reenforce each other. Remind students that the intensity is proportional to the square of the total field, not to the sum of the squares of the individual fields.

B. Point out that if the screen is far away the two rays are nearly parallel, then show that the difference in distance traveled is $d\sin\theta$, where d is the slit separation and θ is the angle the rays make with the forward direction. Explain the condition $d\sin\theta = m\lambda$ for a maximum of intensity and the condition $d\sin\theta = (m+\frac{1}{2})\lambda$ for a minimum.

C. Explain that the two interfering waves must be coherent to obtain an interference pattern. The phase difference at the observation point must be constant over the observation time. Explain why two incandescent lamps, for example, do not produce a stable interference pattern. The light is from many atoms and the emission time for a single atom is about 10^{-8} s. The phase difference changes in a random way over times that are short compared to the observation time. Point out that in this case the intensities add. If you did not use a laser in the demonstration, explain the role of the single slit in front of the double slit.

III. The intensity.

A. Take the two fields to be $E_1 = E_0\sin\omega t$ and $E_2 = E_0\sin(\omega t+\phi)$, where $\phi = (2\pi/\lambda)d\sin\theta$. This is easily shown by remarking that $\phi = k\Delta d$, where $k = 2\pi/\lambda$ and $\Delta d = d\sin\theta$ (derived earlier).

B. Explain how the fields can be represented on a phasor diagram. If you did not cover Chapter 36 explain that a phasor has a length proportional to the amplitude and makes the angle ωt or $\omega t+\phi$ with the horizontal axis. Its projection on the vertical axis is proportional to the field. Sum the phasors to obtain the total field. Show that the amplitude E of the total field is $2E_0\cos(\phi/2)$. Plot the intensity $4E_0^2\cos^2(\phi/2)$ as a function of ϕ. Point out that $\phi = 0$ produces a maximum, that maxima occur at regular intervals, and that the minima are halfway between adjacent maxima.

C. Show that the intensity at a maximum is 4 times the intensity due to one source alone. Remark that the average intensity over the pattern is $2I_0$, just as for two incoherent sources and that no energy is gained or lost over the pattern.

D. Note the width of each maximum, at half the peak, is given by $\sin\theta_{1/2} = \lambda/4d$. The smaller λ/d the sharper the maximum. Near the central maximum, where $\sin\theta \approx \tan\theta \approx \theta$, the linear spread on the screen is $y \approx (\lambda/4d)D$, where D is the distance from the slits to the screen.

E. It is also worth noting that since $\sin\theta = m\lambda/d \leq 1$ for a maximum, the smaller λ/d, the more maxima occur.

IV. Thin film interference.

A. Cut a 1 to 2 mm slit in a 2" square piece of aluminum and insert it in the film gate of a 35 mm projector. Let the beam impinge on a soap bubble to show the effect.

B. Consider normal incidence on a thin film of index n_1 in a medium of index n_2 and suppose the medium behind the film has index n_3. Explain that a wave reflected at the interface with a medium of higher index undergoes a phase change of π. As an analogy use Fig. 15 of Chapter 17, showing mechanical waves on a string. If $n_1 < n_2 < n_3$, waves reflected at both surfaces undergo phase changes of π. Consider all other possibilities and then specialize to a thin film of index n, in air. Give the conditions for maxima and minima for both the reflected light and the transmitted light, assuming near normal incidence. Note that the wavelength in the medium must be used to calculate the phase change on traveling through the medium. Define optical path length and point out its importance for thin film interference.

C. Broaden the discussion qualitatively by including non-normal incidence. Note that for some angles conditions are right for destructive interference of a particular color while at other angles conditions are right for constructive interference of the same color. Also note that these angles depend on λ. Hence the soap bubble colors.

D. If time permits, discuss Newton's rings. Use a plano-convex lens and a plane sheet of glass together with a laser. Use a diverging lens to spread the beam.

SUPPLEMENTARY TOPIC

The Michelson interferometer. This is an excellent example of an application of interference effects. Set up a hallway demonstration and give a brief explanation.

SUGGESTIONS

1. Another point to mention in the discussion of the double slit pattern: the amplitude of the wavelets fall off as 1/r and are not quite the same at the screen. Show this is a negligible effect for the patterns considered here.

2. In the discussion of coherence, give a more detailed explanation of the single slit placed between the source and the double slit. See questions 5, 11, and 27.

3. Questions 6, 12, and 13 are good tests of understanding. Use them in the discussion or ask students to answer them for homework.

4. Problems 6 through 15 deal with the basics of interference. Also consider problem 18. Use problems 19, 23, and 25 to test for understanding of the derivation of the double slit equation.

5. Coherence is covered in problem 28.

6. Assign problems 29, 31, and 34 in connection with the double slit interference pattern.

7. Use problems 38, 39, and 44 to help with the discussion of thin films. Problem 51 is a nice test of understanding of the phase change on reflection.

8. Film

    The Michelson Interferometer, S8, color, 4 min. The Kalmia Company (see Chapter 4 notes for address). Part of the Miller Demonstrations in Physics series. Reviewed TPT 14:253 (1976).

9. Demonstrations

    Double slit interference: Freier and Anderson O14, 5, 9; Hilton O7a. Thin film interference: Freier and Anderson O115 - 18; Hilton O7b, c, d.

    Michelson interferometer: Freier and Anderson O119.

10. Film

    Light Waves, Educational Materials and Equipment Company, P.O. Box 17, Pelham, NY 10803. Apple II, TRS-80 I, III, & W, IBM PC. Simulations of Young's experiment with user selected parameters. Students can view either a graph of the intensity or a simulated

intensity pattern. Useful for lecture demonstrations. Reviewed TPT
September 1984.

11. Laboratory

MEOS Experiment 13-4: <u>Interference and Diffraction</u>. Students observe
double slit patterns of water waves in a ripple tank, sound waves,
microwaves, and visible light. In each case, except water waves,
they measure and plot the intensity as a function of angle, then use
the data to calculate the wavelength. A microcomputer can be used to
take data and plot the intensity of a visible light pattern.

MEOS Experiment 13-6: <u>The Michelson Interferometer</u>. An
interferometer is used to measure the wavelengths of light from
mercury and a laser and to find the index of refraction of a glass
pane and air. Good practical applications.

Chapter 41   DIFFRACTION

<u>BASIC TOPICS</u>

I.   Qualitative discussion of single slit diffraction.

A.   Shine coherent monochromatic light on a single slit and project
     the pattern on the wall. Point out the broad central bright
     region and the narrower, less bright regions on either side, with
     dark regions between. Also point out that light is diffracted
     into the geometric shadow.

B.   Explain that the Huygens' wavelets not only spread into the
     shadow region but that they arrive at any selected point with a
     distribution of phases and interfere to produce the pattern.
     Explain that for quantitative work this chapter deals with
     Fraunhofer diffraction, with the screen far from the slit.

C.   Draw a single slit with a plane wave incident normal to it. Also
     draw parallel rays from equally spaced points within the slit,
     all making the same angle θ with the forward direction. Point out
     that all wavelets are in phase at the slit. The first minimum can
     be located by selecting θ so that, at the observation point, the
     ray from the top of the slit is 180° out of phase with the ray
     from midslit. All wavelets then cancel in pairs. Show that this
     leads to $a\sin\theta = \lambda$, where a is the slit width. Point out that
     this value of θ determines the width of the central bright region

and that this region gets wider as the slit width narrows. Use
sinθ ≈ tanθ ≈ θ to show that the width of the central region is
2Dλ/a. Use a variable width slit or a series of slits to
demonstrate the effect.

D. By dividing the slit into fourths, eighths, etc. and showing that
in each case the wavelets cancel in pairs if θ is properly
selected, find the locations of other minima. Show that $a\sin\theta = m\lambda$ for a minimum.

E. Qualitatively discuss the intensity. Draw a phasor diagram
showing 10 or so phasors representing wavelets from equally
spaced points in the slit. Show that each wavelet at the
observation point is out of phase with its neighbor by the same
amount. First show the phasors with zero phase difference (θ =
0), then show them for a larger value of θ. Show that they
approximate a circle at the first minimum and then, as θ
increases, they wrap around to form another maximum, with less
intensity than the central maximum. Point out that as θ increases
the pattern has successive maxima and minima and that the maxima
become successively less intense.

II. The intensity.

A. Draw a diagram showing 10 or so phasors along the arc of a circle
and let φ be the phase difference between the first and last. See
Fig. 12. Use geometry to show that $E = E_0[\sin\alpha]/\alpha$, where $\alpha = \phi/2$.
Point out that the intensity can be written $I = I_m[(\sin\alpha)/a]^2$. By
examining the path difference for the rays from the top and
bottom of the slit, show that $\alpha = (\pi a/\lambda)\sin\theta$. Explain that these
expressions give the intensity as a function of angle θ.

B. Sketch the intensity as a function of θ (see Fig. 11) and show
mathematically that the expression just derived predicts the
positions of the minima as found earlier.

III. Multiple slit patterns.

A. Consider the double slit arrangement discussed in the previous
chapter. Point out that the electric field for the light from
each of the slits obeys the equation developed for single slit
diffraction and these two fields are superposed. The result for
Fraunhofer diffraction is $I(\theta) = I_m(\cos\beta)^2[(\sin\alpha)/\alpha]^2$, the

product of the single slit diffraction equation and the double slit interference equation.

B. Sketch I vs. θ for a double slit and point out that the single slit pattern forms an envelope for the double slit interference pattern.

C. Qualitatively describe the pattern produced as the number of slits is increased. Principle maxima occur whenever the path difference for rays from two adjacent slits is an integer multiple of the wavelength: dsinθ = mλ. These are flanked by secondary maxima and the interference pattern has the single slit diffration pattern as its envelope.

D. If time permits, consider in detail the case of 3 slits, then 4. Quantitatively show that the principle maxima become narrower as the number of slits increases. Make multiple slits using razor blades and a lamp blackened microscope slide, then use a laser to show the patterns, in order of increasing number of slits. Finish with a commercial grating.

IV. X-ray diffraction (optional).

A. Explain that x rays are electromagnetic radiation with wavelength on the order of 10^{-10} m (1 Å). Point out that crystals are regular arrays of atoms with spacings on that order and so can be used to diffract x rays.

B. Consider a set of parallel crystalline planes and explain that reflection of the incident beam occurs at each plane, with the angle of reflection equal to the angle of incidence. Draw a diagram like Fig. 26 and state that x-ray diffraction is conventionally described in terms of the angle between the ray and the plane, rather than the normal to the plane. Show that waves reflected from the planes interfere constructively if 2dsinθ = mλ.

C. Explain that for a given set of planes intense diffracted waves are produced only if waves are incident at an angle θ that satisfies the Bragg condition, given above. Measurements of these angles can be used to investigate the crystal structure. The same ideas will be used to discuss the diffraction of electron waves by crystals in Chapter 44.

SUPPLEMENTARY TOPICS

1. Diffraction from a circular aperture. This topic is important for its application to diffraction patterns of lenses and the diffraction limit to the resolution of objects by a lens system. If you intend to discuss the resolving power of a grating, the Rayleigh criterion for a circular aperture should be covered first since it is easier to present and understand. You can demonstrate the Rayleigh criterion by drilling two small holes, closely spaced, in the bottom of a tin can. Place the can over a light bulb and let students view it from various distances. See problem 18. Also use red and blue filters to show the dependence on wavelength.

2. Resolving power of a grating. This topic is also of practical interest. Carefully distinguish between dispersion and resolving power.

SUGGESTIONS

1. To test for understanding of the derivation of the single slit diffraction equation, ask questions 16 and 20. Basic relationships are explored in problems 1 through 6 and 8. Problem 11 might help with understanding the derivation.

2. Following the discussion of the equation for the double slit pattern, ask question 24. Characteristics of the pattern are explored in problems 35, 36, 38, and 40.

3. Pick some of problems 43, 44, 46, 47, 48, 50, and 51 in conjunction with gratings.

4. Computer projects
    Assign some of the interference exercises from Chapter 22 of the second volume of the calculator supplement. A computer is used to plot the intensity pattern for various situations including the case when the screen is not far from the sources. If desired, Eureka can be used to carry out the integrations and plot the pattern.
    To illustrate the importance of coherence, assign problem 6, section 22.1 of the calculator supplement. A random number generator is used to select the phase difference and the average of a number of patterns with randomly selected phase differences is calculated.

5. Films
    Joseph Fraunhofer: Diffraction, 16 mm, color, 16 min. Office of Instructional Media, Rensselaer Polytechnic Institute, Troy, NY

12181. This historic film by Professor Leitner illustrates a series of diffraction experiments that were performed by Fraunhofer circa 1820. Reviewed AJP 44:116 (1976).

Interference and Diffraction, 16 mm, b/w, 19 min. Modern Learning Aids (see Chapter 2 notes for address). This PSSC film, although over 20 years old, provides an excellent visual illustration of the phenomena of interference and diffraction through the use of a ripple tank. Reviewed AJP 32:62 (1964).

Shadow of a Hole, S8, color, 3.5 min. American Association of Physics Teachers (see Chapter 2 notes for address). A changing diffraction pattern is shown as the observation screen is moved away from the observer. Reviewed TPT 15:564 (1977) and AJP 46:197 (1978).

Diffraction - Single Slit; Diffraction - Double Slit; Resolving Power, S8, color, 4 min. each. The Kalmia Company (see Chapter 4 notes for address). These excellent loops are part of the Franklin Miller Demonstrations in Physics series. Reviewed TPT 14:253 (1976).

6. Computer program

Physics Simulations III: Diffraction, Kinko's Service Corporation (see Chapter 4 notes for address). Macintosh. Program shows intensity plots for single slits, double slits, and other apertures.

7. Demonstrations

Single slit diffraction: Freier and Anderson O12, 3, 6, 7; Hilton O7a.

Multiple slit diffraction: Freier and Anderson O110, 13; Hilton O7a; g.

Diffraction by circular and other objects: Freier and Anderson O121 - 23; Hilton O7g.

Diffraction by crystals: Freier and Anderson O114.

8. Laboratory

MEOS Experiment 13-4: Interference and Diffraction. See Chapter 40 notes.

MEOS Experiment 13-5: Diffraction Gratings. A grating spectrometer is used to find the wavelengths of the visible lines of helium. The patterns of gratings with various slit spacings are observed. Alternatively, a laser may be used to project the lines. Also see BE Experiment 45: A Study of Spectra with the Grating Spectrometer.

MEOS Experiment 14-4: Bragg Diffraction with Microwaves. Small steel balls imbedded in styrofoam form a simple cubic, body centered cubic, or face centered cubic lattice. Microwaves impinge on the balls and the scattered intensity is measured as a function of scattering angle. The data is used to determine the lattice structure and to compute the lattice constant and ball separation. a large scale working replica of an x-ray crystallography experiment.

Chapter 42   RELATIVITY

BASIC TOPICS

I.   Introduction
  A.   Consider a wave on a string and remind students that its speed relative to the string is given by $v_w = \sqrt{F/\mu}$, where F is the tension and μ is the linear mass density. Explain that, according to non-relativistic mechanics, an observer running with speed v_o with the wave measures a wave speed of $v_w - v_o$ and an observer running against the wave measures a wave speed of $v_w + v_o$. Remark that these results are not valid for light (or fast moving waves and particles). The speed of light in a vacuum is found to be the same regardless of the speed of the observer (or the speed of the source).
  B.   Remark that this fact has caused us to revise drastically our idea of time. If, for example, two observers moving at high speed with respect to each other time the interval between two events they obtain different results.
  C.   Explain that special relativity is a theory which relates measurements taken by two observers who are moving with respect to each other. Although it sometimes seems to contradict everyday experience, it is extremely well supported by experiment.
  D.   State the postulates: the laws of physics are the same for observers in all inertial frames, the speed of light in a vacuum is the same for all directions and in all inertial frames. Remind students what an inertial frame is. Explain that the laws of physics are relationships between measured quantities, not the quantities themselves. Newton's laws and Maxwell's equations are examples. State that relativity has forced us to revise Newton's second law but not Maxwell's equations.

II. Time measurements.

A. Explain the term _event_ and note that three space coordinates and one time coordinate are associated with each event. Explain that each observer may think of a coordinate system with clocks at all places where events of interest occur and that the clocks are synchronized. Outline the synchronization process involving light. State that the coordinate system and clock used by an observer are at rest with respect to the observer and may be moving from the viewpoint of another observer.

B. State that two observers in relative motion cannot both claim that two events are simultaneous. To illustrate show Fig. 5. The meteorites strike simultaneously in the frame of the lower ship, as evidenced by light flashes from the events reaching midship simultaneously, but they do not strike simultaneously in the frame of the upper ship.

C. Explain the light flasher used to measure time, in principle. Consider a flasher at rest in one frame, take two events to be a flash and the subsequent reception of reflected light back at the instrument, then remark that the time interval is $\Delta t_0 = 2D/c$, where D is the separation of the mirror from the flash bulb. Consider the events as viewed in another frame, moving perpendicularly to the light ray, and show the interval is $\Delta t = 2D/c\sqrt{1-\beta^2}$ or $\Delta t = \Delta t_0/\sqrt{1-\beta^2}$, where $\beta = v/c$. This is also written $\Delta t = \gamma \Delta t_0$, where γ $(= 1/\sqrt{1-\beta^2})$ is called the Lorentz factor. State that $\beta < 1$ and $\gamma > 1$.

D. Remark that Δt_0 is the _proper time_ interval; both events occur at the same place in the frame in which it is measured. Point out that Δt is larger than Δt_0. Explain that the same result is obtained no matter what clocks are used for the measurement (as long as they are accurate and each is at rest in the appropriate frame). Ask students to identify a frame to measure the proper time interval for a ball thrown from third to first base. Note that $\Delta t \approx \Delta t_0$ if $v \ll c$.

E. State that time dilation has been observed by comparing clocks carried on airplanes to clocks remaining behind and by comparing the average decay time of fast moving fundamental particles to their decay time when at rest. You might want to discuss the twin paradox here.

III. Length measurements.

A. Point out the problem with measuring the length of an object that is moving relative to the meter stick: the postion of both ends must be marked <u>simultaneously</u> (in the rest frame of the meter stick) on the meter stick. If the speed v of the object is known another way can be used to measure its length: put a mark on a coordinate axis along the line of motion of the object, then measure the time Δt_0 taken by the object to pass the mark. The length is given by $L = v\Delta t_0$. Note that Δt_0 is a proper time interval.

B. Explain that the length of the object, as measured in its rest frame is $L_0 = v\Delta t$, where Δt is the time interval measured in that frame. Substitution of $\Delta t = \gamma\Delta t_0$ leads to $L = L_0/\gamma$. State that L_0, the length as measured in the rest frame of the object, is called the <u>proper length</u>. Since $\gamma > 1$, all observers moving with respect to the object measure a length that is less than the proper length. The same result is obtained no matter what method is used to measure length. Note that $L \approx L_0$ if $v \ll c$.

IV. The Lorentz transformation.

A. Consider two reference frames: S′ moving with speed v in the positive x direction relative to S. Remark that the coordinates of an event as measured in S are written x,y,z,t while the coordinates as measured in S′ are written x′,y′,z′,t′. Write down the Lorentz transformation for the coordinate differences of two events: $\Delta x' = \gamma(\Delta x - v\Delta t)$, $\Delta y' = \Delta y$, $\Delta z' = \Delta z$, $\Delta t' = \gamma(\Delta t - v\Delta x/c^2)$. Remark that these equations reduce to the Galilean transformation if $v \ll c$: $\Delta x' = \Delta x - v\Delta t$, $\Delta y' = \Delta y$, $\Delta z' = \Delta z$, $\Delta t' = \Delta t$.

B. Explain that the transformation equations can be solved for Δx and Δt, with the result $\Delta x = \gamma(\Delta x' + v\Delta t')$, $\Delta t = \gamma(\Delta t' + v\Delta x'/c^2)$. From the viewpoint of an observer in S′, S is moving in the negative x′ direction, so the two sets of equations are obtained from each other when v is replaced by −v.

C. Discuss some consequences of the Lorentz transformation equations:

1. Take $\Delta t' = 0$, $\Delta x' \neq 0$ and show that $\Delta t = \gamma v\Delta x'/c^2$ ($\neq 0$). If two events are simultaneous and occur at different places in S′, then they are not simultaneous in S. Similarly, take $\Delta t = 0$, $\Delta x \neq 0$ and show $\Delta t' = -\gamma v\Delta x/c^2$ ($\neq 0$).

2.   Time dilation. Consider two events which occur at the same place in S and show that $\Delta t' = \gamma \Delta t$. Point out that Δt is proper time interval. Also show that the events do not occur at the same place in S': $\Delta x' = -\gamma v \Delta t$. Work the same problem for two events that occur at the same place in S'.

3.   Length measurement. Suppose the object is at rest in S' and the meter stick is at rest in S. Marks are made simultaneously in S on the meter stick at the ends of the object. Thus $\Delta t = 0$. Show that $\Delta x' = \gamma \Delta x$ and point out that $\Delta x'$ is the proper length. Work the same problem with the object at rest in S and the meter stick at rest in S'.

4.   Causality. Consider two events, the first of which influences the second. For example a particle is given an initial velocity along the x axis and collides with another particle. Remark that t_2 (the time of the collision) must be greater than t_1 (the time of firing). Take $\Delta t = t_2 - t_1$ and $\Delta x > 0$, then show that the Lorentz transformation predicts $\Delta t'$ is positive for every frame for which $v < c$. The collision cannot happen before the firing in any frame moving at less than the speed of light.

5.   Velocity transformation. The notation changes in the text: u now represents the velocity of frame S' relative to S, v and v' represent the velocity of a particle, as measured in S and S', respectively. Divide the Lorentz equation for Δx by the Lorentz equation for Δt to show that the particle velocity in S is $v = (v' + u)/(1 + uv'/c^2)$. Show this reduces to the Galilean transformation $v = v' + u$ for $u \ll c$. Take $u = v' = c$ and show that $v = c$. If $v' < c$ then $v < c$ for all frames. Have students read or reread Section 4-9.

V.   Relativistic momentum and energy.

A.   Explain that the non-relativistic definition of momentum must be generalized if momentum is to be conserved in collisions involving particles moving at high speeds. State that the proper generalization is $\underline{p} = \gamma m \underline{v}$, where γ is the Lorentz factor for the rest frame of the particle. Remark that $\underline{p}$ is unbounded as the particle speed approaches the speed of light.

B.   Remark that γm is sometimes referred to as the relativistic mass of the particle; m is then called its rest mass and is represented by m_0. In this text m is used for the rest mass and is called simply the mass.

C. Explain that Newton's second law in the form $\underline{F} = dp/dt$ is valid but that it is not the same as $\underline{F} = m\underline{a}$ for high speed particles.

D. Remark that the definition of energy must be changed if the work-energy theorem is to hold for particles at high speeds. Write $dW = Fdx = (dp/dt)vdt = (dp/dv)(dv/dt)vdt = (dp/dv)vdv = \gamma^3 mvdv$ for the work done by the net force as a particle moves along the x axis. This can easily be integrated to show that the work done equals the change in the quantity $E = \gamma mc^2$. Take the limit as v/c becomes small and show that E can then be approximated by $mc^2 + \frac{1}{2}mv^2$. Thus the correct relativistic definition of the kinetic energy is $K = E - mc^2 = (\gamma-1)mc^2$. Point out that the particle has energy mc^2 when it is at rest and remark that mc^2 is called the rest energy.

E. Explain that rest mass and rest energy are not conserved in many interactions involving fundamental particles but that total energy is; rest energy can be converted to kinetic energy and vice versa.

F. Derive $E^2 = (pc)^2 + (mc^2)^2$ and remark that $E = pc$ for a massless particle, such as a photon.

SUPPLEMENTARY TOPIC

The Doppler effect for light. The expression for the frequency transformation can be derived easily by considering the measurement of the period in two frames. Suppose an observer in S obtains T for the interval between successive maxima at the same place. This a proper time interval and the interval in another frame S′ is γT. If S′ is moving parallel to the wave, however, the two events do not occur at the same place in S′ and γT is not the period in that frame. An observer in S′ must wait for a time $|\Delta x′|/c$ longer before the next maxima is reached at the place of the first. Thus $T′ = \gamma T + |\Delta x′|/c$ or since $\Delta x′ = -\gamma Tv/c$, $T′ = \gamma T(1+v/c) = T\sqrt{(1+\beta)}/\sqrt{1-\beta}$. Thus $\nu′ = \nu\sqrt{1-\beta}/\sqrt{1+\beta}$. If S′ is moving perpendicularly to the wave the two events occur at the same place in both frames and $T′ = \gamma T$, so $\nu′ = \nu/\gamma$.

SUGGESTIONS

1. Use questions 1 and 2 to broaden the discussion of inertial frames.

2. Simultaneity and time dilation are the issues in questions 10, 11, 12, and 14. Ask them to test understanding. Also assign problems 7 and 20.

3. Length contraction is the issue in question 13. Ask it to test understanding. Also assign problems 11, 13, and 16.

4. Use questions 21, 22, 23, 24, 25, and 26 to broaden the discussion of rest mass. Cover some of them in lecture, assign some of the others.

5. Assign problems 17 and 18 in support of the discussion of the Lorentz transformation. Assign problems 26 and 28 in connection with the relativistic velocity transformation and problems 33 and 34 in connection with the relativistic Doppler effect. Problems 38, 43, 46, and 53 cover relativistic energy and momentum. If you covered cyclotrons in Chapter 30, assign problem 57.

6. Computer project
   Chapter 23 of the calculator supplement contains some problems dealing with the Lorentz and velocity transformations. Eureka can be programed to quickly carry out a Lorentz transformation so students can investigate interesting situations and think about measurements as viewed from various frames.

7. Computer programs
   Physics Simulations I: Einstein, Kinko's Service Corporation (see Chapter 4 notes for address). Macintosh. The screen is split to shown the views of events as seen in two frames which are moving relative to each other. Clocks time intervals between events. Use to demonstrate time dilation, length contraction, twin paradox.
   Intermediate Physics Simulations: Relativistic motion, R.H. Good (see Chapter 15 notes for address). Apple II. User adjusts the velocity (in two dimensions) of a moving clock, which ticks at uniform intervals and lays down a marker at each tick. The screen shows the time in the observer's frame and in the rest frame of the clock. Use to demonstrate time dilation, length contraction, twin paradox.

Chapter 43   QUANTUM PHYSICS - I

BASIC TOPICS

I.   Introduction.
     A. Explain that this chapter deals with some of the fundamental
         results of quantum mechanics. The first few sections describe
         experimental results that can be understood only if light is
         regarded as made up of particles. Remark that interference and
         diffraction phenomena require waves for their explanation.
         Reconciliation of these opposing views will be discussed in the
         next chapter.
     B. Explain that the energy of a photon is related to the frequency
         of the wave through $E = h\nu$ and the momentum of a photon is
         related to the wavelength of the wave through $p = h/\lambda$. The energy
         density in a wave is $nh\nu$, where n is the photon concentration.
         Show these equations predict $p = E/c$, the classical relationship.
         Recall the discussion of the Poynting vector in Chapter 38.
         Explain that Planck's constant is a constant of nature and
         pervades quantum mechanics. Give its value (6.63×10^{-34} J·s) and
         calculate the photon energy and momentum for visible light, radio
         wave, and x-rays.

II. The photoelectric effect.
     A. Sketch a schematic of the experimental setup. Explain that
         monochromatic light is incident on a sample. It is absorbed and
         the energy goes to electrons, some of which are emitted. The
         energy of the most energetic electron is found by measuring the
         stopping potential V_0.
     B. Point out that the stopping potential is independent of the light
         intensity. As the intensity is increased, more electrons are
         emitted but they are not more energetic. Show a plot of the
         stopping potential as a function of frequency and point out that
         the relationship is linear and that as the frequency is increased
         the electrons emitted are more energetic. Also state that
         electrons are emitted promptly when the light is turned on. If
         the radiation energy were distributed throughout the region of a
         wave it would take a noticeable amount of time for an electron to
         accumulate sufficient energy to be emitted, since an electron has

a small surface area. This argument can be made quantitative (see Sample Problem 3).

C. Give the Einstein theory. Electromagnetic radiation is concentrated in photons, with each photon having energy $h\nu$. The most energetic electrons after emission are those with the greatest energy while in the material and, in the interaction with a photon, receive energy $h\nu$. If the light intensity is increased without changing the frequency, there are more photons and hence more electrons emitted, but no single electron can receive more energy. Furthermore, the electron receives energy immediately and need not wait to absorb the proper amount.

D. Show that this analysis leads to $h\nu = \phi + K_{max}$, where ϕ is the work function, the energy needed to remove the most energetic electron from the material. It is characteristic of the material. Remark that $K_{max} = eV_0$ and that the Einstein theory predicts a linear relationship between V_0 and ν and predicts a minimum frequency for emission: $h\nu = \phi$. Remark that the emitted electrons have a distribution of speeds if $h\nu > \phi$ because they come from states with different energies.

III. The Compton effect.
A. Note that in the explanation of the photoelectric effect a photon is assumed to give up all its energy to an individual electron. The photon then ceases to exist. Explain that a photon can also transfer only part of its original energy in an interaction with an electron. Since a lower energy means a lower frequency, the scattered light has a longer wavelength than the incident light. Show Fig. 5. Stress that the experimental data can be explained by considering the interaction to be a collision between two particles, with energy and momentum conserved. Relativistic expressions must be used, however, for energy and momentum.

B. Discuss the experiment. Light is scattered from electrons in matter and the intensity of the scattered light is measured as a function of wavelength for various scattering angles.

C. Remark that the situation is exactly like a two dimensional collision between 2 particles. Write down the relativistic expressions for the momentum and energy of a particle with mass (the electron) and remind students of the rest energy. Assume the electron is initially at rest and that the photon is scattered through the angle φ. The electron leaves the interaction at an angle θ to the direction of the incident photon. Write down the equations for the conservation of energy and the conservation of momentum in two dimensions. Write down the momentum and energy of the photon in terms of the wavelength and solve for the change on scattering of the wavelength.

D. Note that the change in wavelength is independent of wavelength and that the change is significant only for short wavelength light, in the x-ray and gamma ray regions. Also state that the theoretical results successfully predict experimental data. The width of the curves is due chiefly to moving electrons for which Δλ is slightly different for scattering from different electrons and the peak near Δλ = 0 is due to scattering from more massive particles (atoms as a whole). Stress that the particle picture of light accounts for experimental data.

IV. Cavity radiation and the quantization of energy.

A. Describe a radiation cavity as a hollow block of material with a small hole to the outside. The material is kept at a uniform constant temperature and the electromagnetic radiation in the cavity is studied by observing some that leaks out of the hole. Explain that the interior walls absorb and emit radiation and that the distribution of energy among the various wavelengths depends on the temperature but is independent of the material, the size of the cavity, and the shape of the cavity.

B. The quantity of interest is the spectral radiancy S(λ), defined so that S(λ)dλ is the radiation rate per unit area for electromagnetic energy in the wavelength range from λ to λ+dλ. Explain that the electromagnetic energy in the cavity is in thermal equilibrium with the material and the spectral distribution should be predicted when thermodynamics is applied to the radiation. Show Fig. 8, which compares experimental results with classical theory. Point out that the experimental curve reaches a peak in the infrared or red and falls off on either side. As the temperature increases the peak becomes sharper and moves toward the blue end of the spectrum. Classical theory does not predict a peak.

C. Give the Planck law and explain that it fits the data at all wavelengths. Describe the assumptions used by Planck to derive the law: radiation is emitted and absorbed by atoms in thc walls and that the energy absorbed or emitted is quantized. Furthermore the radiation energy in the cavity is also quantized, in units of $h\nu$, where ν is the frequency of the radiation. The law is derived by assuming the quanta of radiation are in thermodynamic equilibrium with the cavity walls.

D. Point out that classically electromagnetic radiation can have any value of energy. Quantum mechanically this is not true but since h is so small the discreteness of the energy values is important only at the atomic level.

E. Briefly discuss the correspondence principle: quantum mechanical results reduce to classical results when the situation is such that classical results are valid. If you presented the Planck expression for $S(\lambda)$ then show it reduces to the classical expression if $\hbar c/\lambda kT \ll 1$.

V. Line spectra and the hydrogen atom.

A. Use a commercial hydrogen tube to show the visible hydrogen spectrum. Since the intensity is low you will not be able to project this but you can purchase inexpensive 8"x10" sheets of plastic grating material, which can be cut into pieces and passed out to the students. Point out Fig. 10. Explain that the discrete lines they see have frequencies described by the empirical Balmer formula $\nu = Rc[(1/2)^2 - (1/n)^2]$, where $R = 1.1 \times 10^7$ m^{-1} was an experimentally determined constant. The red line is given by the formula with n = 3 and the bluest visible line corresponds to n = 6. Lines with n > 6 can be detected with instruments but cannot be seen by the eye. Mention that other series of lines are known and that they are described by a similar formula with the 2 replaced by another integer. Point out that classical physics predicts a continuous spectrum. Since the electron is accelerating as it goes around the nucleus, it should radiate continuously and fall into the nucleus. Stress that it does not, the atom is stable, and the spectrum is discrete.

B. Explain that an electron bound in an atom, molecule, or macroscopic matter can have an energy which is one of a discrete set of values, not a continuum. It radiates or absorbs by changing its state and the frequency of the radiation is given by $h\nu = E_f - E_i$. Demonstrate that radiation from hydrogen can be explained if the electron can have only discrete energies, given by $-hRc/n^2$, where n = 1, 2, 3, ... Point out Fig. 13.

C. Optional. Use classical mechanics to derive expressions for the energy and angular momentum as functions of the orbit radius for an electron in hydrogen. Write down Bohr's hypothesis: $L = nh/2\pi$ and show this leads to the quantization of energy. Also derive the expression for the possible frequencies of radiation emitted from hydrogen. Note that the theory gives an expression for R in terms of fundamental constants and the calculated value is in excellent agreement with the empirical value.

SUGGESTIONS

1. Emphasize that the energy in a light beam is the product of the number of photons and the energy of each photon. Assign some of problems 8, 9, 10, and 11. Ask questions 2, 3, 4, and 5 as part of a discussion of photon properties. Also assign problem 13.

2. After discussing the photoelectric effect, ask questions 10, 11, and 14. Assign problem 24. You may also wish to go over problem 28 after discussing the Compton effect.

3. After discussing the Compton effect, ask questions 16, 17, and 18. Also assign problem 31. If you include the recoil of the electron in the discussion, assign problem 29, 30, or 31.

4. To help in understanding the Planck radiation law, assign problem 48 or go over it in class, then ask the students to do problem 47.

5. After discussing the hydrogen spectrum, ask questions 23, 24, and 29. Also assign problems 58 and 63. If you have emphasized the terms binding energy and excitation energy, assign problem 68.

6. Problem 76 is a good review of the Bohr theory. Also consider problem 69.

7. Question 30 is a good review question. Ask it when the chapter has been studied.

8. Chapter 26 of the calculator supplement deals with some of the topics of this chapter. Root finding programs are used to solve the equations for the photoelectric and Compton effects. Students may be interested, for example, in seeing how the Compton lines broaden when the electrons are not initially at rest. Assign some of these exercises as homework or set aside some laboratory time for a more detailed investigation. Eureka or student generated root finding

programs can be used.

9. Demonstrations

    Photoelectric effect: Freier and Anderson MPb1; Hilton A4a, b.

    Thompson and Bohr models of the atom: Hilton A5a, b.

10. Computer program

    Atoms and Matter, Focus Media (see Chapter 39 notes for address). A series of programs that simulate various modern experiments. One plots radiative intensity vs. frequency for a blackbody at a temperature chosen by the user. This display can be used to illustrate the lecture. The tutorial material can be used by the students. Reviewed TPT December 1986.

11. Laboratory

    MEOS Experiment 14-2: The Photoelectric Effect. Students investigate the characteristics of various photocells, then use a plot of stopping potential vs. frequency to determine Planck's constant. A mercury source and optical filters are used to obtain monochromatic light of various frequencies.

    MEOS Experiment 14-3: Analysis of Spectra. A spectroscope is used to obtain the wavelengths of hydrogen and helium lines. Hydrogen lines are compared with predictions of the Balmer equation.

Chapter 44   QUANTUM PHYSICS - II

BASIC TOPICS

I. Matter waves.

    A. Explain that electrons and all other particles have waves associated with them, just as photons have electromagnetic waves associated with them. The particle energy and wave frequency are related by $E = h\nu$, the particle momentum and the wavelength are related by $p = h/\lambda$. Calculate the wavelengths of a 1 eV electron and a 35 m/s baseball.

    B. By way of example, state that electrons of appropriate wavelength (≈ 1 Å) are diffracted by crystals and the angular positions of the scattering maxima can be found using Bragg's law, just as for x rays.

II. The wave function.

A. State that a matter wave is denoted by $\Psi(\underline{r},t)$ and that $|\Psi|^2$ gives the probability density for finding the particle near $\underline{r}$ at time t. Similarly E^2 is proportional to the probability density for finding a photon. In the limit of a large number of particles $|\psi|^2$ is proportional to the particle concentration. Explain that, at the atomic and particle level, physics deals with probabilities. What can be analyzed is the probability for finding a particle, not its certain position.

B. Explain that, for a particle confined to the region between 0 and L on the x axis, $\Psi_n(x,t) = \psi_{max}\sin(n\pi x/L)f_n(t)$, where n = 1, 2, ..., gives possible wave functions. These are standing waves and vanish at x = 0 and x = L. Explain that $f_n(t)$ is a function of time with magnitude 1. The wavelength of one of the traveling waves in the standing wave is $\lambda = 2L/n$ and the particle momentum has magnitude nh/2L. Use $E = p^2/2m$ to show that $E_n = n^2h^2/8mL^2$ gives the quantized energy levels. Explain that confinement of the particle leads to energy quantization and that energy is quantized for any bound particle.

C. Explain that the particle is certainly between x = 0 and x = L, so $\int |\Psi_n|^2 dx = 1$. The wave function is said to be __normalized__ if it obeys this condition.

D. Use the particle confined to a one dimensional box as an example and explain that $\psi_n^2 dx = \psi_{max}^2 \sin^2(n\pi x/L)dx$ gives the probability that the particle will be found between x and x + dx when it is in the state with the given wave function. Sketch several of the probability density functions and point out that there are several places where the probability density vanishes.

E. Explain that experimentally the probability can be found, in principle, by performing a large number of position measurements and calculating the fraction for which the particle is found in the designated segment of the x axis. Since a position measurement changes the state of the particle, it must be restarted in the same state each time.

III. The hydrogen atom.

A. Graph the potential energy as a function of distance from the proton for the electron in a hydrogen atom and explain the electron is bound. Give the expression for the energy levels in terms of the principal quantum number.

B. Give the ground state wave function and obtain the expression for the probability density. Show that the volume of a spherical shell with thickness dr is $4\pi r^2 dr$ and define the radial probability density as $P(r) = 4\pi r^2 |\psi(r)|^2$. Sketch $P(r)$ for the ground state and point out there is a range of radial distances at which the electron might be found. Contrast this with the Bohr model. Locate the most probable radius and the average radius.

IV. Barrier tunneling.

A. Show Fig. 13 and explain that a wave function penetrates a finite barrier. It is oscillatory (in position) outside the barrier, where E > V, and exponential inside, where E < V.

B. Explain that the particle has a probability of being found on either side of the barrier. Contrast to the behavior of a classical particle.

C. Write down Eqs. 20 and 21 for the transmission coefficient and explain that this measures the probability of transmission through the barrier. Remark that transmission is small for high, wide barriers and becomes larger as the barrier height decreases and the barrier width narrows.

V. The uncertainty principle.

A. Point out that a different answer might result each time the position of the electron is measured and, in the limit of a large number of measurements, the distribution of results follows closely the probability density function. There is an uncertainty in the result and the uncertainty can be defined similarly to the standard deviation of a collection of experimental results. Similar statements can be made about momentum measurements. Explain that the uncertainties in position and momentum are both determined by the particle wave function. If the electron is placed in a state for which the uncertainty in position is small then the uncertainty in momentum is large and vice versa.

B. As an example, explain that for a particle in a one dimensional infinite square well the uncertainty in position is about L and the uncertainty in momentum is about 2p = nh/L. Remark that the uncertainty in position can be reduced by reducing L and that this increases the uncertainty in momentum. Show that $(\Delta x)(\Delta p) = nh$.

    C.   State the uncertainty principle: $(\Delta x)(\Delta p_x) \geq h$ and give similar expressions for the other cartesian components. Explain that Δx and Δp_x can be changed by changing the state of the particle, but that their product is always greater than h.

VI.   Waves and particles.

    A.   Photons, electrons, and other quantum entities have both particle and wave properties and these complement each other. Explain that particle properties are detected in some experiments and wave properties in others.

    B.   Consider electrons incident on a double slit and explain that an interference pattern is obtained even if only one electron is in the system at any time. Now suppose a particle detector is placed at each slit and assume each electron is detected as it passes through one slit or the other. Graph the intensity pattern at the screen and remark that interference effects are missing. Explain that the act of detecting a particle collapses the wave function so it is transmitted through one slit only (the slit at which the particle was detected).

SUGGESTIONS

1.   In the discussion of wave-particle duality, include questions 1, 3, 5, and 6. After the photoelectric effect and particle diffraction have been studied, ask question 25.

2.   Include questions 2, 10, and 12 in a discussion of de Broglie wavelength. Also assign problems 10 and 16.

3.   After discussing a particle confined between rigid walls, ask questions 16 and 19. Assign problems 21, 23, and 25.

4.   To emphasize probability densities, ask question 18 and assign problem 29.

5.   Following the discussion of the uncertainty principle, ask questions 18, 29, 30, and 31. Assign problem 42.

6.   As questions 22 and 24 in connection with tunneling. Also assign problems 37 and 38.

7.   Discuss problems 33 and 34 in connection with the hydrogen atom.

8.   Computer project

    Ask students to solve problem 1 of section 26.2 of the calculator supplement. This problem emphasizes the probabilistic nature of quantum mechanics by showing how to use a random number generator to find the average position and the uncertainty in position for a particle in a box. Chapter 26 of the calculator supplement contains other exercises that illustrate wave functions and the uncertainty principle.

9. Film

    Matter Waves, 16 mm, b/w, 28 min. Modern Learning Aids (see Chapter 2 notes for address). A comparison of electron and optical diffraction patterns are shown in a modern version of the original experiment that showed the wave nature of the electron. Reviewed AJP 33:63 (1965).

10. Demonstrations

    Electron diffraction: Hilton 13b.

11. Laboratory

    MEOS Experiment 14-5: Electron Diffraction. The Sargent-Welch electron diffraction apparatus is used to investigate the diffraction of electrons by aluminum and graphite.

Chapter 45   ALL ABOUT ATOMS

BASIC TOPICS

I. Hydrogen atom states.

    A. Explain that electron wave functions and energies for the electron in a hydrogen atom can be found by solving the Schrodinger equation. The important point is that wave functions and energy levels are determined by the potential energy function, through the Schrodinger equation. It is not necessary to write down the equation. As an example, state that the potential energy function for hydrogen is $U = e^2/4\pi\epsilon_0 r$ and the condition $\psi \rightarrow 0$ as $r \rightarrow \infty$ is applied. Explain that states for hydrogen are classified using 4 quantum numbers.

    B. Explain that the principal quantum number n gives the energy: $E_n = -(me^4/8\epsilon_0^2 h^2)(1/n^2)$, where n can take on the integer values 1, 2, 3, ... Remark that the energy is quantized (as it is for all bound particles) and explain the meaning of the negative sign. Show a diagram of the lowest few levels, to scale. Traditionally each value of n is said to label a shell and the shells are named

K, L, M, N, ... in order of increasing n.

C. State that the magnitude of the orbital angular momentum is given by $L = \sqrt{\ell(\ell+1)}\,\hbar$, where $\hbar = h/2\pi$. Note that L is quantized. The orbital quantum number ℓ can take on the values 0, 1, 2, ..., n-1 for a given n. These are said to label the subsells, which are named s, p, d, f, g, h, ... in order of increasing ℓ. Go over the allowed values of ℓ for n = 1, 2, and 3. Emphasize that n = 2, ℓ = 0 and n = 2, ℓ = 1 states, for example, have the same energy but different angular momenta.

D. State that the z component of the angular momentum is given by $L_z = m_\ell \hbar$, where m_ℓ = 0, ±1, ±2, ..., ±ℓ. m_ℓ is called the magnetic quantum number. The z axis can be in any direction, perhaps defined by an external magnetic field. Point out that the angle θ between the angular momentum vector and the z axis is given by $\cos\theta = m_\ell/\sqrt{\ell(\ell+1)}$. The smallest value of θ occurs when $m_\ell = \ell$ and it is not zero. Explain that the angles $\underline{L}$ makes with the x and y axes cannot be known. Discuss this in terms of the precession of $\underline{L}$ about the z axis.

E. Explain that the electron and some other particles have intrinsic angular momentum, as if they were spinning on axes. The magnitude of the electron spin angular momentum is $\sqrt{3/4}\,\hbar$ and the z component is either $-\frac{1}{2}\hbar$ or $+\frac{1}{2}\hbar$ (there are two possible states).

F. List all the states for n = 1, 2, and 3. Group them according to n and remark that all states with the same n have the same energy, all states with the same ℓ have the same magnitude of orbital angular momentum, and all states with the same m_ℓ have the same z component of orbital angular momentum. Remark that states with different values of n, ℓ, and m_ℓ have different wave functions. Include m_s and note that the number of states doubles.

G. Hydrogen atom wave functions. Remind students of the definition of the radial probability density and show graphs of P(r) for n = 1, ℓ = 0 and n = 2, ℓ = 0 states (Fig. 8). Write down expressions for P(r) and explain that all ℓ = 0 wave functions are spherically symmetric. Note that the average radius and most probable radius both increase with n. If you wish to broaden the discussion a little, write down expressions for the radial probability density associated with an n = 2, ℓ = 1 state and point out it is a function of the spherical coordinate θ. Note that the average radius and most probable radius decrease with increasing ℓ for the same n.

II. Magnetic moments.

A. Explain that the electron has a magnetic dipole moment because of its orbital motion and write down $\mu_\ell = -(e/2m)L$ and $\mu_{\ell z} = -(e/2m)L_z = -(e\hbar/2m)m_\ell$. Give the value of the Bohr magneton ($\mu_B = e\hbar/2m = 9.28 \times 10^{-24}$ J/T). Remind students that because of its motion the electron experiences a torque in an external magnetic field and produces its own magnetic field (provided $m_\ell \neq 0$).

B. State that the spin magnetic moment is $\mu_{sz} = -2m_s\mu_B$. Stress the appearance of the factor 2. The electron produces a magnetic field and experiences a torque in a magnetic field because of this moment.

C. Remark that the energy of an electron is changed by $-\mu_z B$ when an external field is applied in the positive z direction. Thus states with the same n but different m_ℓ have different energies in a magnetic field. State that this is called the Zeeman effect.

D. Briefly describe the Stern-Gerlach experiment. Explain that a magnetic dipole in a <u>non-uniform</u> magnetic field experiences a force and that $F_z = \mu_z \partial B/\partial z$ for a field in the z direction that varies along the z axis. Atoms with different values of m_ℓ experience different forces and arrive at different places on a screen. That discrete regions of the screen receive atoms is experimental evidence for the quantization of the z component of angular momentum.

E. To emphasize the practical, qualitatively explain NMR and its use in diagnostic medicine.

III. Atom building and the periodic table.

A. Give the "rules" for atom building.

1. The 4 quantum numbers n, ℓ, m_ℓ, m_s can be used to label states. They have the same restrictions on their values as for hydrogen. Remark that wave functions and energies are different for electrons with the same quantum numbers in different atoms. Also remark that the energy depends on ℓ.

2. No more than one electron can have any given set of quantum numbers. This is a general principle of quantum mechanics, called the Pauli exclusion principle.

3. As another proton is added to the nucleus and another

electron is added to the region outside, the electron goes into the vacant state with the lowest energy. This produces the ground state of the atom.

B. Explain that for a given n, the subshells are filled in order of increasing ℓ. Point out that electrons with low ℓ have a greater chance of being near the nucleus, where the potential energy is lower, than electrons with higher ℓ. Electrons that are not near the nucleus experience a force that is reduced by the screening of inner electrons.

C. Explain that as more protons are added to the nucleus the electron wave functions pull in toward regions of low potential energy. This means that a shell may not be completed before the next shell is started. For example, a 5s state is generally lower in energy than a 4d state, in different atoms. It also accounts for the fact that all atoms have nearly the same size.

D. Show a periodic table. Point out the inert gas atoms and explain they all have filled shells. Point out the alkali metal and alkaline earth atoms and state they have one and two electrons, respectively, outside closed shells. Remark that electrons in partially filled shells are chiefly responsible for chemical activity. Point out the atoms in which d and f states are being filled and finally those in which p states are being filled.

IV. X rays and the numbering of the elements.

A. Explain that x rays are produced by firing energetic electrons into a solid target. Show Fig. 16 and point out the continuous part of the spectrum and the sharp lines. Also point out that there is a sharply defined minimum wavelength to the x-ray spectrum. Explain that the continuous spectrum results because the electrons lose some or all of their kinetic energy in close (decelerating) encounters with nuclei. This energy appears as photons and $\Delta K = h\nu$. Explain that a photon of minimum wavelength is produced when an electron loses all its kinetic energy. Derive the expression for the minimum wavelength in terms of the original accelerating potential and point out it is independent of the target material.

B. Explain that the line spectrum in Fig. 16 appears because incident electrons interact with atomic electrons and knock some of the deep lying electrons out of the atoms. Electrons in higher levels drop to fill the holes, emitting photons with energy equal to the difference in energy of the initial and final atomic levels. The K_α line is produced when electrons drop from the L (n = 2) shell to the K (n = 1) shell and the K_β line is produced when electrons drop from the M (n = 3) shell to the K shell. Show Fig. 18.

C.   Show Fig. 20 and state that when the square root of the frequency for any given line is plotted as a function of the atomic number of the target atom, the result is nearly a straight line. Argue that the innermost electrons have an energy level scheme close to that of hydrogen but with an effective nuclear charge of (Z-b)e, where b accounts for screening by electrons close to the nucleus. Z is the number of protons in the nucleus, the atomic number. Use the expression for hydrogen energy levels and, for K_α put n =2 for the initial state, n = 1 for the final state, then show that $\sqrt{\nu}$ is proportional to (Z-b).

D.   Remark that this relationship was used to position the chemical elements in the periodic table, independently of their chemical properties. This technique was particularly important for positioning elements in the long rows of the periodic table; they contain many elements with similar chemical properties. Today the technique is used to identify trace amounts of impurities in materials.

V.   The laser.

A.   List the characteristics of laser light: monochromatic, coherent, directional, can be sharply focused. See the text for quantitative comparisons with light from other sources.

B.   Explain the mechanism of light absorption: an incident photon is absorbed if hν corresponds to the energy difference of two electron states of the material and the upper state is initially empty. An electron makes the jump from the lower to the upper state. Explain spontaneous emission: an electron spontaneously (without the aid of external radiation) makes the transition from one state to a lower state (if that state is empty) and a photon with hν equal to the energy difference is emitted. Emphasize that in most cases the electron remains in the upper state for a time on the order of 10^{-9} s but that there are metastable states in which the electron remains for a longer time ($\approx 10^{-3}$ s). Explain stimulated emission: with the electron in an upper state an incident photon with the proper energy can cause it to make the jump to a lower state. The result is two photons of the same energy, moving in the same direction, with waves having the same phase and polarization. Remark that laser light is produced by a large number of such events, each triggered by a photon from a previous event. Hence all photons are identical. Explain that metastable states are important since the electron must remain in the upper state until its transition is induced. Compare with light produced by random spontaneous transitions.

C. Explain that, in thermodynamic equilibrium, upper levels are extremely sparsely populated compared to the ground state. To obtain laser light the population of an upper level must be increased. A laser must be pumped. If you covered Section 21-7, write down $n_x = Ce^{-E_x/kT}$ and show the connection with the Maxwell distribution law.

D. Use Fig. 27 to describe the three level laser. First describe the equilibrium distribution of electrons among the states, then describe pumping from the ground state to the highest level and the fast decay to a metastable state. Describe the distribution when the population has been inverted. Finaly describe the stimulation of emission and the build up of the number of photons with the same energy and phase.

E. Discuss the helium-neon laser. First give the energy level scheme and describe the pumping and stimulation processes. Discuss the role of the walls and mirror ends. Go over the four characteristics of laser light discussed earlier and tell how each is achieved.

SUGGESTIONS

1. To emphasize the difference between the Bohr and quantum models of the hydrogen atom and to get students to think about wave functions, ask questions 1, 2, and 13.

2. To test for understanding of the angular momentum quantum numbers, assign problems 3, 5, 7, and 8.

3. Assign problems 7 and 8 in connection with orbital angular momentum. To stress the connection between angular momentum and magnetic moment, also assign problems 10 and 11. To discuss the Stern-Gerlach

experiment in more detail, include questions 14 and 17. Also assign problems 29 and 32.

4. Hydrogen wave functions and radial probability densities are covered in problems 20, 22, and 26. Ask one or more of these.

5. Use problem 40 to test for understanding of the Pauli exclusion principle. To emphasize the role played by spin in the building of the periodic table, ask questions 18 and 20. Also include question 22 in the discussion of the periodic table. Assign problem 39.

6. The existence of a minimum wavelength in the continuous x-ray spectrum provides an argument for the particle nature of light. Either discuss this or see if the students can devise the argument. Assign question 24 and problem 46.

7. After discussing characteristic x-ray lines, ask questions 26, 28, and 31.

8. Ask questions 33 and 34 to see if students understand the properties of laser light. Ask questions 38 and 39 to test for understanding of the laser mechanism. Also assign problems 60 and 64.

9. Film:
    Introduction to Lasers, 16 mm, color, 17 min. Encyclopaedia Britannica Educational Corporation (see Chapter 17 notes for address). This film discusses the development and uses of the laser. Featured are three of the principal developers. The phenomenon of "lasing" is illustrated through animated sequences and laboratory demonstrations. Reviewed AJP 42:525 (1974).

10. Demonstrations
    Zeeman effect: Freier and Anderson MPc1; Hilton A20a.
    X-ray apparatus: Hilton A7a, b, c.
    Lasers: Hilton A12.

11. Computer program
    Animation Demonstration: Electron Waves in an Atom, Conduit (see Chapter 17 notes for address). Compares classical electron orbits and quantum wave patterns. Shows quantization of orbits by applying boundary conditions and simulates radiative transitions. Reviewed TPT November 1986

Chapter 46   CONDUCTION OF ELECTRICITY IN SOLIDS

<u>BASIC TOPICS</u>

I.   Electron energy bands and conduction.
   A.   Explain that energy levels for electrons in solids are grouped
     into bands with the levels in any band being nearly continuous
     and with gaps of unallowed energies between. Remark that bands
     are produced when atoms are brought close together. Wave
     functions then overlap and extend throughout the solid. Show Fig.
     3 and remark that low energy bands are narrow since the wave
     functions are highly localized around nuclei and overlap is
     small. High energy bands are wide because overlap is large.
     Sample Problem 1 gives an estimate of the separation of levels in
     a band.
   B.   Remind students that since the Pauli exclusion principle holds
     the lowest total energy is achieved when electrons fill the
     lowest states with one electron in each state. Thus at T = 0 K
     all states are filled up to a maximum energy.
   C.   Explain that the electric current is zero when no electric field
     is present because states for which the velocities are **v** and −**v**,
     for example, have the same energy. If one is filled then so is
     the other. Thus the average velocity of the electrons vanishes. A
     current arises in an electric field because the electrons
     accelerate: they tend to make transitions within their band to
     other states such that the changes in their velocities are
     opposite to the field. Point out that a filled band cannot
     contribute to the current because there are no empty states to
     accept electrons.
   D.   Explain that insulators and semiconductors have just the right
     number of electrons to completely fill an integer number of bands
     and that, in the lowest energy state, all bands are either
     completely filled or completely empty. For metals, on the other
     hand, the highest occupied state is in the middle of a band.
     Metals always have partially filled bands. Compare Figs. 4 and 5.
     Remark that the energy of the most energetic electron in a metal
     is called the Fermi energy. Identify the valence and conduction
     bands for an insulator.
   E.   Explain that as the temperature is raised from T = 0 K a small

fraction of the electrons in the valence band of an insulator or semiconductor are thermally excited across the gap into the conduction band. For a semiconductor the gap is small (about 1 eV) and at room temperature both bands can contribute to the current. The conductivity, however, is still small compared to that of a metal. For an insulator the gap is large (more than 5 eV), so the number of promoted electrons is extremely small and the current is insignificant for laboratory fields. Explain that silicon and germanium are the only elemental semiconductors although there are many semiconducting compounds. Carbon is a prototype insulator, with a gap of 5.5 eV. Compare with silicon, which has a gap of 1.1 eV. Resistivities of metals and semiconductors are compared in Table 1.

II. Metallic conduction.

A. Write down the Fermi-Dirac probability function p(E), given by Eq. 1, and state that it gives the thermodynamic probability of a state with energy E being occupied. Show that for T = 0 K, p(E) = 1 for $E < E_F$ and p(E) = 0 for $E > E_F$. To give a numerical example, calculate the probabilities of occupation for states 0.1 and 1 eV above the Fermi energy, then 0.1 and 1 eV below, at room temperature. Graph E vs. p(E) for T = 0 and for T > 0. See Fig. 6. Also show the graph for a still higher temperature and point out that the central region (from p = .9 to p = .1, say) widens. This quantitatively describes the thermal excitation of electrons to higher energy states. Remark that the Fermi-Dirac probability function is valid for any large collection of electrons, including the collections in metals, insulators, and semiconductors.

B. Define the state distribution function n(E) and the particle distribution function $n_0(E)$. Explain that $n_0(E) = n(E)p(E)$ and that the total electron concentration in a metal is given by n = ∫n(E)p(E)dE. In principle, this equation can be solved for the Fermi level as a function of temperature. State that for nearly free electrons in a metal $n(E) = (35.5m^{3/2}/h^3)E^{1/2}$ and that the Fermi level is given by $E_F = (0.121h^2/m)n^{2/3}$, where n is the free electron concentration. Evaluate the expression for copper and show that E_F is about 7 eV above the lowest free electron energy.

C. Explain that the acceleration caused by an electric field does not continue indefinitely because the electrons are scattered by atoms of the solid. As a result, the electron distribution distorts only slightly. Some states with energy slightly greater than E_F and velocity opposite the field become occupied while some states with energy slightly less than E_F and velocity in the direction of the field become vacant. Electrons with energy E_F have speeds v_F given by $E_F = \frac{1}{2}mv_F^2$ but the average speed (the drift speed) is considerably less because most electrons can be paired with others moving with the same speed in the opposite direction.

D. Explain that a steady state is reached and that the drift velocity is then proportional to the applied electric field. This explains Ohm's law. Steady state is approached exponentially with a characteristic time called the relaxation time. Mention that the relaxation time is determined by collisions. In a rough way, if there are few collisions per unit time then the relaxation time is long. Ask students to review section 28-6. Write down Eq. 2 for the resistivity and explain that τ is the relaxation time. A long relaxation time means a low resistivity. As the temperature increases the number of collisions per unit time increases and the relaxation time becomes smaller. This explains the increase with temperature in the resistivity of a metal.

III. Semiconductors.
   A. When electrons are promoted across the gap they contribute to the current in an electric field. The valence band becomes partially filled and electrons there also contribute to the current. It is usually convenient to think about the few empty states in this band rather than the large number of electrons there. These are called holes and behave as if they were positive charges. In contrast to electrons, holes drift in the direction of E. Compare the carrier concentrations of metals and semiconductors. See Table 1.
   B. Explain the different signs for the temperature coefficients of resistivity, also given in Table 1. Explain that for both metals and semiconductors near room temperature the relaxation time decreases with increasing temperature. For metals the electron concentration is essentially constant but for semiconductors n

increases dramatically with temperature as electrons are thermally promoted across the gap. This effect dominates and the resistivity of an intrinsic semiconductor decreases with increasing temperature.

C. Explain that the proper kind of impurities (donors) can increase the number of electrons in the conduction band and other kinds (acceptors) can increase the number of holes in the valence band. They produce n and p type semiconductors, respectively. By considering the number of electrons in their outer shells explain why phosphorous is a donor and aluminum is an acceptor. Point out that wave functions for impurity states are highly localized around the impurity and so do not contribute to the conductivity. Go over Sample Problem 5, which shows that only a relatively small dopant concentration can increase the carrier concentration enormously. Doped semiconductors are used in nearly all semiconducting devices.

IV. Semiconducting devices.

A. Show a commercial junction diode and draw a graph of current vs. potential difference (Fig. 12). Include both forward and reverse bias. Explain that it is a rectifier, with high resistance for current in one direction and low resistance for current in the other direction. Demonstrate the I-V characteristics by placing the diode across a variable power supply and measuring the current and potential, as described in note IIC for Chapter 28.

B. Describe a p-n junction and explain the origin of the electric field in the transition region and the origin of the contact potential. Stress that the field is due to uncovered impurity atoms, positive donors on the n side and negative acceptors on the p side. Remark that diffusion of carriers leaves a small depletion zone, nearly devoid of carriers, straddling the metallurgical junction.

C. Describe a diffusion current as one which arises because particles diffuse from regions of high concentration toward regions of low concentration. Explain that this motion results from the random motion of the particles. More particles leave a high concentration region simply because there are more particles there, not because they are driven by any applied force. State that the diffusion current for both electrons and holes in an unbiased p-n junction is from the p to the n side, against the contact electric field. Point out that the drift current is from the n toward the p side and that the diffusion and drift currents cancel when no external field is applied. Point out the depletion zone and the currents on Fig. 11.

D. Draw a circuit with a battery across a p-n junction, the positive terminal attached to the n side. Explain that this is reverse bias. The internal electric field is now larger, the barrier to diffusion is higher, and the reverse current is extremely small. See Fig. 14. Also explain that the width of the depletion zone is increased by application of a reverse bias.

E. Draw the circuit for forward bias. The internal electric field is now smaller, the barrier to diffusion is lower, and the current increases dramatically. See Fig. 14. The depletion zone narrows.

F. Explain how diodes are used for rectification and how light emitting diodes work.

G. Optional. Explain the mechanism by which the gate voltage of a MOSFET controls current through the channel. Remove the covers from a few chips and pass them around with magnifying glasses for student inspection.

SUGGESTIONS

1. To test for understanding of the conduction process, ask questions 5 and 7.

2. Use questions 2, 4, and 6 to help in the discussion of the distribution of electrons among the states. Also assign problems 6, 9, and 18. After discussing holes, assign problem 17.

3. After discussing the differences between conductors, semiconductors, and insulators, ask questions 11, 12, 18, and 23. Also ask question 21.

4. Use questions 19 and 20 in a discussion of doping. Also assign problems 33 and 34.

5. After discussing p-n junctions, ask questions 30 and 32. Also assign problem 37.

6. Computer project:
Ask students to use Eureka or their own root finding program to carry out calculations of the electron concentration in the conduction band and hole concentration in the valence band of both intrinsic and

doped semiconductors. Then ask them to calculate the contact
potential for a p-n junction with given dopant concentrations. See
Section 28.1 of the calculator supplement.

Chapter 47   NUCLEAR PHYSICS

BASIC TOPICS

I.   Nuclear properties.
   A.   Explain that the nucleus of an atom consists of a collection of
        tightly bound neutrons, which are neutral, and protons, which are
        positively charged. The proton has the same magnitude of charge
        as the electron. Define the term nucleon and state that the
        number of nucleons is called the mass number and is denoted by A,
        the number of protons is called the atomic number and is denoted
        by Z, and the number of neutrons is denoted by N. Point out that
        A = Z + N. Remark that nuclei with the same Z but different N are
        called isotopes. The atoms have the same chemical properties and
        the same chemical symbol. Show Fig. 5 or a wall chart of the
        nuclides. Refer to Table 1 when discussing properties of
        nuclides.
   B.   Explain that one nucleon attracts another by means of the strong
        nuclear force and that this force is different from the
        electromagnetic force. It does not depend on electrical charge
        and is apparently the same for all pairs of nucleons. It is
        basically attractive; at short distances (a few fm) it is much
        stronger than the electrostatic force between protons, but it
        becomes very weak at larger distances. Two protons exert
        attractive strong forces on each other only at small separations
        but they exert repulsive electric forces at all separations.
        Because of the short range, a nucleon interacts only with its
        nearest neighbors via the strong force. Because the nucleus is
        small, the much stronger nuclear force dominates and both protons
        and neutrons can be bound in stable nuclei.
   C.   Show Fig. 4 and point out the Z = N line and the stability zone.
        Explain why heavy nuclei have more neutrons than protons. Also
        explain that unstable nuclei are said to be radioactive and
        convert to more stable ones with the emission of one or more

particles. Show Fig. 5 and point out the stable and unstable nuclei.

D. Explain that the surface of a nucleus is not sharply defined but nuclei can be characterized by their mean radii and these are given by $R = R_0 A^{1/3}$, where $R_0 = 1.1$ fm (1 fm = 10^{-15} m). Stress how small this is compared to atomic radii. Show that this relationship between R and A leads to the conclusion that the mass densities of all nuclei are nearly the same. Show that the density of nuclear matter is about 2×10^{17} kg/m^3.

E. Explain that the mass of a nucleus is less than the sum of the masses of its constituent nucleons, well separated. The difference in mass is accounted for by the binding energy through $E_B = \Delta m c^2$, where Δm is the magnitude of the mass difference. The binding energy is the energy which must be supplied to separate the nucleus into well separated particles, at rest. Generalize this equation to the case of a nucleus with Z protons and N neutrons: $E_B = Z m_p c^2 + N m_n c^2 - m c^2$. This expression is used implicitly in Sample Problem 3. Show Fig. 6 and point out that there is a region of greatest stability, near iron. For heavier nuclei the binding energy per nucleon falls slowly but nevertheless does fall. For lighter nuclei the binding energy per nucleon rises rapidly with increasing mass number. The high mass number region is important for fission processes, the low mass number region is important for fusion processes.

F. State that nuclear masses are difficult to measure with precision so binding energies are usually expressed in unified atomic mass units: 1 u = 1.66×10^{-27} kg. Also state that tables usually give atomic rather than nuclear masses and so include the mass of the atomic electrons. Show that the electron masses cancel in the expression for the binding energy.

G. Explain that nuclei have discrete energy levels. An excited nucleus can make a transition to a lower energy state with the emission of a photon, typically in the gamma ray region of the spectrum. Explain that a nucleus may have intrinsic angular momentum and a magnetic moment. Spins are on the order of $\hbar$, like atomic electrons, but moments are much less than electron moments because the mass of a nucleon is much greater than the mass of an electron.

204

II. Radioactive decay.

A. Explain that nuclei may be either stable or unstable and those which are unstable ultimately decay to stable nuclei. Decay occurs by spontaneous emission of an electron (β^-), a helium nucleus (α), a positron (β^+), or larger fragments. The resulting nucleus has a different complement of neutrons and protons than the original nucleus.

B. Explain that decay is energetically favorable if the total mass of the products is less than the original mass. Define a decay symbolically as $X \rightarrow Y + b$, where X is the original nucleus, Y is the daughter nucleus, and b is everything else. Point out that charge, number of nucleons, and energy are all conserved. Define the disintegration energy by $Q = (m_X - m_Y - m_b)c^2$. Note that an appropriate number of electron rest energies must be added or subtracted so that atomic masses may be used. Note also that Q must be positive for spontaneous decays and Q appears as the kinetic energy of the decay products.

C. Explain that each radioactive nucleus in a sample has the same chance of decaying and that the decay rate ($R = -dN/dt$) is proportional to the number of undecayed nuclei present at time t: $-dN/dt = \lambda N$. This has the solution $N = N_0 \exp(-\lambda t)$, so the decay rate is given by $R = R_0 \exp(-\lambda t)$. Define the term half life and show that $\tau = (\ln 2)/\lambda$. Go over Sample Problem 4, show Fig. 8, and point out the half life. Remark that R decreases by a factor of 2 in every half life interval.

D. Discuss α decay. Write down Eq. 9 and explain that the daughter nucleus has 2 fewer neutrons and 2 fewer protons than the parent. Go over Sample Problem 6 to show that α decay is energetically favorable for ^{238}U. Show Fig. 9 and explain that the deep potential well is due to the strong attraction of the residual nucleus for the nucleons in the α particle, while the positive potential is due to Coulomb repulsion. The two forces form a barrier to decay. Explain that the α particle can tunnel through the barrier. Its wave function does not go to zero at the inside edge but has a finite amplitude in the barrier and on the outside. There is a non-zero probability of finding the α particle on the outside. High, wide barriers produce a small probability of tunneling and a long half life while low, narrow barriers produce the opposite effect. Note the wide range of half lives that occur in nature.

E. Discuss β decay. Explain that a neutron can transform into a
proton with the emission of an electron and a neutrino and that a
proton can transform into a neutron with the emission of a
positron and a neutrino. Mention the properties of a neutrino:
massless, neutral, weakly interacting. Only protons bound in
nuclei can undergo β decay but both free and bound neutrons can
decay. These transformations lead to decays such as the ones
given in Eqs. 10 and 11. Explain that the energy is shared by the
decay products and the electrons or positrons show a continuous
spectrum of energy up to some maximum amount (see Fig. 10).
Explain that neutron rich nuclides generally undergo β^- decay
while proton rich nuclides generally undergo β^+ decay. This is a
mechanism for bringing the nucleus closer to stability. Carefully
discuss the addition of electron rest energies to the Q value so
that atomic masses can be used. In particular, show that in β^-
decay there is no excess electron mass but in β^+ decay there is
an excess of 2 electron rest masses.

F. Define the units used to describe radioactivity and radiation
dosage: curie, roentgen, and rem.

SUPPLEMENTARY TOPICS

Radioactive dating.
Nuclear models.

SUGGESTIONS

1. Nuclear constitution is covered in problems 5, 8, and 11. Nuclear
radius and density are covered in problems 4, 6, and 12.

2. Include questions 2, 3, 5, and 13 in the discussion of the strong
interaction and its role in nuclear binding. Problems 8, 9, 10, 13,
19, and 21 illustrate nuclear stability and binding. Be sure to
include problem 10 if you intend to discuss fission (Chapter 48).

3. Problems 26 and 27 cover basic half-life calculations. Problems 28
through 41 involve half-life calculations drawn from many interesting
applications. Assign some of them.

4. Following the discussion of α decay, students should be able to
answer questions 15 and 22 in their own words. The disintegration

energy and barrier height are covered in problems 49 and 50. Problem 48 asks students to take into account the recoil of the residual nucleus. Problem 51 shows why alphas are emitted rather than well separated nucleons.

5.   After discussing β decay, ask questions 23, 24, and 25. Also assign one or more of problems 52, 53, and 55. Problem 56 shows that β particles do not exist inside nuclei before decay occurs. The β decay discussion can be broadened somewhat by including the recoil of the nucleus. See problem 61.

6.   Film:

The Discovery of Radioactivity, 16 mm or 3/4" videocassette, color, 15 min. International Film Bureau (see Chapter 6 notes for address). The experiments and discoveries of Roentgen, Curie, Becquerel, Elster, Rutherford, and Geitel are featured.

7.   Demonstrations

Geiger counter: Freier and Anderson MPa2.

Radioactivity: Hilton A15, A16, A18.

8.   Laboratory

Many of the following experiments make use of a Geiger tube and scalar.

BE Experiment 47: The Characteristics of a Geiger Tube describes how students can systematically investigate the plateau and resolving time of a Geiger tube. They also learn how to operate a scalar. Consider prefacing the other experiments either with this experiment or with a demonstration of the same material.

BE Experiment 48: The Nature of Radioactive Emission. Statistical fluctuations in the counting rate for a long life time source provide a demonstration of the statistical nature of radioactivity. Students also study variations in the counting rate as the source-counter separation is increased.

MEOS Experiment 14-7: Half-Life of Radioactive Sources. A Geiger counter and scalar are used to measure the decay rate as a function of time for indium, cesium 137, and barium 137. For the first and last, the data is used to compute the half life. Other sections explain how to use a microcomputer to collect data and make the calculation and how to use a emanation electroscope to collect data. A neutron howitzer or minigenerator is required to produce radioactive sources.

BE Experiment 50: Measurement of Radioactive Half-Life. Nearly the same as MEOS 14-7. The generation of sources with short half lives is discussed.

MEOS Experiment 14-6: <u>Absorption of Gamma and Beta Rays</u>. The particles are incident on sheets of aluminum and the number which pass through per unit time is counted. Students make a logarithmic plot of the counting rate as a function of the thickness of the aluminum and determine the range of the particles.

BE Experiment 49: <u>Properties of Radioactive Radiation</u>. Essentially the same as MEOS 14-6 but cardboard and lead as well as aluminum absorbers are used. Students can compare the relative absorbing power of these materials.

Chapter 48   ENERGY FROM THE NUCLEUS

<u>BASIC TOPICS</u>

I. The fission process.
    A. Refer back to the binding energy per nucleon vs. A curve (Fig. 47-6). It suggests that a massive nucleus might split into two or more fragments nearer to iron, thereby increasing the total binding energy. Each fragment is more stable than the original nucleus. This is the fission process.
    B. Remark that many massive nuclei can be rendered fissionable by the absorption of a thermal neutron. Such nuclei are called fissile. Give the example $^{235}U + n \rightarrow {}^{236}U^* \rightarrow X + Y + bn$. Explain that a thermal neutron (≈ 0.04 eV) is absorbed by a ^{235}U nucleus and together they form the intermediate fissionable $^{236}U^*$ nucleus. This nucleus splits into 2 fragments (X and Y) and several neutrons. The sequence of events is illustrated in Fig. 3. Point out ^{236}U on Fig. 47-6. The disintegration energy for one possible fission event is calculated in Sample Problem 1.
    C. Explain that different fission events, starting with the same nucleus, might produce different fragments. The fraction of events that produce a fragment of a given mass number A is graphed in Fig. 2. Point out that fragments of equal mass occur only rarely. Explain that the parent nucleus is neutron rich, the original fragments are neutron rich, and that the original fragments expel neutrons to produce the fragments X and Y. These

generally decay further by β emission.

D. Show Fig. 4 and explain that the parent nucleus starts in the energy well near r = 0. The incoming neutron must supply energy to start the fission process. The required energy is slightly less than E_b since tunneling can occur. Point out the energy Q released by the process. Point out Table 2 and explain that E_n is the actual energy supplied by an incoming thermal neutron. Point out nuclides in the table for which fission does not occur.

E. Write out several fission modes for ^{235}U (see the Student Study Guide) and note that on average more than one neutron is emitted. Explain that some neutrons come promptly while others come from later decays (the delayed neutrons). Point out that the average mode yields Q ≈ 200 MeV, of which 190 MeV or so appears as the kinetic energy of the fission fragments and 10 MeV goes to the neutrons.

II. Fission reactors.

A. Note that to have a practical reactor the fission process must be self sustaining, once started. Also, there must be a way to control the rate of the process and to stop it, if desired.

B. To be self sustaining, a chain reaction must occur: neutrons from one fission event are used to trigger another. The neutrons emitted from a typical fission event share about 5 to 10 MeV energy and they must be slowed to thermal speeds to be useful. Some sort of moderator, often water, is used.

C. Explain that on average about 2.5 neutrons are produced per fission event. Describe in detail what happens to them. Some leak out of the system, some of the slowed neutrons are captured by ^{238}U, some are captured by fission fragments, and the rest start fission in ^{235}U. Fig. 5 shows some typical numbers.

D. Explain the terms critical, subcritical, and supercritical. Note that the control rods, which absorb slow neutrons, are used to achieve criticality. Point out that without the delayed neutrons, control would not be possible since time is needed to move the rods into or out of the reactor.

E. Use Fig. 6 to describe the essential features of a nuclear power plant. Apart from the fact that the fission process is used to heat water or generate steam, this schematic could apply to any power plant. Remark on the special problems attendant on nuclear plants.

III. Fusion.

    A. Return to Fig. 47-6 and remark that if two low mass nuclei are combined to form a higher mass nucleus the binding energy is increased considerably. The energy is transformed to the kinetic energy of the resulting nucleus and any particles emitted. In order to carry out the fusion process, the nuclei must be given sufficient energy to overcome the electrostatic repulsion of their protons. They can then approach each other closely enough for the attraction of the strong force to bind them. For ^{3}He the height of the barrier is about 1 MeV. Since tunneling is possible, fusion can occur at somewhat smaller energies.

    B. To achieve a large number of fusion events, hydrogen or helium gases must be raised to high temperatures. Even at the temperature of the sun only a small fraction of the nuclei have sufficient energy to overcome the Coulomb barrier. Go over Fig. 9.

    C. Discuss fusion in the sun. Remark that the core of sun is 35% hydrogen and 65% helium by mass. Outline the principal proton-proton cycle: 2 protons fuse to form a deuteron, a positron, and a neutrino. A deuteron fuses with a proton to form ^{3}He and two ^{3}He nuclei fuse to form ^{4}He and two protons. Remark that 6 protons are consumed and two are produced for a net loss of 4. The two positrons are annihilated with electrons to produce photons. Note that the process can be simplified to $4p \rightarrow \alpha + 2\beta^+ + 2\nu$ and the Q value is computed from the mass difference between the alpha particle and the 4 protons. To this is added the energy of annihilation of the positrons.

    D. Calculate the energy released. Show that Q = 26.7 MeV and note that the neutrinos take about 0.5 MeV with them when they leave the sun. Point out that the fusion process produces about 20 million times as much energy per kg of fuel as the burning of coal.

    E. If time permits, discuss helium burning. Use the solar constant to calculate the rate at which the sun converts mass to energy. Speculate on the future of the sun. Also mention the carbon cycle, which is essentially the same as the proton-proton cycle. Carbon acts as a catalyst.

F. Discuss controlled thermonuclear fusion. Explain that deuteron-deuteron and deuteron-triton fusion events are being studied. Point out that high particle concentrations at high temperatures must be maintained for sufficiently long times in order to make the process work. Discuss some means for doing this: the tokamak for plasma confinement by magnetic fields, inertial confinement, and laser fusion. State that the right combination has not yet been achieved but work continues.

SUGGESTIONS

1. After explaining the basic fission process, test understanding with questions 4, 5, 6, and 8. Use question 17 to discuss critical size. Also assign problem 4.
2. Following the discussion of the fission reactor, ask questions 10, 11, 12, 13, and 15. To help students understand the role of a moderator, assign problem 27.
3. Following the discussion of the basic fusion process, ask question 21.
4. To help students understand fission and fusion processes as energy sources, assign problems 8, 29, 34, and 49.
5. Film
   Fusion: The Ultimate Fire, 16 mm, color, 14 min. BFA Educational Media (see Chapter 30 notes for address). Various types of fusion research are presented. An explanation of the fundamental concepts of fusion is aided by animation. Includes visits to fusion labs.
6. Demonstrations
   Chain reaction: Freier and Anderson MPa1.

Chapter 49   QUARKS, LEPTONS, AND THE BIG BANG

BASIC TOPICS

I. The particle "zoo".
   A. Show a list of particles already familiar to students. Include the electron, proton, neutron, and neutrino, then add the muon and pion. Explain that many other particles have been discovered in cosmic ray and accelerator experiments. To impress students

with the vast array of particles and the enormous collection of data, make available to them a Review of Particle Properties paper, published roughly every 2 years in Reviews of Modern Physics.

B. Explain that many new particles are discovered by bombarding protons or neutrons with electrons or protons and show bubble chamber pictures, such as Fig. 3. State that the pictures show tracks of charged particles in a strong magnetic field, hence the curvature. Remind students that the radius of curvature can be used to find the momentum of a particle if the charge is known. Indicate the collision point and emphasize that the new particles were not present before the collision: the original particles disappear and new particles appear. In most cases the total rest energy after the collision is much greater than the total rest energy before. Kinetic energy was converted to mass.

C. Mention that a few particles seem to be stable (electron, proton, neutrino) but most decay spontaneously to other particles. Point out decays on a bubble chamber picture. Explain the statistical nature of decays and remind students of the meaning of "half life". Examples: $n \rightarrow p + e^- + \nu$, $\pi^+ \rightarrow \mu^+ + \nu$.

D. Explain that for each particle there is an antiparticle with the same mass. A charged particle and its antiparticle have charge of the same magnitude but opposite sign. Their magnetic moments are also opposite. A particle and its antiparticle can annihilate each other, the energy (including rest energy) being carried by photons or other particles produced in the annihilation. Example: $e^+ + e^- = \gamma + \gamma$. Antiparticles (except the positron) are denoted by a bar over the particle symbol. Some uncharged particles (such as the photon) are their own antiparticles. Our part of the universe is made of particles; perhaps other parts are made of antimatter.

II. Particle properties.

A. Spin angular momentum. Remind students that many particles have intrinsic angular momentum. Explain that the magnitude is always an integer or half integer times $\hbar$. Remark that particles with half integer spins are called fermions while particles with integer spins are called bosons. Remind students of the Pauli exclusion principle and its significance, then state that fermions obey the principle while bosons do not. Give examples: electrons, protons, neutrons, and neutrinos are fermions, photons, pions, and muons are bosons. Remark that spin angular momentum is conserved in particle decays and interactions. An odd number of fermions, for example, cannot interact to yield bosons only.

B.   Charge. Remind students of charge quantization and charge conservation. Even if the character and number of particles change in an interaction, the total charge before is the same as the total charge after. Example: $n \rightarrow p + e^- + \nu$.

C.   Momentum and energy. Explain that energy and momentum are conserved in decays and interactions. Give masses and rest energies for the particles in the list of part I. Give the expressions for relativistic energy and momentum in terms of particle velocity and go over Sample Problem 1.

D.   Forces. Remark that all particles interact via the force of gravity and all charged particles interact via the electromagnetic force. The force of gravity is too weak to have observable influence at energies presently of interest. Remark that there are two additional forces, called strong and weak, respectively. Remind students of the role played by the strong force in holding a nucleus together and the role played by the weak force in beta decay. These topics were covered in Chapter 47.

E.   Leptons and hadrons. State that particles which interact via the weak force but not the strong are called leptons while particles which interact via the strong force (as well as the weak force) are called hadrons. List the leptons (electron, muon, tauon, and their neutrinos) and explain that a different neutrino is associated with each of the other leptons. Remark that the neutrino that appears following muon decay is not the same as the neutrino that appears following beta decay. Neutrinos are labelled with subscripts giving the associated lepton: ν_e, ν_μ, and ν_τ.

F.   Baryons and mesons. Remark that some hadrons (proton, neutron) are fermions and are called baryons while others (pion) are bosons and are called mesons. Explain that a baryon number of +1 is assigned to each baryon particle, a baryon number of −1 is assigned to each baryon antiparticle, and a baryon number of 0 is assigned to each meson. Then baryon number is conserved in exactly the same way charge is conserved: the total baryon number before a collision or decay is the same as the total baryon number after. This conservation law (and conservation of energy) accounts for the stability of the proton, the baryon with the smallest mass. There is some speculation that baryon number is not strictly conserved and that protons may decay to other particles, but the half life is much longer than the age of the universe. Some physicists are trying to observe proton decay.

G. Strangeness. Explain that another quantity, called strangeness, is conserved in strong interactions. Neutrons and protons have S = 0, K^- and Σ^+ have S = -1. A particle and its antiparticle have strangeness of opposite sign. Conservation of strangeness allows $\pi^+ + p \to K^+ + \Sigma^+$ but prohibits $\pi^+ + p \to \pi^+ + \Sigma^+$, for example.

III. Modern theories.
A. Quarks and the eight-fold way. Remark that the properties of hadrons can be explained if we assume they are made up of particles (called quarks). List the quarks and their properties (Table 6). Particularly note the fractional charge and baryon number. Baryons are constucted of three quarks, antibaryons of three antiquarks, and mesons of a quark and antiquark. Show that uud has the charge, spin, and baryon number of a proton and udd has the charge, spin, and baryon number of a neutron. Give the quark content of the spin 1/2 baryons (Fig. 6a) and explain Fig. 4a. Point out that the strange quark accounts for the strangeness quantum number. Give the quark content of the spin 0 mesons (Fig. 6b) and explain Fig. 4b. Mention the charm, bottom, and top quarks and point out they lead to other particles.

B. Explain that the existence of internal structure allows for excited states: there are other particles with exactly the same quark content as those in the figure but they are different particles because the quarks have different motions. The additional energy results in greater mass. Contrast this with the leptons, which have no internal structure. Quarks and leptons are believed to be truly fundamental.

C. Messenger particles (optional). Explain that particles interact by exchanging other particles. Electromagnetic interactions proceed by exchange of photons, for example. Also explain that energy may not be conserved over short periods of time but this is consistent with the uncertainty principle. State that the strong interaction proceeds by the exchange of gluons by quarks and the weak interaction proceeds by the exchange of Z and W particles by quarks and leptons. The interaction that binds nucleons in a nucleus is the same as the interaction that binds quarks in a hadron. In the former case gluons are exchanged between quarks of different nucleons, in the latter they are exchanged between quarks of the same hadron.

D. Explain that quarks are conserved in strong interactions. Either the original quarks are rearranged to form new particles or quark-antiquark pairs are created, then both the original and the new quarks are rearranged. This accounts for conservation of strangeness. Example: $K^+ \rightarrow K^0 + \pi^+$ ($u\bar{s} \rightarrow d\bar{s} + u\bar{d}$). A $d\bar{d}$ pair is formed. The d quark couples to the $\bar{s}$ quark to form a K^0 and the $\bar{d}$ quark couples to the u quark to form a π^+. Contrast this with the weak interaction, which can change one type quark into another. Illustrate with beta decay, in which a d quark is converted to a u quark.

E. Explain that quarks have another property, called color. It can have three values, usually termed red, yellow, and blue. Color produces the gluon field, much as charge produces the electromagnetic field: baryons interact via the strong interaction because quarks have color. Be sure students understand that "color" in this context has nothing to do with the frequency of light. The condition that all particles be color neutral limits the strongly interacting particles to baryons (three quarks - one red, one yellow, and one blue) and mesons (a quark-antiquark pair with both members the same color).

IV. The big bang.
A. Evidence. Remind students of the doppler shift for light and state that spectroscopic evidence convinces us that on a large scale matter in the universe is receding from us and that belief in the cosmological principle leads us to conclude the universe is expanding. Write down Hubble's law and give the Hubble parameter: 17×10^{-3} m/(s·ly). Show that this implies a minimum age for the universe of about 18×10^9 y (see Sample Problem 8).
B. Discuss the microwave background radiation and state that

physicists believe it was generated about 500,000 years after the big bang, when the universe became tenuous enough to allow photons to exist without being quickly absorbed.

C. Remark that in the early universe the temperature was sufficiently high that the exotic particles now being discovered (and others) existed naturally. We need the results of high energy physics to understand the early universe.

D. Go over the chronological record given at the end of section 14. Point out that in the very early universe the forces we know today evolved from a single unified force. Then baryons and mesons were formed, followed by the light nuclides, then atoms. Heavy nuclides are created chiefly in supernova explosions.

SUGGESTIONS

1. To test for understanding of the conservation laws and the stability of particles, ask questions 10, 11, 12, 13, 14, and 19. Problems 2, 8, 10 (or 11), 12, 13, 15, 16, 17, and 20 each deal with one or more of the conservation laws. Assign several. Discuss question 18 in connection with problem 14.

2. To help clarify particle properties and classifications, ask questions 4, 8, 27, 28 and 29.

3. Problems 23, 24, 26, 27, and 28 provide excellent illustrations of the quark model. Assign a few of them.

4. Questions 15, 21, 22, 24, and 35 deal with the fundamental forces. Ask several of them.

5. Include questions 36, 37, 39, and 41 in discussions of cosmology. Also assign problems 30 and 38. Assign problem 31 or 32 in connection with the red shift. If you discussed the relativistic Doppler shift in connection with Chapter 42, assign problem 33.

6. Demonstrations
    Show nuclear emulsion plates, available from Brookhaven National Laboratory, Fermilab, and other high energy laboratories.
    Elementary particles: Hilton A23.

7. Laboratory
    MEOS Experiment 14-8: <u>Nuclear and High Energy Particles</u>. A dry ice and alcohol cloud chamber is used to observe the tracks of alpha and beta particles as well as the tracks produced by cosmic rays. A magnet is used to make circular tracks.

SECTION THREE
ANSWERS TO SELECTED QUESTIONS

In this section of the manual, answers are given to many of the end-of-chapter questions suggested in the Lecture Notes. Roughly a fourth of the questions in the extended version of FUNDAMENTALS OF PHYSICS are answered. Questions which contain references to discussions elsewhere are not answered here and usually only one or two of a set of similar questions are answered. The answers given cover the main points; detailed answers are not given where these are long and involved.

Some of the questions highlight common misconceptions of students. It is often worthwhile to assign some of these questions prior to or along with problems. In some instances, questions and problems are linked in the SUGGESTIONS sections of the Lecture Notes.

Many of the questions concern applications of the theory to areas not discussed in the text. These are an excellent source of lecture and recitation section material or might form the basis for student projects and term papers. In some cases, reference is made in the question to material in the literature and the references, in turn, give bibliographies or suggest a series of experiments.

Some instructors, in an attempt to have students think qualitatively about physical phenomena, ask questions similar to those in the text as examination questions. In some cases, both qualitative and quantitative aspects can be combined as separate parts of the same question.

Chapter 1

1. If a great many measurements show the same systematic variation, the variation might be attributed to the standard and a new standard sought. Remark on the variation of the solar day as measured by a cesium clock. When the current mean solar day was used as a standard, systematic variations on the order of 10^{-8} s occured in time measurements. The standard was changed to a particular mean solar day, then to a cesium clock.

2. Ideally, comparison with secondary standards should be easy to carry out with high precision. The standard should be indestructible, easy to store, safe to use, and convenient in size.

6. No. The old definition is no longer used.

7. The bar expands and contracts with changes in temperature. If two length measurements of the same object at the same temperature are made using the standard at two different temperatures, different results are obtained. To avoid this, the bar is taken as the standard only when it is at a certain temperature. No inconsistencies occur as long as the measurement of temperature is well defined.

10. (a)

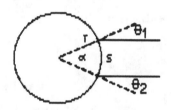

Simultaneously at two points on the earth's surface find the angle between the vertical and the line to a star. Use these angles and the arc length to calculate the radius of the earth. For the case shown $\alpha = \theta_1 + \theta_2$ and $r = s/\alpha$, where α is in radians.

(b)

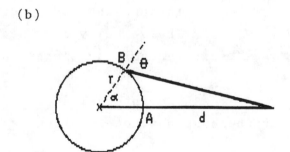

When the sun is overhead at A, sight to the center of the sun from B and use d = $r \sin\theta/\sin(\theta-\alpha)$. The radius r is known from part (a) and $\alpha = s/r$ is in radians.

(c) Measure the angular separation of the sun's edge across a diameter, then use the known distance to the sun to calculate its radius.

11. (a) Measure the total thickness of 100 sheets, then divide by 100.

(b) Use the interference fringe technique described in the text for measuring the length of a master machinist's gage block.

(c) Measure the mass of one atom using a mass spectrometer, measure the mass of a sample of the material, and take the ratio to find the number of atoms in the sample. Measure the volume of the sample, compute the volume per atom, and take the cube root. This assumes the atoms are closely packed. If the atomic configuration is known (from x-ray analysis, say), geometry can be used to find the interatomic spacing.

19. An atomic standard is certainly more accessible, less easily destroyed, less variable, and more reproducible than the present standard kilogram. It is difficult, however, to make high precision comparisons of atomic and macroscopic masses. Techniques are improving with technology, however.

Chapter 2

4. The meanings are different. Consider a round trip and point out that the average speed is zero by the first definition but non-zero by the second.

7. No. When v = constant, $x = x_0 + vt$ for any interval of duration t. The average velocity is $(x-x_0)/t = v$, the same as the instantaneous velocity.

8.

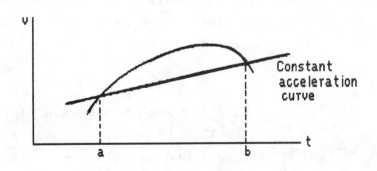

No in general. Graph v(t) for a variable acceleration. Pick an interval (a,b) and draw the straight line through the end points. The velocity for the variable acceleration is greater for every t in the interval so the average velocity is greater than for the constant acceleration case. Other situations can be considered.

10. (a) Yes. Point out that an object thrown straight up has zero velocity at its highest point but that it has a non-zero acceleration. If it did not it would remain at the highest point.

(b) Yes. Point out that a body rounding a curve at constant speed has a non-zero acceleration because the direction of its velocity vector is changing with time.

(c) No. A constant vector has a constant magnitude. Point out that each component of the constant vector is constant.

11. Yes. A ball thrown upward has constant acceleration, if air resistance can be neglected. At the top of its trajectory it stops, then starts down.

12. Yes. If the velocity and acceleration remain in the same direction, the speed increases regardless of changes in the magnitude of the acceleration.

Chapter 3

1. The beginning and end points are the same, so the displacements are the same. Emphasize that the path between is not relevant.

2. No. Draw one vector and a circle, centered at its head, with a radius equal to the magnitude of the second vector. Point out that the vectors from the tail of the first to points on the circle give all possible resultants as the orientation of the second vector changes. None of these is zero. It is possible to find many sets of three vectors which add to zero. Vectors of a set form the sides of a triangle.

3. No. The terms in $A^2 = A_x^2 + A_y^2 + A_z^2$ are all positive, so if one or more components do not vanish, A cannot vanish. Also point out that A = 0 implies that all components vanish.

6. Explain that 3 vectors that sum to zero form a triangle and therefore define a plane.

7. Explain that units are attached to the components, not to the unit vectors. This allows the use of the same unit vectors for quantities with different units. Also if a unit vector is defined by $\underline{A}/A$, the units cancel.

Chapter 4

3. For a given initial speed the length of the jump depends on the angle of take-off and so does the height of the jump. So different heights mean different initial angles and different lengths of jump.

4. They both fall the same distance in the same time. Over the same horizontal distance, however, the water falls a greater distance because the horizontal component of its velocity is much less than that of the electron.

5.   The minimum speed is at the highest point on the trajectory, where
     the vertical component of the velocity vanishes. Remind students that
     the horizontal component is constant. If the projectile remains above
     the firing point, it has its maximum speed at firing. If it goes
     below the firing point it has its maximum speed at the end of its
     flight.

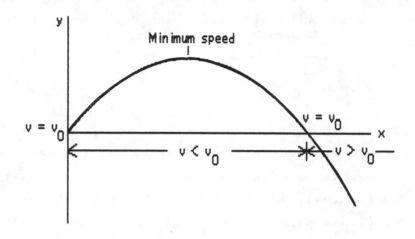

11.   In each case the density of the air is less in the first situation
      than in the second, air resistance is less, and the projectile
      travels farther.

12.   When air resistance is taken into account the firing angle for
      maximum range (with the landing point at the same elevation as the
      firing point) is somewhat greater than 45°.

13.   No. The initial velocity of the projectile and the acceleration due
      to gravity determine a plane. The projectile remains in that plane
      throughout its flight. If air resistance is significant and a wind is
      blowing perpendicularly to the plane of $\underline{v}_0$ and $\underline{g}$, then the projectile
      will move out of the plane.

14.   If the direction of the velocity changes with time the acceleration
      does not vanish, regardless of whether the magnitude of the velocity
      changes or not. In rounding a curve the direction of the velocity
      changes so the acceleration cannot vanish. Point out that a
      projectile moves on a curved path.

Chapter 5

1. When the train decelerates, the upper part of the body continues forward if the force exerted on it (by muscles of the lower back) is not great enough to give it the same deceleration as the train. The lower body is restrained by the frictional force of the seat and floor, unless the force required is too great. If it is, you slide off the seat. The situation when the train accelerates is similar. The back of the seat now provides the necessary force. When the train rounds a curve you will tend to go straight or round a curve of greater radius unless the force of the seat and floor is sufficient to hold you in the same curve as the train.

2. It is easiest for a passenger to maintain balance if the line through his feet is parallel to the acceleration of the car. When the car is starting or stopping the acceleration is in the forward or backward direction. When the car is traveling at constant speed the acceleration is sideways when the car rounds a bend. Even on a straight path the car may rock a little from side to side.

5. (a) The body moves at constant speed if the resultant force changes with time such that it is always perpendicular to the velocity. If the resultant force is constant in direction, the object cannot move with constant speed.

    (b) The velocity can be instantaneously zero, even though the acceleration is not. An example is a ball thrown straight up, at the top of its trajectory.

    (c) The sum of the two forces cannot vanish since the body is accelerating and these are the only forces acting on it.

    (d) The two forces can have any orientation relative to each other (except that they cannot be of equal magnitude and oppositely directed).

13. The horizontal component of the resultant force on the horse is the net result of the wagon pulling back and the ground pushing forward. The horse can accelerate if he pushes back on the ground with a force greater than that of the wagon on him. The ground then pushes forward with sufficient force. The horizontal component of the resultant force on the wagon is due to the horse and to the ground. If this does not vanish, the wagon accelerates. Point out that the two forces in an action-reaction pair act on different bodies.

14 (a) Yes.

    (b) Yes. The propeller pushes air back and the air pushes the propeller forward.

(c) Yes.

(d) Yes.

(e) No. Three different bodies are involved.

15. (a) Not true. Mass is the inertial property of an object. It is a measure of the acceleration of the object when it is subjected to a 1 N force. Weight is the force of the earth on the object.

(b) True.

(c) True only if the gravitational field acting on the body remains the same as the mass is changed.

(d) Not true. Variations in weight come about because the local gravitational field varies. The mass remains the same no matter what the local field.

22. Yes. The resultant of a vertical and horizontal force can never vanish, so the mass accelerates.

24. None. The acceleration of the body is in the direction of the total force acting on it. The velocity, on the other hand, can be in any direction whatsoever.

34. The tension is the least when the elevator is descending with increasing speed. The magnitude of the tension is then less than the weight of the object. The tension is the greatest when the elevator is descending with decreasing speed. It is then greater than the weight of the object. When the elevator is at rest or moving with constant speed the tension in the chord equals the weight of the object.

35. The forces on the woman are those due to gravity and to the spring balance. The spring balance shows zero (minimum) if both the elevator and woman are in free fall. It shows the maximum when the elevator and woman are accelerating upward.

37. If the pulley is frictionless, the elevator is in free fall.

Chapter 6

1. At first polishing removes some microscopic roughness from the surfaces. There are fewer "bumps" where the surfaces can catch on each other or where welds can form. When the surfaces are highly polished a large fraction of the atoms of one surface are close to atoms of the other surface and the surfaces tend to adhere. The materials tend to form a single sample with the atoms of one attracting atoms of the other.

2. Both coefficients can have values greater than 1. For an object resting on a horizontal surface, this means the horizontal force required to start it moving is greater than its weight and, once moving, the force of friction is greater than its weight. For an example, see problem 5.

4. Yes, by pushing with a force that has an upward vertical component as well as a horizontal component. The magnitude of the force that starts the crate is then less than $\mu_s W_c$, where W_c is the weight of the crate. The magnitude of the force you can generate without slipping is greater than $\mu_s W$, where W is your weight.

6. An external force must act on her. Such a force cannot be exerted by the ice so she cannot walk or roll off the ice. She can, however, throw a shoe. She exerts a force on the shoe and it exerts an equal and opposite force on her. Once started, she would continue to move were it not for air resistance. Throwing her arm out causes the rest of her body to move for a short time but this is of limited value since she must soon pull back on her arm and this stops the motion.

7. The coefficient of static friction is greater than the coefficient of kinetic friction so you should push just hard enough to prevent slipping.

11. Since $m = (4\pi/3)\pi r^3$ and $A = 4\pi r^2$, Eq. 17 yields $v_t = (2r/3)^{1/2}$, where r is the radius of the drop. Larger drops fall faster than small drops.

12. Initial speeds may be greater than the terminal speed. The baseball then approaches the terminal speed from above, rather than below.

15. If the road is banked the normal force of the road on the car has a component that is radially inward toward the center of the curve. It provides some or all of the centripetal force required for the car to round the curve. If the road were not banked drivers would need to rely on the force of friction to supply the required force. If it cannot, the car slides. Emphasize that the centripetal force needed to round a curve must be supplied by some agent.

25. The force of friction supplies the centripetal force required for the coin to go around with the record. As the record speeds up, the force required becomes larger. When the upper limit for static friction is reached, friction can no longer supply a sufficiently large force and the coin flies off.

Chapter 7

3. Generally there are forces of friction acting in all machines and
    some of the work done by the operator ends up as thermal energy. For
    some machines (the lever and multiple pulleys, for example) some work
    goes into the potential energy of the machine parts. This is
    recoverable but, in practice, this energy rarely is recovered in a
    useful way. Machines are used because a small force applied at the
    input produces a large force at the output. The work put in always
    equals or exceeds the work obtained.

4. Suppose the winning team pulls the rope through its hands and does
    not move. It does positive work on the rope and the rope does a small
    amount of negative work on the team (same magnitude force but smaller
    displacement). Suppose the losing team holds onto the rope. It does
    negative work on the rope and the rope does an equal amount of
    positive work on it. If the losing team drags its feet, the earth
    does negative work on them. Look at the system composed of the two
    teams and the rope. The net work done by external forces is negative.
    But the center of mass is accelerating toward the winning team and
    kinetic energy is increasing. The net work done at the hands of the
    winning team is internal work and must be considered to account for
    the energy.

9. You have done positive work mgh and the earth has done negative work
    −mgh. It is important to specify the force being considered when work
    is discussed.

15. While the ball is moving upward, the force of gravity does negative
    work on it and if this is the only force acting the ball loses
    kinetic energy and slows down. The ball stops when the magnitude of
    the work equals the initial kinetic energy. On the way down the force
    of gravity does positive work. If this is the only force acting the
    ball gains kinetic energy. Since the magnitude of the work done when
    the ball falls equals the magnitude of the work done when the ball
    rises, the ball returns to the throwing point with speed equal to its
    initial speed. If air resistance is present it does negative work
    throughout the trip. The ball does not rise as far and it returns
    with less kinetic energy and a slower speed than it had initially.

Chapter 8

1. It is transformed almost wholly to thermal energy in the brake while the wheel rotation is being stopped and at the road, where the tires skid. Once the wheels stop rotating no more work is done in the brakes.

3. Energy from somewhere increased the mechanical energy of the object. For example, the object may originally have had stored elastic potential energy (like a compressed spring) that was released on impact or an explosion may have occured.

4. It appears as potential energy of the counterweight, thermal energy in the hydraulic fluid, electric energy in the lines behind the motor.

6. The same amount of energy must be transferred, with or without an air bag. With an air bag the transfer occurs over a larger distance than without, so the average force is less.

9. The kinetic energy of the water at the bottom of the falls equals the sum of its kinetic and potential energies at the top. If the kinetic energy at the top is large the water has much more energy at the bottom than it would have if it merely dropped from rest. In a collision with a rock the water might bounce straight up. It would then reach a greater height than it originally had.

20 The minimum force required to turn away from the wall is $F_1 = mv^2/d$. This causes the car to turn in a circle with radius d. Slowing down to a stop before reaching the wall requires a force $F_2 = \frac{1}{2}mv^2/d$. Both forces must arise from friction between the tires and the road. Since $F_1 > F_2$, hitting the brakes seems advisable.

21. The potential energy stored in the spring appears as kinetic energy of the molecules freed in the reaction that dissolved the spring.

Chapter 9

1. No. The center of mass of a doughnut, for example, is in the hole. The center of mass of a table or desk with sufficiently massive legs lies under the top. The center of mass of a person bending from the waist is not within the body.

2. Suppose a cone is generated by positioning a large number of thin triangular prisms so their bases are diameters of a circle. The center of mass of each prism is one third of the way up but the mass density of the cone they generate is greater near the apex than near the base. Mass must be moved down to achieve a uniform density and

this lowers the center of mass.

4. The center of mass of the earth's atmosphere is quite close to the center of the earth.

5. Yes, if his body bends around the bar.

16. After being thrown, the external forces acting on the firecracker are the force of gravity and the normal force of the ice on each piece after it lands. After the toss but before the first piece lands, the firecracker's center of mass follows the parabolic trajectory of a projectile. As each piece lands, the normal force of the ice causes the vertical component of the center of mass velocity to decrease until, when all pieces have landed, that component vanishes. Meanwhile the horizontal component remains constant throughout. When all the pieces have landed, the firecracker's center of mass moves across the ice. For the man-firecracker system, the net force is the same as for the firecracker alone. After throwing the firecracker, the force of gravity on the man is canceled by the normal force of the ice on him. The center of mass in initially projected straight upward by the act of throwing since the ice can can exert only a normal force. It rises, then falls along a vertical line with acceleration g. When the fragments hit, the center of mass velocity decreases until it vanishes, when the last fragment has landed. It then remains zero. The center of mass of the man-firecracker system is then stationary.

Chapter 10

7. The terms _elastic_ and _inelastic_ describe collisions in which kinetic energy is conserved and not conserved, respectively. In a collision between two helium atoms momentum is conserved whether or not kinetic energy is conserved.

12. There is a change in energy associated with the permanent deformation of the balls. This is chiefly a potential energy due to the changed configuration of the atoms. In addition, the motions of the atoms change on collision and the internal energy increases.

14. The velocity of B after the collision is given by $v_{Bf} = 2m_A v_{Ai}/(m_A+m_B)$. For the greatest speed, $m_A/(m_A+m_B)$ should be as large as possible or $m_A \gg m_B$. For the greatest kinetic energy, $m_B m_A^2/(m_A+m_B)^2$ should be as large as possible. This function has its maximum value for $m_A = m_B$. If $m_A \gg m_B$, then $v_{Bf} \approx 2v_{Ai}$ and A

continues on with little decrease in speed. If $m_A \ll m_B$, then A rebounds backward with $v_{Af} \approx -v_{Ai}$. In the first case B acquires the greatest speed it can, whereas in the second case it acquires the greatest momentum. Initially all the kinetic energy is vested in A. This is the most that B can acquire and it does so only if A stops. For this to happen $m_B = m_A$.

Chapter 11

1. Yes. The units in $\omega = \omega_0 + \alpha t$ and $\theta = \omega_0 t + \frac{1}{2}\alpha t^2$ are consistent if θ is in degrees, ω is in deg/s, and α is in deg/s^2. Remind students that $s = \theta r$, $v = \omega r$, and $a_T = \alpha r$ are valid only if θ is in radians, ω is in rad/s, and α is in rad/s^2. The latter relations follow directly from the definition of the radian. If θ is given in degrees, then $s = (\pi/180)\theta r$, for example.

11. If $\alpha = 0$, the point has radial acceleration $\omega^2 r$ but $a_T = 0$. If $\alpha \neq 0$, then the point has both radial and tangential acceleration: $a_R = \omega^2 r$, $a_T = \alpha r$. For $\alpha = $ constant, $\omega = \omega_0 + \alpha t$ and the radial component (but not the tangential component) changes with time, so both the magnitude and direction of <u>a</u> change with time.

13. The linear speeds of points on the rims of the two gears are the same, so $R_1\omega_1 = R_2\omega_2$.

14. No. The rotational inertia depends on the <u>distribution</u> of mass, not just the total mass present. For a given fixed axis there are points where the mass can be concentrated and the rotational inertia will be the same. They are points such that the distance to the axis equals the radius of gyration. See problem 56. According to the parallel axis theorem the radius of gyration must be greater than the distance from the center of mass to the axis so the point of concentration cannot be at the center of mass.

16. For both disks $I = \frac{1}{2}MR^2$. The disk with the greater density has the smaller radius and the smaller rotational inertia.

17. Apply a known torque about the chosen axis, measure the angular acceleration, and calculate I using $\tau = I\alpha$. In practice you must correct for frictional torque in the bearings. It is usually convenient to use a constant known torque, measure the angular displacement for a given time and use $\theta = \frac{1}{2}\alpha t^2$ to find α.

18. The hoop has the largest rotational inertia since all its mass is at a large distance from the axis. On the other hand, most of the mass comprising the prism is relatively close to the axis so this object

228

has the smallest rotational inertia.

19. When the stick is pivoted at the wooden end its rotational inertia is greater than when it is pivoted at the steel end. The more dense steel is further from the axis of rotation in the first configuration. Thus, for the same torque, the angular acceleration is less in the first configuration.

Chapter 12

4. If both cylinders are uniform they have the same rotational inertia and reach the bottom at the same time. After the hole is drilled, the brass cylinder has the larger rotational inertia and takes longer than the wooden cylinder to reach the bottom.

8. Conservation of energy yields $\frac{1}{2}v_1^2[m + I/R_1^2] = \frac{1}{2}v_2^2[m + I/R_2^2]$. Since $R_2 > R_1$, $v_2 > v_1$.

17. If the net external torque vanishes, a decrease in rotational inertia results in an increase in angular velocity. Example: a spinning skater who drops her arms. If torque is applied in the direction of the angular velocity and the rotational inertia does not change, then the angular velocity increases. Example: a wheel on an accelerating bicycle.

18. Angular momentum is conserved since the net external torque vanishes. Since the student's rotational inertia decreases, his angular velocity increases.

Chapter 13

2. Yes. First suppose $\underline{F} = 0$ and $\underline{\tau} = 0$ about some axis. Since the torque about one point differs from the torque about another by $\underline{R} \times \underline{F}$ (where $\underline{R}$ is the displacement of the second point from the first), it follows that $\underline{\tau} = 0$ about every axis. Use $\underline{F} = m\underline{a}_{cm}$ and $\tau = I\alpha$ to show $\underline{a}_{cm} = 0$ and $\alpha = 0$ about any axis. The given condition is sufficient. Now suppose $\underline{a}_{cm} = 0$ and $\alpha = 0$ about some axis. Use $\underline{F} = m\underline{a}_{cm}$ and $\tau = I\alpha$ to show $\underline{F} = 0$ and $\underline{\tau} = 0$ about that axis. Again $\underline{F} = 0$ means that $\underline{\tau} = 0$ about every axis, so the condition is necessary.

3. No. There is a net force (due to gravity) acting on it.

4. Yes. At the bottom of its swing, the net force and net torque are both zero.

5. If the center of gravity does not coincide with the center of mass and furthermore the line between the two is not vertical, gravity exerts a torque about the center of mass of the ball and it spins.

9. Turn a bicycle over and apply forces to the the axle and rim of the front wheel so that the center of mass does not move. Turn on a phonograph and watch it come up to speed. Push down on the accelerator when the car is on ice. Explain the case when two tangential forces, equal in magnitude but opposite in direction act at the rim of disk. The disk then has an angular acceleration but its center of mass does not accelerate.

11. Neglect the force of friction at the wall and suppose the ladder makes the angle θ with the horizontal. Then if the man is a distance x up the ladder the force of friction at the ground must be $f = (wx + \frac{1}{2}W\ell)/\ell\tan\theta$ for the ladder to remain in equilibrium. Here w = weight of the man and W = weight of the ladder. As x increases f must increase to hold the ladder, so the danger of slipping increases as the man climbs.

19.

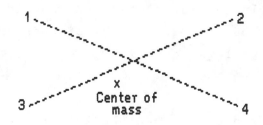

If the forces at the tires are vertical and the center of gravity is in the vertical plane through 1-4, then the forces exerted by the tires at 2 and 3 are equal. Other wise they are not. Similarly, if the center of gravity is in the vertical plane through 2-3, then the forces exerted by the tires at 1 and 4 are equal. Otherwise they are not. This is all that can be deduced from the conditions of equilibrium. The problem is indeterminate unless further information (the elastic properties of the tires and springs) is known and the tilt of the car is taken into account. One point of this question is to illustrate that not all equilibrium problems are solvable using the equations of equilibrium alone.

Chapter 14

3. Take the y axis to be positive in the downward direction, with the origin at the lower end of the spring when it is unstretched. When weight W is added, the spring stretches to y_0, where $ky_0 = W$. Now exert an additional force $\underline{F}$, down. The spring stretches to y, where $ky = W + F$ or, using the first result, $k(y-y_0) = F$. Recognize that $y-y_0$ is the displacement from the new equilibrium point and note that it is multiplied by the same spring constant k to yield the force. This answer emphasizes the determination of the spring constant by

stretching the spring. The spring length must stay within its elastic limit.

5. The extension of the spring is $\Delta y = mg/k$, so $k/m = g/\Delta y$. This is the square of the natural angular frequency so the period is $T = 2\pi/\omega = 2\pi\sqrt{\Delta y/g}$. The natural frequency depends on the ratio k/m and not on k and m separately.

8. The period and force constant do not depend on the amplitude. The period depends on k/m and the force constant is a property of the spring. The total mechanical energy is given by $\frac{1}{2}kA^2$, the potential energy when the displacement is a maximum and the velocity vanishes. The total energy can also be written $\frac{1}{2}mv_{max}^2$, the kinetic energy when the displacement is zero. So $v_{max} = \sqrt{k/m}\,A$ and it is proportional to A.

9. Since $v_{max} = A\omega$, you could double the amplitude (by doubling both the initial displacement and initial speed) or double the frequency (without changing the amplitude). In the later case, k/m should be made 4 times as large. This means changing the spring or mass and might not be allowed by the wording of the question.

Chapter 15

15. Viewed on a large scale, the moon travels around the sun in a nearly elliptical path, pulled by the gravitational force of the sun. Its orbit is nearly the same as that of the earth and even if there were no gravitational attraction between the moon and the earth they would travel close together. Superposed on this motion is its motion around the earth. This motion is on a much smaller scale.

17. The moon and sun pull on both the earth and its oceans. What is important is the <u>difference</u> between the gravitational field of the sun or moon at a point on the earth's surface (the water) and the earth's center. This difference is much greater for the moon than for the sun because the moon is closer to the earth. A calculation is required to show this. The ratio of the gravitational force of the sun on the earth to that of the moon on the earth is $(m_s/m_m)(r_{em}/r_{es})^2 \approx 175$. The ratio if the center-surface difference for the two gravitational fields is $(m_s/m_m)(r_{em}/r_{es})^3 \approx 0.4$.

Chapter 16

20. The ice displaces a volume of water with weight equal to its own
weight. When it melts, it produces an amount of liquid with the same
weight and hence with a volume equal to the volume of water
originally displaced. The water level does not change. When ice
containing sand floats, it displaces a volume of water with weight
equal to the weight of the ice and sand. When it melts the ice
portion produces a volume of liquid with weight equal to its own
weight. But the sand sinks showing that it alone displaces a volume
of water with weight less than its weight. In all, less water is
displaced after melting than before so the water level falls. If the
ice contains air, the level rises when it melts.

22. As the air pressure increases the ball rises slightly. It drops
slightly as the air is pumped out. To simplify the problem, consider
a cylinder of length ℓ, floating upright with length h submerged. If
p_a is the air pressure on its top, then $p_a + \rho_a g(\ell-h)$ is the air
pressure at the water level, and $p_a + \rho_a g(\ell-h) + \rho_w gh$ is the water
pressure on its bottom. The net force on the cylinder is $mg + p_a A -$
$[p_a + \rho_a g(\ell-h) + \rho_w gh]A$. This vanishes so $h/\ell = (\rho-\rho_a)/(\rho_w-\rho_a$, where
ρ is the density of the cylinder. As the air pressure increases, ρ_a
increases and h decreases. Since ρ_a is so much smaller than either ρ
or ρ_w, the effect is very slight.

26. It is not necessarily true. For stability all that is required is
that when the object is tilted the net torque (due to the force of
gravity and the buoyant force) tends to right the object. Two
examples of stability are shown. The center of gravity is marked by
X, the center of buoyancy is marked by ·. For the tilt shown the
center of buoyancy must be to the right of the center of gravity,
after the tilt. Remark that the center of buoyancy may shift relative
to the object as the object tilts.

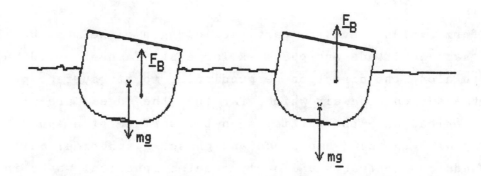

37. Consider the view from
the rest frame of the
boats, as shown. The
water speed v far from
the boats is the same
as the boat speed in the
frame of the water. The
region between the boats
narrows from point 1 to

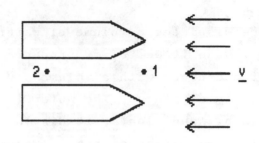

point 2. Since vA remains constant, the water speed at 2 exceeds v.
Therefore the pressure drops and the boats are sucked together. A
similar effect, due to air, occurs between cars. It is most
noticeable when a large truck passes slowly by your car with both car
and truck going at high speed.

Chapter 17

5. For any given value of t all these functions increase indefinitely
with x. Such disturbances are impossible to produce since they
require that the source have infinite amplitude at a time infinitely
far in the past.

8. For a two dimensional circular wave the power crossing every circle
centered on the source must be the same, so the power per unit length
of circumference must decrease as $1/r$ and the amplitude must decrease
as $1/\sqrt{r}$. In practice the amplitude decreases more strongly with r
because energy is converted to thermal energy by frictional forces.

13. Even when the string is straight it is still moving and has kinetic
energy. At maximum displacement the string is not moving and all the
energy is potential energy, stored as elastic energy in the stretched
string. At intermediate displacements the energy is partially kinetic
and partially potential.

Chapter 18

5. To form a plane wave the particles along a plane must be moved with
the same amplitude and phase. A large stiff sheet moved back and
forth along the direction perpendicular to its surface produces waves
in the surrounding air which, far from the edges, are plane. To form
a spherical wave the particles on the surface of a sphere must be
moved with the same amplitude and phase. A spherical balloon which
expands and contracts uniformly creates spherical waves in the

surrounding air.

12. The valves and slide change the length of the air column in the instrument and thus change the resonant frequencies of the instrument.

13. With no valves the bugler is limited to the natural resonant frequencies of the instrument. Notes are produced by blowing air through the lips into the instrument. The lips are closed by the low pressure behind the moving air and are pulled open by the pressure wave reflected at the far end of the bugle. Different notes can be sounded by changing the response of the lips to the column of vibrating air. This is done chiefly by changing the tension in the lips, pursing and relaxing them.

20. Pitch, loudness, and tone quality are all subjective characteristics of sound. Pitch roughly corresponds to frequency but the relationship also depends on the intensity. Two notes, near each other in frequency, might be ordered with either one at the higher pitch, depending on the relative intensity. Loudness corresponds roughly to intensity but, as a look at Fig. 6 reveals, there are low frequency notes which are inaudible and higher frequency notes, played at the same intensity, which can be heard and so are louder. Tone quality is determined chiefly by the mixture of overtones present, their relative amplitudes, when they reach peak amplitude, and when they die out (relative to the fundamental).

Chapter 19

3. Energy is transferred from the thermometer to the ice. Since the enclosure is evacuated the transfer takes place via electromagnetic radiation. In this case, the thermometer radiates more energy to the ice than the ice radiates to the thermometer.

12. (a) The sun's temperature can be found by examining the radiation from it. One method is to measure the radiancy and apply Stefan's law. Another is to find the wavelength for which the spectral radiancy is a maximum and calculate the temperature of a blackbody which gives the same peak.

(b) The temperature of the upper atmosphere can be inferred from the drag on a satellite. Atmospheric force on the satellite depends on the density of the atmosphere which, in turn, depends on the temperature. It is also possible to measure the distribution of the speeds of charged particles in the upper atmosphere. The distribution is temperature dependent, as explained in Chapter 21.

(c) The thermometer for this application must be quite small.
Thermocouples are usually used. These are made of two fine wires of
different metals, joined together at each end. There is a potential
difference, end to end, which depends on the temperature difference
between the ends. This instrument can be made in the form of a
hypodermic needle and inserted into the insect.

(d) Most of the light from the moon is directly reflected sunlight.
Some sunlight is absorbed, however, and reradiated in the infrared.
This is the portion which heats the moon and analysis of the spectrum
gives the temperature.

(e) The property measured should be insensitive to pressure and show
wide variation in the range from -15° C to +15° C. The resistivities
of metals meet the requirements. The resistance of a platinum sample
is usually measured.

(f) The resistance of a semiconductor (usually GaAs) is measured.
This shows great variation with small changes in temperature in the
region around 4 K. It does not change phase or undergo structural
changes from room temperature down.

13. Many gases give the same temperature reading (the ideal gas
temperature) in the limit of small concentration. They differ in
convenience of measurement and range of usefulness, however. The gas
cannot undergo a phase change in the pressure and temperature range
of interest, so a low boiling point is desirable. Since the
temperature limit as $p_{tr} \rightarrow 0$ must be evaluated, it is desirable that
the curve shown in Fig. 6 have as small a slope as possible.
Generally helium is picked as the gas which best satisfies these
requirements. In addition, the gas and the container material must be
picked to minimize diffusion of the gas into the container walls. If
platinum is used for the container, helium also satisfies this
requirement.

Chapter 20

2. Any adiabatic process in which work is done on or by the system
results in a temperature change. Use a piston to compress a gas in a
well insulated container.

4. Yes. Heat added to ice at the melting point causes the formation of
liquid water at the same temperature. In the definition of heat the
temperature difference referred to is between the environment and the

system. The environment must be at a higher temperature to add heat to the system but the temperature of the system need not change while heat is being added.

9.   If heat ΔQ is transferred the temperature of the hotter body drops by $\Delta T_h = -\Delta Q/m_h c_h$, where m_h is its mass and c_h is its specific heat. The temperature of the colder body increases by $\Delta T_c = \Delta Q/m_c c_c$. These temperature changes are not the same unless $m_h c_h = m_c c_c$.

22.   The outside air is much colder than the inside air and heat flows from the air inside to the wall, through the wall, to the air outside. The rate of heat flow per unit temperature difference is greater for flow through the wall than for heat transfer at either surface. If the temperature difference between the inside air and the inside wall surface is small or zero, the energy leaving the surface through the wall is replaced at a lower rate than it leaves and the temperature of the surface decreases. The decrease reduces the rate at which energy leaves the surface and increases the rate at which energy is transferred from the air to the wall. Steady state is reached when the two rates are the same. There is then a temperature difference. At the outside surface the temperature of the wall increases until the rate at which energy passes through the wall equals the rate at which energy is transferred to the air. There is then a temperature difference between the surface and the air. Mention that the thermal conductivity of air is quite small and energy gets to the wall on the inside and is removed from the wall on the outside chiefly by convection. The convection rate is to be compared with the conduction rate through the wall.

27.   The internal energy is increased if heat flows from the environment to the system or if the environment does positive work on the system. If a gas in a well insulated container is compressed, its internal energy increases but no heat flows. If the gas expands adiabatically against a piston, it does work and its internal energy decreases. No heat flows.

Chapter 21

3.   If the concentration is reduced by removing molecules there will be a point where the assumption of equal numbers of molecules moving in all directions fails. Temperature and pressure are then not well defined and we must consider the motion of individual molecules to calculate the force on the container walls, for example. On the other

hand, as the concentration is increased the gas becomes non-ideal because the sum of the molecular volumes is too large a fraction of the container volume and because the potential energy of molecular interaction is too large a fraction of the total energy. In general real gases do not meet the requirements to be ideal. However, for many gases, particularly those with small, weakly interacting molecules, there is a fairly broad range of concentrations for which kinetic theory works and for which the gas is nearly ideal. Fig. 19-6 shows some of these gases. For some gases, say a plasma of charged particles (which interact strongly over large distances), such a concentration range may not exist.

10. We consider equal time intervals which are so long that during any one of them a large number of molecules strike the walls. The number of collisions with the walls is not precisely the same for each interval but it nearly the same and as the interval becomes large the fractional deviation from the average becomes small. Except for these small fluctuations the force exerted by the molecules on the walls is constant. It is important that the interval be small compared to the mean time between collisions of a single molecule. It is if the gas concentration is sufficiently high.

46. If heat is added at constant volume, all of the additional energy becomes part of the internal energy of the system. The temperature rises. If the same amount of heat is added at constant pressure, part of the additional energy is lost to the environment (a piston, say), against which the system does work. Less energy is added to the pool of internal energy so the temperature rise is less and the specific heat is greater.

48. During expansion, when a molecule bounces off a _moving_ piston, it does work on the piston and loses kinetic energy. If no energy enters the system, then internal energy decreases and, since the temperature is proportional to the internal energy, it also decreases.

Chapter 22

6. When one surface is close to the another, bonds are formed by the attraction of molecules to each other across the boundary. When the bodies move relative to each other, work is done to break the bonds. This energy is supplied by the agent pushing the bodies past each other. The molecules in the same body also exert forces on each other and when one molecule near the surface is pulled out of position to

form a bond, other molecules, deep inside the material, are also pulled away from their normal positions. When the bond is broken, the energy is distributed among the various out-of-place molecules and eventually spreads to a great many, if not all, the molecules of the body. The temperature of the material increases. The reverse process does not occur since that would require the energy to return to a relatively few molecules near the surface, where it would be used to move surface molecules and finally to provide the kinetic energy of the other body.

11. Work cannot be calculated using a p-V diagram if the process is irreversible. In fact, pressure is undefined. Work is done, however. Molecules strike the moving piston and exert a force on it.

12. Reversible heat transfer can be approximated. Use a series of heat reservoirs (large chunks of material with large specific heats, at uniform temperatures), at slightly different temperatures. One reservoir should be at the initial temperature and one at the final temperature. Start with the reservoir at the initial temperature and place the system in contact with successive reservoirs, in order of increasing temperature, until the final temperature is reached. The process can be reversed by reversing the order in which the reservoirs are used. The smaller the difference in temperature from one reservoir to the next, the closer the process to being reversible.

15. For any heat engine, the greater the ratio of work obtained to heat added, the greater the efficiency. If all the heat added were converted to work, with no heat rejected or retained as internal energy, the engine would be 100% efficient. This is prohibited by the second law of thermodynamics unless the temperature of the cold reservoir is 0 K. The third law asserts that it is impossible to achieve exactly 0 K.

16. All reversible heat engines have the same efficiency and irreversible engines have lower efficiencies, so reduction of efficiency occurs when there are departures from reversibility: compression and expansion strokes are not quasi-static or heat transfer is accomplished by contact with an environment at a greatly different temperature than that of the working substance. There is also heat loss: not all of the heat leaving the boiler enters the working substance.

17. Use $e = 1 - T_c/T_h$ to show that $de/dT_c = -1/T_h$ and $de/dT_h = T_c/T_h^2$.

238

Since $T_c < T_h$, the efficiency increases more rapidly with a decrease in T_c than with an increase in T_h.

29. For the earth $\Delta S_e = \Delta Q/T_e$ and for the sun $\Delta S_s = -\Delta Q/T_s$, where ΔQ is the heat absorbed by the earth. Since $T_s > T_e$, it follows that $|\Delta S_e| > |\Delta S_s|$ and $\Delta S = \Delta S_e + \Delta S_s > 0$.

Chapter 23

1. Start with neutral spheres and, while they are touching each other, bring the charged rod rod close to one of them. The positive charge on the rod attracts negative charge to the surface near the rod, some of it coming from the other sphere where it leaves positive charge behind. Now separate the spheres and finally remove the rod. The sphere which was nearest the rod is negatively charged while the other sphere is positively charged. Since charge is conserved, the magnitudes of the charges are the same. The spheres need not be the same size.

2. Use two identical spheres and charge one of them (by touching it to the glass rod or by induction). Then remove the rod and touch the spheres together. Since they are identical, they will carry identical charge.

3. The rod polarizes the cork dust so that the surface of a dust particle nearest the rod is charged oppositely to the rod and the surface away from the rod is charged with the same sign charge as the rod. Since the electric force decreases with distance, the near side of the dust is attracted more strongly than the far side is repelled and dust jumps to the rod. When in contact, charge flows and the dust obtains a net charge of the same sign as the charge on the rod. The dust may then be repelled.

11. The electrons in the metal exert forces on each other. Electrons already at the end tend to repel other electrons, for example. Flow stops when the net force on each electron, due to charge on the insulator and to other electrons in the metal, vanishes.

14. (a) No. Charge separation occurs in the object, with negative charge closer to the rod than positive charge. Since the force between two charges decreases with their separation, the result is attraction.
    (b) Yes. Charge separation in a neutral rod would cause attraction. For repulsion the net charge must be positive.

20. The magnitudes of the two forces are both proportional to the product of the charges and are therefore equal.

Chapter 24

3.   When the test charge is placed at the point, it will repel the ball
      and the ball will move to a new position, further from the test
      charge. The test charge experiences a force which is smaller than it
      would experience if the ball had not moved. The magnitude of $\underline{F}/q$ is
      less than the magnitude of the field at the point, in the absence of
      a test charge.

5.   Wherever there is an electric field its direction is not ambiguous.
      It can be found, for example, by evaluating the vector sum of the
      fields due to all charges. On the other hand, if two field lines
      cross, each would give a different direction for the field at the
      point of crossing. We conclude field lines do not cross.

6.   At a point which is far from the charge distribution compared to the
      distance between charges, the field must become the field of a single
      charge, equal to the net charge in the distribution. For net positive
      charge, this field is radially outward from the charge.

7.   At any instant the acceleration of a charge is along the field line
      through its position at that instant. Stress that it is the field
      line due to all _other_ charges. Since the particle starts from rest,
      it starts moving along the field line through its initial position.
      If the line is straight, it continues along that line. If the line
      bends, the particle does not follow it once the particle acquires a
      non-vanishing velocity.

13.   The lines of force shown are for the total field, created by both
      charges. The total field is large near either one of the charges
      because the field of that charge is large. The force on a charge,
      however, is due to the field of all other charges, not its own field.
      To find the force on the lower charge in Fig. 5 only the field due to
      the upper charge should be used.

Chapter 25

5.   (a) _All_ the charges. (b) Equal. Charges q_3 and q_4 do not contribute
      to the net flux through the surface.

7.   (a) No change occurs since the amount of charge inside the two
      surfaces is the same. The shape of the surface is immaterial.
      (b) No change. The size of the surface is immaterial.
      (c) No change. The position of the charge is immaterial as long as
      it remains inside.

(d) The flux is now zero. There is zero charge inside the surface.

(e) No change from the original situation. The charge inside contributes the same flux as before while the charge outside contributes zero flux.

(f) The flux is now that due to the two charges and is $(q_1+q_2)/\epsilon_0$.

8. The integral has the same value if $\underline{E}$ is the total field, due to charge both inside and outside the surface, or if $\underline{E}$ is the field due to the charge inside only. The flux due to charge outside vanishes, although the field does not not.

10. If the surface encloses zero net charge, the integral for the flux vanishes but $\underline{E}$ need not be zero anywhere on the surface. Think of the surface divided into many small elements $\Delta\underline{S}$. For some elements $\underline{E}\cdot\Delta\underline{S}$ is positive while for others $\underline{E}\cdot\Delta\underline{S}$ is negative. The sum may vanish even though none of the individual terms are zero. Consider a closed surface with a single charge outside. $\underline{E}$ does not vanish anywhere but the integral is zero. The converse is true, however. If $\underline{E} = 0$ everywhere on the surface then the flux through the surface vanishes and the net charge enclosed must be zero.

13. $\underline{E} = 0$ inside a spherical balloon with a spherically symmetric charge distribution on its surface, provided no other charge is present. For a sausage shaped balloon with a uniform charge distribution on its surface, $\underline{E} \neq 0$ inside. Note that the flux through any closed surface wholly inside either balloon vanishes. Also note that the proof that $\underline{E} = 0$ inside a uniform distribution depends on the symmetry of the situation.

25.

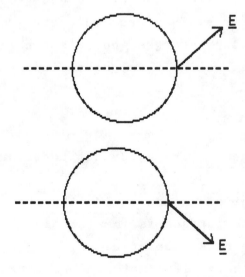

Imagine an electric field with a tangential component. Rotate the sphere by 180° around the dotted line. The field rotates to the direction shown in the second diagram. The charge distribution, however, is the same before and after the rotation because it is spherically symmetric, so the field must be the same. Only a radial field, not a field with a tangential component, remains the same after a rotation about any axis. Also show that the magnitude of $\underline{E}$ is the same at every point on a sphere centered at the center of the distribution.

Chapter 26

1. Yes. The potential at any point (not occupied by a point charge) can be given any value whatsoever. If a point is said to be a +100 V instead of 0, the potential at all points would be given values +100 V higher than previously and the potential difference between any two points would remain the same. Only potential _differences_ have physical meaning.

13. No. $\underline{E}$ must be known for all points along some path joining the point where V = 0 to the point of interest. Then the path integral that defines the potential difference can be evaluated.

14. E can be estimated using $|\Delta V/\Delta d|$, where Δd is the distance between two equipotential surfaces whose potentials differ by ΔV. The surfaces are closer together on the left side of the figure so the field has a larger magnitude there.

18. Charge is not necessarily distributed uniformly over the surface of a conductor. In fact the charge per unit area is large where the radius of curvature is small and vice versa. If, on the other hand, the magnitude of the field is uniform over the surface then the charge distribution is uniform, as shown by $E = \sigma/\epsilon_0$.

21. $\underline{E}$ = 0 in that region. Note that V must be constant throughout a volume, not just along a line or on a surface, for $\underline{E}$ to vanish.

23. The electric field must be zero in that region. Remove all charge from the region and completely surround it with a good conductor. The field is zero, not only in the conductor, but in the cavity.

24. In one possible arrangement the charges are at the vertices of an equilateral triangle and $q_1 = q_2 = -2q_3$.

Chapter 27

1. Charge is conserved. All charge removed from one plate goes to the other. Neither the battery nor the wire collect charge. The plates were originally neutral, so they end up with charge of equal magnitude and opposite sign, regardless of their size or shape.

2. Yes. As a simple example, consider two spherical shells, one inside the other. The outer shell is $(q/4\pi\epsilon_0)(1/b - 1/a)$ higher in potential than the inner shell. Here a is the radius of the inner shell and b is the radius of the outer shell.

242

6.   (a) The capacitance increases. To see this, use $C = \epsilon_0 A/d$.
      (b) The new capacitance is $C = \epsilon_0 A/(d-t)$, where d is the plate
      separation and t is the slab thickness. The effect is small if
      $t \ll d$.
      (c) The capacitance is doubled.
      (d) The charge on the larger plate spreads out into an area slightly
      larger than that of the smaller plate. For the same charge, the field
      and the potential difference are smaller (the field lines are further
      apart) and the capacitance is larger. This is a small effect since
      most of the charge on the larger plate is opposite charge on the
      smaller plate.
      (e) The capacitance is reduced by slightly more than 50%.
      (f) There is no effect. The charge also doubles.
      (g) The capacitance increases. Think of the situation as a collection
      of capacitors, wired in parallel, each succeeding one with a slightly
      smaller separation than the one before.

14.   In a uniform field the polarization is also uniform so the dielectric
      acts like a body with charge distributed only on its surface. The
      total charge is zero and since the field is the same at each surface,
      the new force is zero.

18.   The charge remains the same. It cannot change once the battery is
      removed. Because the dielectric is polarized with positive
      polarization charge on its surface nearest the negative plate and
      negative polarization charge on its surface near the positive plate,
      the field between the plates is reduced. The potential difference,
      which is the path integral of the field, is also reduced. A smaller
      potential difference for the same charge means a larger capacitance.
      The energy stored is reduced since $U = \frac{1}{2}q^2/C$. Negative work was done
      by the agent which moved the dielectric into place (the field pulled
      the slab in, the agent had to pull back on the slab to keep it from
      accelerating).

19.   The potential difference and electric field remain the same since
      they are maintained by the battery. The sum of the charge on either
      plate and the polarization charge on the nearby surface of the
      dielectric is the same as the charge on the plate before the slab was
      inserted. These charges create the field and the field is the same.
      Since the polarization charge and the plate charge have opposite
      signs, the charge on the plate must increase. An increase in the
      charge for the same potential difference means a greater capacitance.

Since $U = \frac{1}{2}CV^2$, the energy stored increases by $\frac{1}{2}(\kappa-1)C_0V^2$. The battery does work $(\Delta q)V = (\Delta C)V^2 = (\kappa-1)C_0V^2$, so the agent does work $-\frac{1}{2}(\kappa-1)C_0V^2$. The agent again had to hold back on the slab to keep it from accelerating.

Chapter 28

1. In the steady state, no charge builds up anywhere and the rate at which charge enters the volume equals the rate at which it leaves. The net current out of the surface vanishes, so the integral vanishes. This is true no matter what circuits lie within the surface. Gauss's law holds and since the charge within the surface does not change with time, the electric flux through the surface does not change either.

2. Whatever the convention, it must describe all possible situations. In particular, if positive and negative charge move in opposite directions the current densities must be in the same direction and the currents must have the same sign. If they move in the same direction, the current densities must be in opposite directions and the currents must have opposite signs. This stems from an important property of nature. Positive charge moving in one direction and negative charge moving in the opposite direction are both brought about by the same electric field or emf and they produce the same effects (magnetic field, for example). For most applications (excepting the Hall effect, for example), the current may be considered to be composed of positive charge moving in one direction or negative charge moving in the opposite direction. With this limitation, it is possible to take the electron to be either positive or negative and, in a separate convention, to take the current to be in the direction of electron flow or in the opposite direction. The laws of physics must then be written to conform to the convention and describe nature.

5. (a) In the steady state, no charge builds up anywhere and the current into the one corner of the cube must be the same as the current passing through any plane completely through the cube, regardless of its position and orientation.

(b) Near the sides of the cube the current density is parallel to the sides, while near the body diagonal connecting the leads it is parallel to the body diagonal. In going from a side toward the body diagonal, the current density changes direction. Most of the charge

flows near the body diagonal so the current density also changes
magnitude across the plane. The plane cannot be oriented so that $\underline{J}$ is
uniform on it.

(c) Yes.

(d) Yes.

6. (a) Doubling V doubles the electric field in the wire and so doubles
the acceleration of the electrons between collisions. The drift speed
is therefore doubled.

(b) If ℓ is doubled, with the same potential difference, the electric
field is halved and so is the drift speed.

(c) There is no effect as far as the drift speed is concerned. The
electric field is the same if the potential difference between the
ends is the same. The current, however, increases by a factor of 4
since the cross sectional area increases by that factor and the
current density remains the same.

14. Between collisions, the electrons accelerate in the direction
opposite to that of the field. The curved lines are parabolas for a
uniform field.

21. The apparent paradox is resolved by specifying what is held fixed
while R changes. If V is fixed, P decreases as R increases according
to $P = V^2/R$. If i is fixed, P increases as R increases according to $P = i^2 R$.

Chapter 29

1. No. An emf is directed from the negative to the positive terminal,
through the seat, under open circuit conditions. It is independent of
the current direction.

4. To measure the emf, place a high resistance voltmeter across the
terminals, with no other circuit attached. To find the internal
resistance, first place a small resistance across the terminals and
measure both the current and the potential difference across the
terminals. Then use $V = \epsilon - ir$ to calculate r.

6. If R is the resistance of the light bulb, r is the internal
resistance of the battery, and ϵ is the emf, then the current is $i = \epsilon/(R+r)$. If the internal resistance is large then $i = \epsilon/r$,
independently of the bulb resistance. The power dissipated by the
bulb is $i^2 R = \epsilon^2 R/r^2$, which is higher for the high resistance bulb
(25 W). The situation is reversed if the internal resistance is low.
Then the potential difference across the battery is nearly ϵ,

independently of the bulb resistance. The power dissipated by the
bulb is ϵ^2/R, which is lower for the high resistance bulb.

Chapter 30

2. The magnetic field would then be different for a test charge moving
   in different directions. When it is defined as usual, we may think of
   the field as existing independently of the test charge.

5. No. The electron might be traveling parallel or antiparallel to the
   field.

6. No. An electric field may be pointing in the direction opposite to
   the electron's acceleration. For example, the electric field of a
   positive point charge might cause an electron to travel uniformly
   around a circle just as a uniform magnetic field does. The test for a
   magnetic field must start with a test for an electric field, using a
   stationary test charge. After the electric force is subtracted from
   the total force, what remains is the magnetic force.

9. For typical magnetic fields ($\approx$ 1 T), the dominant force is the force
   of the magnetic field on
   the electrons and the
   interaction between
   electrons can be ignored.
   At first they repel each
   other and speed up but
   once they are separated
   even slightly this
   effect can be neglected.
   Look down on the room.

   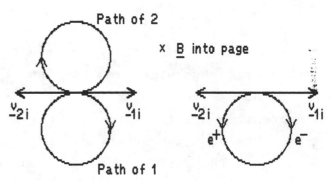

   Each electron travels in a horizontal circle, through a point very
   near the point of release and tangent to its initial velocity. Both
   travel in a clockwise direction so the circles are on opposite sides
   of the line along the initial velocities.
   (b) The positron travels around its orbit in a counterclockwise
   direction while the electron travels in a clockwise direction as
   before. The circles coincide. The charges do not move in circles but
   actually spiral inward as they lose speed in collisions.

14. Yes. The force on the free electrons must be transmitted to the wire
    as a whole; that is, to the ions. This comes about through collisions
    between electrons and ions. If there were no interaction between
    electrons and ions, electrons would pass out of the wire and the wire

would not move.

17. If electrons form the current, they are moving and experience a
     magnetic force. The protons are at rest (macroscopically) and do not
     experience a magnetic force.

Chapter 31

3.   Not necessarily. If $\underline{B}$ is constant in magnitude the same number of
     field lines pass through each unit area perpendicular to the lines.
     The lines do not spread or come together. If the lines are more dense
     in one region than in another, along a line, the magnitude of $\underline{B}$
     cannot be constant along that line.

10. Near each wire the
     lines are circular
     since the field of
     the nearest wire
     dominates. In the
     region between the
     wires the individual
     fields tend to cancel
     and there are few

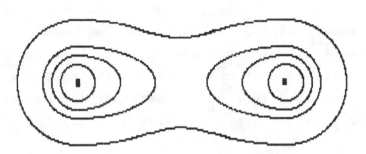

     lines. In the region beyond either wire the individual fields tend to
     reenforce each other and the density of lines is greater. Far away
     from both wires the field is like that of a single wire carrying
     current 2i and the lines are circular again.

15. Yes. The path can be any closed path which forms the boundary of a
     surface. The surface and its boundary may be a mathematical
     construction or may be a physical entity. Fig. 10 shows a case where
     the path is entirely within a conductor.

16. First use the right hand rule: curl the fingers in the direction of
     integration around the path; then the thumb points in the direction
     of positive current. Use the rule to assign the correct sign to each
     of the currents. Then algebraically sum the currents through the
     surface bounded by the path.

Chapter 32

8.   The magnetic field due to the current in the larger loop is left to
     right through the smaller loop as seen in the diagram. Since the
     field is increasing in magnitude, the field of the smaller loop must
     point right to left through that loop. The current must be in the

counterclockwise direction. The coils behave like two magnets with like poles facing each other. They repel each other.

15. Current in the left loop is counterclockwise and is decreasing with time. The field through the right loop is into the page and is decreasing in magnitude. The current induced in the right loop creates a field which is into the page inside the loop and, to create this field, the current must be clockwise.

20. The induced current produces a magnetic field that is out of the page in the interior of the loop. The external field must be into the page since its flux increases as the rod moves. This increase is then opposed by the induced field.

22. It can be translated, without rotation, in any direction. Since the field is uniform the flux through the loop does not change. It can be rotated about any axis parallel to the field, through the center of the loop or elsewhere. Rotation about any other axis will produce a changing flux and an induced emf.

24. Suppose a strip without slots starts as shown and swings into the field. The flux through it increases at first and currents are induced. The induced field must oppose the increase in flux, so the induced currents are counterclockwise. Current flows up the right portion of the strip (in the field) and down the left portion (outside the field). The magnetic force on the current in the right portion retards the motion of the strip. Similarly, as the strip leaves the field on the other side, the magnetic force tends to pull it back. The slots force the current loops to close inside the field, for the most part, and damping is reduced.

Chapter 33

2. Use $\epsilon = -L \, di/dt$. Assume a changing current and calculate the flux and emf for each loop of the coil, using a consistent sign convention. Add the emfs, then solve for L.

4. Wind the coil in layers with the current in opposite directions in successive layers.

7. After the switch is thrown current continues to flow for a long time and charge collects at the switch blades. The potential across the blades increases (it must match $L \, di/dt$ across the inductor) and when it is sufficient to ionize the air, an arc jumps.

11. The derivative of a function can be large at a point where the function vanishes. The function passes through zero with a large

slope. In this case, di/dt is large but i itself is zero. At the instant the switch is closed the electrons start to experience an electric field and they accelerate. At that instant the drift speed is zero.

13. When current passes through the coil the potential difference is iR + Ldi/dt just as for two separate elements. The point here is that the resistance and inductance pictured in Fig. 5 may actually be in the same physical circuit element, a coil for example. When the time constant is calculated, one must use the total inductance and total resistance in the circuit and take all sources into account. The circuit itself forms a loop and so has a nonvanishing inductance. For completeness it should be mentioned that the resistance of many materials depends on the magnetic field so it might matter whether or not the resistance and inductance are physically close. This is not usually an important practical consideration and, in any event, it is accounted for when those properties are measured.

Chapter 34

1. Lay one bar on a table. With their long axes at right angles to each other, pass the other bar along the length of the first. If the force of attraction is strong when the second bar is at either end of the first and weaker when it is at the midpoint, the stationary bar is magnetized. If the force of attraction is fairly uniform, the moving bar is magnetized. This technique works because the magnetic field of a magnetized bar is strong at the ends and weak in the middle.

2. One of the bars is weakly magnetized or unmagnetized. If both were strongly magnetized, repulsion would occur for some orientations.

13. In a paramagnetic substance, the atoms have dipole moments that tend to align with the field. Collisions disorient the atoms and reduce the magnetization. The effect of collisions increases as the temperature increases and the number of dipoles aligned with the field is therefore temperature dependent. In a diamagnetic substance, the part of the dipole moment of interest comes about through a distortion of the electron orbit by the field. A collision does not change the distortion much; the electron still picks up an additional motion around the direction of the field. So the magnetization is relatively insensitive to temperature.

17. When a field $\underline{B}_0$ is applied to a superconductor the total field $\underline{B}$

vanishes inside. Since $\underline{B} = \underline{B}_0 + \underline{B}_M$, where $\underline{B}_M$ may be thought of as the field due to magnetization, it follows that $\underline{B}_M = -\underline{B}_0$. The field produced by magnetization exactly cancels the applied field. Thus the superconductor may be described as a perfect diamagnet. Actually, $\underline{B}_M$ is due to current on the surface of the sample.

Chapter 35

1. When the capacitor is completely discharged ($q = 0$) current is flowing and the capacitor immediately begins to collect charge again. The situation is analogous to a mass on the end of a spring as it passes the equilibrium point. Its displacement is zero but its velocity is not, so it continues on by.

4. (a) The product: $\omega = 1/\sqrt{LC}$.

   (b) The initial charge on the capacitor and the initial current, as well as ω: $Q^2 = q_0^2 + (i_0/\omega)^2$. This can be obtained from $q_0 = Q\cos\phi$, $i_0 = -\omega Q\sin\phi$.

7. Take q to be positive if the upper plate has positive charge on it; take the current to be positive in the clockwise direction, and use $q_0 = Q\cos\phi$, $i_0 = -\omega Q\sin\phi$.

   (a) $q = Q$, $i = 0$ so $\phi = 0$.

   (b) $0 < q < Q$, $i < 0$ so $0 < \phi < \pi/2$.

   (c) $q = 0$, $i < 0$ so $\phi = \pi/2$.

   (d) $-Q < q < 0$, $i < 0$ so $\pi/2 < \phi < \pi$.

   (e) $q = -Q$, $i = 0$ so $\phi = \pi$.

   (f) $-Q < q < 0$, $i > 0$ so $\pi < \phi < 3\pi/2$.

   (g) $q = 0$, $i > 0$ so $\phi = 3\pi/2$.

   (h) $0 < q < Q$, $i > 0$ so $3\pi/2 < \phi < 2\pi$.

13. Capacitance corresponds to the reciprocal of the spring constant; resistance corresponds to the coefficient of the drag force, assumed to be proportional to the velocity; charge corresponds to the displacement of the mass; electric field energy corresponds to elastic energy stored in the spring; magnetic field energy corresponds to the kinetic energy of the mass; current corresponds to the velocity of the mass.

Chapter 36

4.   (a) The emf drives the current. If they did not have the same
      frequency, the phase difference between ε and i would change with
      time. If the voltages across various elements of the circuit summed
      to zero at one time, they would not at a later time.

5.   Potential differences, currents, and emfs are not vectors and, for
      example, do not add as vectors. The direction of a phasor does not
      represent the direction of any physical quantity. Its projection on
      an axis, however, has the same time dependence and phase as the
      physical quantity associated with it. A phasor diagram shows the
      phase relationships of the various physical quantities represented.
      Furthermore, when two quantities with the same frequency but
      different phases are added, the result is the projection of the
      resultant phasor, the "vector sum" of the two phasors. So a phasor
      diagram provides a useful way to add such quantities.

8.   A high frequency indicates a rapid variation of the charge on the
      capacitor and so a large current amplitude. The voltage across the
      capacitor is proportional to the charge, so as the frequency
      increases the current amplitude also increases if the voltage
      amplitude does not change. The capacitive reactance, which multiplies
      the current amplitude to give the voltage amplitude, decreases. A
      high frequency also means a rapid variation of the current so the
      amplitude of the emf generated by the inductor increases with the
      frequency if the current amplitude remains the same. The inductive
      reactance multiplies the current amplitude to give the emf amplitude
      and so increases with frequency.

10.   If the phase of the voltage is between 0 and 180° greater than the
      phase of the current, the voltage is said to lead the current. Plot
      both the voltage and current as functions of time and look at a peak
      of one function and the nearest (in time) peak of the other. The peak
      of the voltage occurs at an earlier time than the peak of the
      current. If the phase of the voltage is between 0 and 180° less (or
      between 180° and 360° more) than that of the current, the voltage is
      said to lag the current. It reaches its peak later than the nearest
      current peak.

11.   (a) $X_L > X_C$ means $\omega^2 LC > 1$. For fixed ω, the product LC should be
      relatively large.

      (b) For fixed L and C, ω should be relatively large.

16.   Write $P_{av} = (\varepsilon_{rms})^2 R/Z^2$ and differentiate with respect to R, C, or L,

holding ε_{rms} constant, then evaluate the derivative for the given values of R, C, L, and ω. To find how the power factor changes, differentiate R/Z. For the circuit of Sample Problem 4:

(a) The sign of dP_{av}/dR is negative so P_{av} decreases with increasing R. The power factor increases and, since $\phi = -29.3°$ for the circuit, ϕ increases.

(b) The sign of dP_{av}/dC is positive so P_{av} increases with increasing C. The power factor increases and ϕ increases.

(c) The sign of dP_{av}/dL is positive so P_{av} increases with increasing L. The power factor increases and ϕ increases.

Chapter 37

3. Even if a laboratory electric field is made to change at an extremely fast rate, only a small magnetic field is produced. Consider a uniform field that changes at 10^{10} V/m·s in a circle with a 1 cm radius. The magnetic field at the rim is about 5×10^{-10} T and is not easily measurable. On the other hand, if a uniform magnetic field in the same circle varies at 1 T/s, the electric field at the rim is about 5×10^{-3} V/m and is easily measurable.

5. Take $\underline{A}$ to be into the page. In the integral $\oint \underline{B} \cdot d\underline{s}$, $d\underline{s}$ is in the clockwise direction. The question can be answered by finding the sign of $d\Phi_E/dt$ and hence of $\underline{B} \cdot d\underline{s}$. It can also be answered by finding the direction of the displacement current and observing that the lines of $\underline{B}$ around i_d are the same as the lines around a true current.

(a) $\Phi_E > 0$ and $d\Phi_E/dt < 0$ so the integral is negative and $\underline{B}$ points in the counterclockwise direction. Here i_d is opposite to $\underline{E}$.

(b) $\Phi_E > 0$ and $d\Phi_E/dt < 0$ so the integral is negative and $\underline{B}$ points in the counterclockwise direction. Here i_d is along $\underline{E}$.

(c) $\Phi_E < 0$ and $d\Phi_E/dt > 0$ so the integral is positive and $\underline{B}$ points in the clockwise direction. Here i_d is into the figure.

(d) $d\Phi_E/dt = 0$ and, since the electric field has cylindrical symmetry, $\underline{B} = 0$. Here $i_d = 0$.

7. The displacement current density is in the direction of $d\underline{E}/dt$. Since the line of $\underline{E}$ does not change, it is in the direction of $\underline{E}$ if E is increasing and in the opposite direction if E is decreasing. For the situation shown, i_d is into the page (left diagram) or left to right (right diagram). The lines of $\underline{B}$ form circles around the direction of the displacement current. Point the right thumb in the direction of i_d (or $d\underline{E}/dt$), then the fingers curl in the direction of $\underline{B}$.

Chapter 38

10. Through Faraday's law the electric field in the wave is related to the rate of change of the magnetic field. From one viewpoint the changing magnetic field induces the electric field. Through the Ampere-Maxwell law the changing electric field can be said to induce the magnetic field. Thus the fields induce each other. Of course, the ultimate sources of the fields are accelerating charges.

15. Faraday's law describes the mechanism by which the changing magnetic field induces the electric field. The displacement current term in the Ampere-Maxwell law is needed to complete the cycle: it describes the mechanism by which the changing electric field induces the magnetic field.

17. No. Light carries momentum and, when it is absorbed, the conservation law demands that the momentum be transferred to the absorbing medium.

28. First use two sheets with their polarization axes perpendicular to each other. Light transmitted by the second sheet is polarized along its axis and is not transmitted by the second sheet. Now add the middle sheet with its axis in any direction not parallel to that of either of the other sheets. Since the polarization direction of light transmitted by the first sheet is not perpendicular to the axis of the middle sheet, some light is transmitted through it. The polarization direction of the light transmitted by the middle sheet is along the axis of that sheet and is not perpendicular to the axis of the last sheet. Hence some light is transmitted through the last sheet.

29. You need a good source of plane polarized light with known direction of polarization. Shine the light on the sheet and turn the sheet until the transmitted intensity is a minimum (zero for a good source), then mark the direction perpendicular to the original polarization direction. As a source use a marked Polaroid sheet or use reflected light. See section 7. If reflected light is used, start with a well collimated beam and experiment with the angle of incidence to find the minimum. This is Brewster's angle, for which the reflected light is completely polarized.

34. The speed of light in a vacuum is the same for all light waves and does not depend on the wavelength, frequency, or state of polarization. It also does not depend on the motion of the observer or source. This is the basis for the special theory of relativity.

Chapter 39

28. The approximation made is that all of the rays from the source to the
    mirror make small angles with the optic axis. Then Eqs. 19 are valid
    and the mirror equation follows. All of the rays cross the axis at
    very nearly the same place and the image is sharp. To see how this
    comes about in detail, use the diagram shown below to obtain $h = r\sin\beta = r\sin(s/r)$, $d = r(1 - \cos\beta) = r[1 - \cos(s/r)]$. Then $\tan\alpha = h/(o-d) = [r\sin(s/r)]/\{o - r[1 - \cos(s/r)]\}$. If α is small, then $\tan\alpha \approx \alpha$ (in radians) and s/r is also small, so $\sin(s/r) \approx s/r$. Then $\alpha \approx s/o$, which is the first of Eqs. 19. A similar analysis can be made to
    show $\gamma \approx s/i$. Finally the exact expression $\alpha + \gamma = 2\beta$ is used to
    obtain the mirror equation.

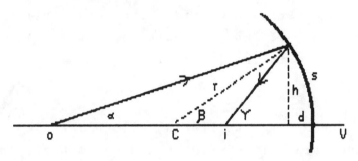

29. Assume the source is real so $o > 0$.
    (a) The image is real if $i > 0$ or $2/r > 1/o$. For a convex mirror ($r < 0$) the image is always virtual. For a concave mirror ($r > 0$) the
    image is virtual for $o < r/2$ and real for $o > r/2$.
    (b) Since $m = -i/o$ is positive for erect images and negative for
    inverted images, all virtual images are erect and all real images are
    inverted.
    (c) Combining the mirror equation and the expression for
    magnification yields $m = -r/(2o-r)$. The image is larger than the
    object if $|2o-r| < |r|$. This occurs for concave mirrors if $o < r$. The
    image is smaller if $o > r$. For convex mirrors the image is always
    smaller than the object.

30. Virtual images can be projected on screens and photographed. The lens
    of the projector or camera causes the diverging rays to converge at
    the final image. Paper at the site of a virtual image will not ignite
    since there is no light there.

36. Converging lens: If $o > 2f$ the image is real, inverted, and smaller

than the object. If f < o < 2f the image is real, inverted, and larger than the object. If o < f the image is virtual, erect, and larger than the object. Diverging lens: the image is always virtual, erect, and smaller than the object.

43. Use $1/f = (n-1)(1/r' - 1/r'')$, where r' is the radius of the first surface struck by the light and r'' is the radius of the second surface. These are positive or negative according to whether the surface is convex or concave when viewed from the side of the incident light. If f is positive the lens is converging; if f is negative the lens is diverging.

(a) The lens is converging. Consider the cases: (i) r' positive, r'' negative; (ii) both r' and r'' positive, with $r'' > r'$; (iii) both r' and r'' negative with $|r'| > |r''|$.

(b) The lens is diverging. Consider the cases: (i) r' negative, r'' positive; (ii) both r' and r'' positive, with $r' > r''$; (iii) both r' and r'' negative, with $|r''| > |r'|$.

Chapter 40

1. In classical theory the oscillating electric field in the wave causes charges in the material to vibrate and to emit electromagnetic radiation. The vibrational frequency of the charges is the same as the frequency of the incoming wave and so is the frequency of the reradiated light. The quantum mechanical description also leads to absorption and reradiation at the same frequency. The speed of the wave is different in the material and, since the frequency is the same, the wavelength is different. For completeness, it should be mentioned that for some materials a small portion of the reradiated light has a frequency which is a multiple of the original frequency.

6. Maxima still occur for $d\sin\theta = m\lambda$ and minima still occur for $d\sin\theta = (m+\frac{1}{2})\lambda$ but now λ is the wavelength of the light in water and is somewhat shorter than the wavelength in air. The pattern is tighter, with the minima closer together, for example.

10. Coherence influences the result only if interference effects take place. For the formation of an image by a lens or mirror system, for example, all waves travel the same optical path length from a point source to its image. They start in phase because they originate as parts of the same spherical wave, so they arrive at the image in phase. Even if the phase of the next wave train is different, the intensity at the image does not change. On the other hand, if the

wave is split (as in thin film reflection, for example), then the different parts may travel different optical path lengths and coherence is important. If the film is thick the wave that is reflected a few times in the film may have originated from a different train than the wave that is reflected only once. If the relative phase of the two waves is changing rapidly, the pattern is washed out.

12. The double slit interference pattern washes out but red and blue single slit diffraction patterns remain, with their intensities superposed. This may happen for two reasons. The red and blue parts of the spectrum come from different sets of atoms or from the same set of atoms at different times. They are incoherent. Even if they are coherent in the sense that the phase constants do not change, the interference pattern fluctuates at the beat frequency. To see this, picture two phasors rotating with different periods. The observer sees the time average of the pattern and this does not show the double slit interference effect.

13. If one slit is covered the intensity is proportional to E_1^2. If both slits are open the intensity is proportional to $(E_1 + E_2)^2 = 4E_1^2$. The constant of proportionality is the same so the intensity decreases by a factor of 4 when one slit is closed.

27. Light waves from the two sources are not coherent. The phase difference is rapidly changing. The eye perceives the time average intensity, which is nearly the same at every illuminated point.

Chapter 41

8. (a) When the wavelength is increased without changing the slit width the pattern spreads out. Any two adjacent minima, for example, have a larger angular separation and the central bright band is wider.
(b) When the slit width is increased without changing the wavelength the pattern becomes more compact. Any two adjacent minima have a smaller angular separation and the central bright band is narrower.

16. The resultant of wavelets r_1 and r_3 is out of phase with the resultant of wavelets r_2 and r_4. The order of adding the wavelets is immaterial.

20. Place the screen far from the slit and parallel to the surface in which the slit is made. θ is the angle between a ray from the slit to the screen and the perpendicular to the screen. It describes the direction of the rays. ϕ gives the phase difference, at the screen,

between wavelets from the top and bottom of the slit. α is the phase of the resultant relative to the phase of the wavelet from the top of the slit. It is $\phi/2$.

24. As Eq. 12 shows, the intensity is the product of the function for single slit diffraction and the function for double slit interference. In a loose sense, either may be said to modulate the other. However, one usually considers the more slowly varying function to be the modulating function. The center-to-center distance between slits must be larger than the width of either one of slits, so the central maximum of the single slit diffraction pattern is, of necessity, wider than any of the interference peaks. One or more of the interference peaks fit inside the central diffraction peak so it is more suitable to say that the interference fringes are modulated by the diffraction pattern than vice versa.

Chapter 42

2. Ocean and air currents and ballistic missile trajectories all show significant effects. Rotation of the earth must be taken into account in launching a space craft. On a smaller scale, precise measurements of weight also show effects. They must be taken into account when the measurements are used (by geologists, for example) to find the local density of the earth.

12. As the equations in Table 3 reveal, if Δx and Δt both vanish then both $\Delta x'$ and $\Delta t'$ also vanish, regardless of the relative velocity of the two frames. Thus the events occur at the same time and at the same place in all frames.

13. The events recorded by the two observers are different so different results should not seem paradoxical. Each observer marks the positions of the ends of the other meter stick at the same time, according to his clocks. These events are not simultaneous according to the clock of the other observer, in the rest frame of the stick being measured. Note that the processes are different because different sticks at rest in different frames are involved. Also note the symmetry. Point out that an observer in S´, watching an observer in S measure the stick in S´, agrees that the result is less than 1 m. S´ sees S mark one end first, then the other end. During the interval the first mark moves toward the position of the second so, according to S´, the marks are closer than 1 m at the time the second mark is made. If the observer in S´ did not know about relativity he

would claim that S did not carry out a length measurement since the making of the marks was not simultaneous.

18. Consider a particle moving with speed c along a line which makes the angle θ with the x' axis. The components of its velocity, in S', are $v_{x'} = c\cos\theta$, $v_{y'} = c\sin\theta$. In frame S, moving with speed u along the x axis of S', $v_x = (v_{x'} + u)/(1 + v_{x'}u/c^2) = c(c\cos\theta + u)/(c + u\cos\theta)$ and $v_y = v_{y'}/\gamma(1 + v_{x'}u/c^2) = c^2\sin\theta/\gamma(c + u\cos\theta)$. The first expression is the appropriate modification of Eq. 23 while the second can be derived by substituting the Lorentz transformation equations into $v_y = \Delta y/\Delta t$. A little straightforward algebra and use of $\cos^2\theta + \sin^2\theta = 1$ gives $v^2 = v_x^2 + v_y^2 = c^2$, independently of u. Thus the speed of the particle is c in every inertial frame.

According to Eq. 33 the momentum of a massless particle moving at a speed different from c vanishes. Since E = pc so does its energy. No change can occur in any interaction so such a particle cannot be observed.

26. The rest mass of a collection of particles is given by $m = E/c^2$, where E is the total energy of the system as measured in a frame for which the total momentum vanishes. The total energy includes the kinetic energies of the individual particles and is greater than the sum of the rest energies of the particles. As the material cools the total kinetic energy decreases and so does the rest mass of the material. A sufficiently sensitive scale would indicate the change.

Chapter 43

10. The existence of a cutoff frequency means that the energy transferred to an electron depends on the frequency of the light. When the frequency is too low, not enough energy is transferred to drive the electron from the material. The fact that the cutoff frequency is independent of the light intensity means that the energy transferred is not proportional to the square of the amplitude as it would be if the classical wave theory were correct. In the photon theory the effect is explained by postulating that energy can be transferred only in units of $h\nu$. These are the photons.

17. The change in wavelength occurs when light is scattered from free or nearly free electrons. Materials differ in the concentration of these electrons and in their energies. The energy transferred in Compton scattering is large compared to the original electron energy so, to a good approximation, the scattering is nearly as if the electrons were

at rest. The concentration of electrons does not influence Δλ; it does influence the number of scattering events which occur, not the result of any single event. So Δλ is nearly the same for all materials.

18. The maximum change in wave length which can occur is about 0.024 Å, independent of the wavelength of the incident light. This change is extremely small compared to the wavelength of visible light (about 5000 Å). The change occurs only in the sixth significant figure and is difficult to detect.

29. Yes. The electron is freed with some kinetic energy.

30. Cavity radiation: oscillators in the cavity walls have energies which are integer multiples of $h\nu$, where ν is the natural frequency; they emit or absorb radiation energy in units of $h\nu$; the number of oscillators with a given energy is given by the usual law of statistical mechanics.

Photoelectric effect: light energy comes in integer multiples of $h\nu$, where ν is the frequency of the light; in the interaction with an electron, energy $h\nu$ is transferred to the electron, which then gives up some of its energy to the material and escapes with the remaining energy in the form of kinetic energy.

Compton effect: light behaves like a collection of particles each having energy $h\nu$ and momentum $h\nu/c = h/\lambda$; the interaction between a photon and an electron can be analyzed as any other 2 particle collision in which energy and momentum are conserved.

Hydrogen atom: the electron can exist only in certain quantum mechanical states. It changes state with the emission or absorption of a photon. The lowest energy state is stable against emission.

Chapter 44

1. Both wave and particle properties are associated with any physical object. The wavelength of the wave is related to the particle momentum; the frequency is related to the particle energy. Some experiments are sensitive to wave properties and some to particle properties.

16. For a particle confined between rigid walls, ψ can be written as the superposition of two traveling waves of equal amplitude, one traveling to the right and the other traveling to the left. For each of these waves, taken separately, particle momentum is well defined ($p_x = \pm nh/2L$) but it is in opposite directions for the two waves. We

interpret this to mean the probability of observing the particle with momentum in the positive x direction is the same as the probability of observing it with momentum in the negative x direction.

18. Quantum mechanics is capable, in principle, of giving exact predictions of the probabilities of the occurrence of events. It is not capable of giving an exact prediction of the outcome of any single event. Quantum mechanics can be used to predict the average outcome and the distribution of results for a large number of identical experiments.

19. Look at the bottom graph of Fig. 9. The probability of finding the particle near either wall is very small. It is not zero since the volume element extends away from the wall and the probability density is slightly greater than zero a small distance from the wall.

29. Since the probability density is zero at two points, the particle will never be found at those points while the it is in that state. It might be found on either side of the points. In addition, it is not possible to find the particle first on one side of a zero and then on the other side, without altering the wave function. Performance of the first position measurement alters the wave function. After the first measurement it might be placed in the original state and the position measured again. This time it might be found on the other side of the zero. To get it into the original state, we had to give it a certain well defined energy and momentum magnitude and we lost control of its postion.

30. In the Bohr theory the electron is assigned a definite orbit, a circle of definite radius, for example. Since the value of r is not uncertain, the radial component of the electron's momentum has large (infinite) uncertainty. If this momentum component is measured many times, with the electron initially placed in the same orbit, the standard deviation of the results will be huge. This is not substantiated by experiment.

Chapter 45

1.   Wave mechanics correctly gives the probability of finding the particle in any chosen volume. The Bohr theory does not. In addition, wave mechanics predicts the probability that the particle makes a transition to another state when it is subjected to an external force.

13. The probability density function gives the probability per unit

volume that the particle is in an infinitesimal volume in the neighborhood of $\underline{r}$ at time t. The wave function $\psi(\underline{r},t)$ is a function such that the probability density is given by $|\psi|^2$. It satisfies a differential equation (the Schrodinger equation) and it is complex, while $|\psi|^2$ is real and positive. Since ψ can be constructed as a linear combination of waves, it can exhibit interference and diffraction effects. The radial probability density, defined by P(r) = $4\pi r^2|\psi|^2$, gives the probability per unit radial distance that the particle is in a spherical shell of infinitesimal thickness dr.

17. The torque is produced by the magnetic field acting on the magnetic moment of the atom: $\underline{\tau} = \underline{\mu} \times \underline{B}$. In the field (assumed to be along the z axis), states with different values of m_ℓ (and hence different values of μ_z) have different energy. The atoms are distributed among these states, so different atoms may have different values of μ_z. One may think of the angular momentum and dipole moment as precessing about the direction of the field, the precession being produced by the torque. The force on the atom is produced by the field gradient: $F_z = \mu_z \partial B/\partial z$. This causes atoms with dipole moments in different directions to follow different trajectories as they cross the field.

20. The lengths of the periods in the periodic table support the need for a fourth quantum number. For example, the chemical properties of the inert gases are similar and these atoms occur for Z = 2, 10, 18, 36, 54, and 86. The numbers of electrons in the various shells can be found using these numbers. When the rules for specifying ℓ and m_ℓ are applied, it is found that each shell contains only half as many states as are needed to produce the periodic table. An additional quantum number, with two possible values, is needed. If there were no physical manifestation of the fourth quantum number, perhaps its need could be eliminated by revising the Pauli exclusion principle. However, a magnetic dipole moment is associated with spin and the exclusion principle is supported by relativity theory.

22. In the ground state, the atoms of the lanthanide series all have two electrons in 6s states and have empty 6p states. The wave functions of the 6s electrons extend the farthest distance from the nucleus of any of the electron wave function for atoms in the series. It is these electrons that interact with neighboring atoms and determine the chemical properties of the elements, so the chemical properties are similar. Atoms of the series differ in the number of 4f electrons. These have wave functions which are more confined than the

6s functions and they do not participate in chemical interactions. The atoms of the series also differ in the number of protons, so the energies of the less shielded inner electrons are reduced from atom to atom through the series. This affects the characteristic x-ray spectrum and the atoms fit on the Mosley plot in their predicted places.

24. The cutoff wavelength is the wavelength of a photon with energy equal to the original kinetic energy of the electron. It is a clue to the photon nature of light because it ties the wavelength (and hence the frequency) of the emitted light to the energy lost by an electron. The electron loses all its kinetic energy in the creation of a photon with the same energy.

26. A characteristic x ray is emitted when a higher energy atomic electron falls to a lower vacant state. The lower state was vacated when its original occupant was knocked out of the atom by an electron incident from outside the atom. The frequencies of characteristic x rays are proportional to the energy differences of the atomic states involved and these, in turn, depend on the atomic number (the nuclear charge). A plot of $\sqrt{\nu}$ vs. Z is nearly linear.

28. Atomic hydrogen does not emit characteristic x rays. If the single electron in a hydrogen atom were removed, the atom could pick up a free electron and one or more photons would be emitted as the electron dropped to the ground state. The energy of the most energetic photon possible is 13.6 eV, too small by a factor of 50 to be an x ray.

38. Without population inversion the intensity of the beam would be small or non-existent. When the population of the upper state is less than the population of the lower state the system is closer to thermodynamic equilibrium and the radiation energy absorbed is more nearly equal to the radiation energy emitted in any time interval. It is only when the populations are inverted that significant light amplification takes place.

39. A metastable state is one for which spontaneous transitions to other states (the ground state, for example) are not highly probable and the electron, once in a metastable state, remains there, on average, for a time which is long compared to occupancy times for "normal" states. For the laser to operate, it must be much more likely for the transition to occur by stimulation than for it to occur spontaneously. Then the wave for the emitted photon is in phase with

the wave for the incident photon. Electrons in metastable states
satisfy this requirement.

Chapter 46

5.   Without the Pauli exclusion principle all electrons would be in low
      lying states, tightly bound to atoms, and the conductivity would be
      extremely small. With the exclusion principle, one or more electrons
      per atom are in states with sufficiently high energy that the
      electrons are loosely bound and can accelerate in an electric field
      to produce a large current.

6.   The state distribution function n(E), when multiplied by dE, gives
      the number of quantum mechanical states per unit volume with energy
      between E and E+dE. The states need not be occupied and, in fact, the
      function is not influenced by whether they are or not. This function
      is different for different systems since, in general, different
      systems have different sets of electron states, with different
      energies. The Fermi-Dirac probability function p(E) gives the
      probability that a state with energy E is occupied at temperature T.
      It ranges from 0 to 1. It is completely independent of the states of
      the system, and the same function is valid for all large collections
      of electrons, regardless of the state distribution function. In fact,
      a particular system may not have a state with the energy E for which
      the function is evaluated. The particle distribution function $n_0(E)$,
      when multiplied by dE, gives the thermodynamic average number of
      electrons per unit volume with energy between E and E+dE. This must
      be the product of the number of states per unit volume in the energy
      range and the probability of occupation: $n_0(E) = n(E)p(E)$. As an
      example, suppose in some small energy range there are 5 states per
      unit volume and each has a 25% chance of being occupied. We then
      expect 1.25 electrons per unit volume to have energy in that range,
      on average.

12.   In each case look at the distribution of electrons among the states
      at T = 0 K and examine the most energetic electrons. If there are a
      large number of vacant states nearby in energy, the material is a
      conductor. If the vacant states are separated in energy from the
      occupied states but a significant number of electrons are promoted
      thermally from occupied states as the temperature increases to room
      temperature, the material is a semiconductor. This may come about
      because the gap between the valence and conduction bands is small or

because there are impurity states in the gap. If a significant number of electrons are not thermally promoted, the material is an insulator.

13. Band theory can be used to predict whether a given material is a conductor, insulator, or semiconductor. It gives us the state distribution function and so is used to predict the number of electrons which contribute to the current when an electric field is turned on. Band theory is used, in conjunction with a known mean free time, to compute the electron drift speed.

21. Collisions with atoms control the temperature dependence of the conductivity of metals. They enter the simple theory through the mean free time. As the temperature increases so do the number of collisions; the mean free time decreases and so does the conductivity. For semiconductors at room temperature and above, the number of electrons in the conduction band and the number of holes in the valence band control the temperature dependence of the conductivity. As the temperature increases the number of electrons thermally promoted across the gap increases and so does the conductivity. The number of collisions also increases, as for metals, but thermal promotion is far more important at these temperatures.

24. As far as electrical conduction is concerned, the net effect of all electrons in a band with one empty state is the same as that of a single positive charge with momentum equal to the total momentum of the electrons. The positive particle is called a hole. Normally holes in the valence band, rather than the conduction band, enter the description of a semiconductor because there are relatively few holes in these bands and all of them have nearly the same energy.

28. It is the internally generated (so-called contact) electric field that provides the barrier to electron and hole flow across the junction and that is of primary importance for the electrical properties of the junction. The field is created by ionized impurities: positive donors on the n side and negative acceptors on the p side. Raising of the barrier by an externally generated field causes the high reverse resistance of the junction while lowering of barrier brings about the low forward resistance.

Chapter 47

2. The strong force is independent of the electrical charge on the particles, the electrostatic force is not. The strong force has a short range and nearly vanishes when the separation of the particles is on the order of 2 fm or more, while the electrostatic force is long range and dies out much more slowly with separation. At separations where both exist, the strong force is much stronger than the electrostatic force. Except at very small separations the strong force is attractive, while the electrostatic force can be attractive or repulsive, depending on the particles involved.

5. All nucleons attract neighboring nucleons via the strong force but the force is weak for nucleons with greater separations. On the other hand, protons repel all other protons which are beyond the range of the strong force. The nucleus becomes more stable if there are more neutrons than protons.

13. The shape of the binding energy curve is due chiefly to saturation of the strong force and to the reduction in binding energy because of the mutual electrostatic repulsion of protons. For small A the strong interaction is not saturated so each new nucleon interacts with all nucleons already there and the binding energy per nucleon increases strongly with A. At large A each new nucleon interacts via the strong force with about the same number of nucleons as the nucleons already there. If no other forces were present the curve showing binding energy per nucleon vs. A would level off. In fact, it decreases as A increases. As A increases so does Z and the increase in electrostatic interactions between protons tends to decrease the binding energy.

22. Look at the potential barrier diagrammed in Fig. 8. The larger the energy of the α particle the narrower the barrier, the greater the probability of tunneling, and the shorter the half-life. Once the α is outside and moving away, its potential energy drops toward zero. The energy of the α while inside becomes part of the disintegration energy and is shared with the recoil nucleus as the particles repel each other. Thus the larger the energy of the α, the greater the disintegration energy.

24. The energy lost by the nucleus in β decay can be measured and this energy defines the upper limit to the β spectrum. For most β decays, however, the β has less energy and physicists must account for the missing energy or declare the principle of energy conservation to be invalid. It was postulated that another particle, a neutrino, is

emitted with the β and that the neutrino carries the previously missing energy. Later experiments, in which interactions involving incident neutrinos were observed, substantiated the existence of these particles.

25. Neutrinos have spin with m_s either $+\frac{1}{2}$ or $-\frac{1}{2}$, like electrons. They obey the Pauli exclusion principle and the Fermi-Dirac probability function is valid for them. Photons, on the other hand, have $m_s = -1$, 0, or +1 and do not obey the Pauli exclusion principle. It is possible to have any number of photons in any given state. The number of photons is not conserved in interactions.

Chapter 48

4. Strictly speaking, Q depends on the masses of the fragments. However, for the overwhelming majority of fission events X and Y are medium mass nuclei. These have nearly the same binding energy per nucleon, so Q does not deviate much from 200 MeV, which is characteristic.

5. The curve is approximately symmetric. For each heavy fragment there is a companion light fragment. The mass numbers of the fragments sum to 236, less the number of emitted neutrons. The heavier fragments have a greater ratio of neutrons to protons and the number of neutrons emitted varies. This destroys the symmetry slightly.

6. The decaying fragments are all neutron rich and become more stable by conversion of a neutron to a proton, with the emission of a β^- particle. β^+ decay would make the fragments even more neutron rich.

15. While the power level is being decreased, the rods are pushed in and k decreases. When the desired level of neutron flux is achieved the rods are adjusted so k = 1 again. For long term sustained operation k = 1 no matter what the power level.

21. No. The number of fusion events per unit volume is given by the product n(K)p(K). Thus the greatest number of events occur for the kinetic energy at which the product of the two curves is a maximum.

Chapter 49

2. Waves associated with high energy particles have greater frequencies and shorter wavelengths than those associated with low energy particles. They can therefore be used to probe smaller regions. Of equal importance, high energy is needed to produce massive particles from the constituents of ordinary matter.

6. Although all neutrinos travel at the speed of light they may have

different momenta. The relativistic expression $p = mv/\sqrt{1-v^2/c^2}$ becomes indeterminate as m approaches 0 and v approaches c. A zero mass particle can have non-vanishing momentum only if it travels at the speed of light. Since E = pc for a zero mass particle, a different momentum leads to a different energy.

8.   Photons are bosons (spin $\hbar$) while neutrinos are fermions (spin $\hbar/2$). Neutrinos participate in the weak interaction while photons are the messenger particles for the electromagnetic interaction. Neutrinos are produced in weak decays, such as beta decays. Photons are produced in electromagnetic transitions of other particles, such as the change in the state of an electron in an atom. Both particles can be detected by observing the process that is the inverse to the producing process: the absorption of photons by electrons in atoms and the change of a proton to a neutron when a neutrino is absorbed.

11.   Of all the charged particles, the electron (and the positron) has the smallest mass. Its decay, if it occurs, must be to a particle with equal or less mass and so would not conserve charge.

13.   Spin angular momentum cannot be conserved by such a decay. The z component of the spin angular momentum of the products must be either $-\hbar$, 0, or $+\hbar$ while the z component of the spin angular momentum of the neutron must be either $-\hbar/2$ or $+\hbar/2$.

28.   Baryons interact via the strong interaction, leptons do not. Baryon number (+1 for baryons, −1 for antibaryons) seems to be conserved in all interactions. There is no analogous "lepton number" that is conserved. Baryons are composed of more fundamental particles (quarks) while leptons are not.

33.   To decay, one of the quarks in a charged pion must change flavor (i.e. u must change to d or $\bar{d}$ must change to $\bar{u}$). This occurs as a result of a weak interaction and requires a relatively long time.

36.   The cosmological principle states that the universe looks the same for two separated observers observing <u>at the same time</u>. The quasars we see existed roughly 13×10^9 years ago, the time it takes light to travel from them to us. An observer presently in that region of space sees no nearby quasars.

40.   To take a simple example, suppose the galaxy is spherical, with each part rotating about the center under the gravitational pull of the other parts. If the mass density is uniform all parts will have the same angular speed. On the other hand, if the density decreases with increasing radius, the angular speed of the outer parts will be less

than that of the inner parts. For real galaxies, the density of luminous matter does decrease from the center to the rim but the angular speed goes not decrease in proportion. We conclude that there must be more mass. The argument can be put on a quantitative basis and generalized to non-spherical galaxies. The angular speed as a function of distance from the center can be used to map the mass distribution.

SECTION FOUR
TRANPARENCY MASTERS

This section of the manual contains a series of diagrams that can be used to make transparencies for an overhead projector. All figures needed to carry out suggested discussions in the Lecture Notes have been included, along with many others. These, we hope, will facilitate lectures and discussions. A smaller set, composed of the most useful transparencies, on acetate and ready for projection, is available from your Wiley representative.

TRANSPARENCY MASTERS

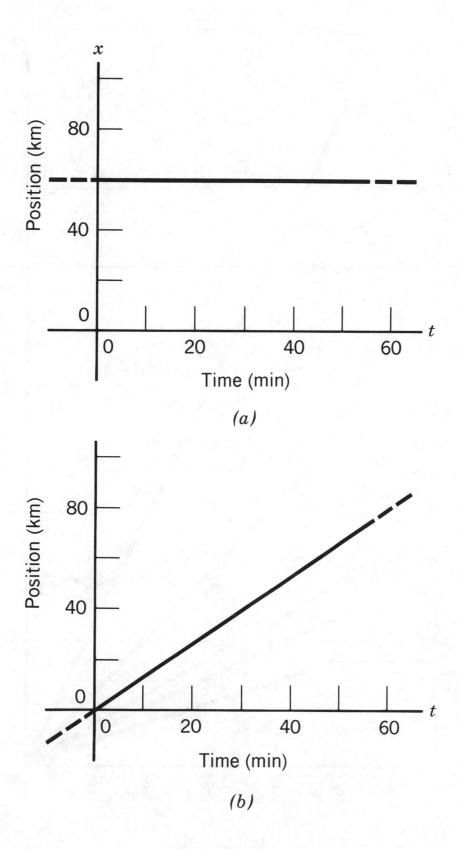

Position as a function of time for an object with (a) zero velocity and (b) constant non-zero velocity.

1

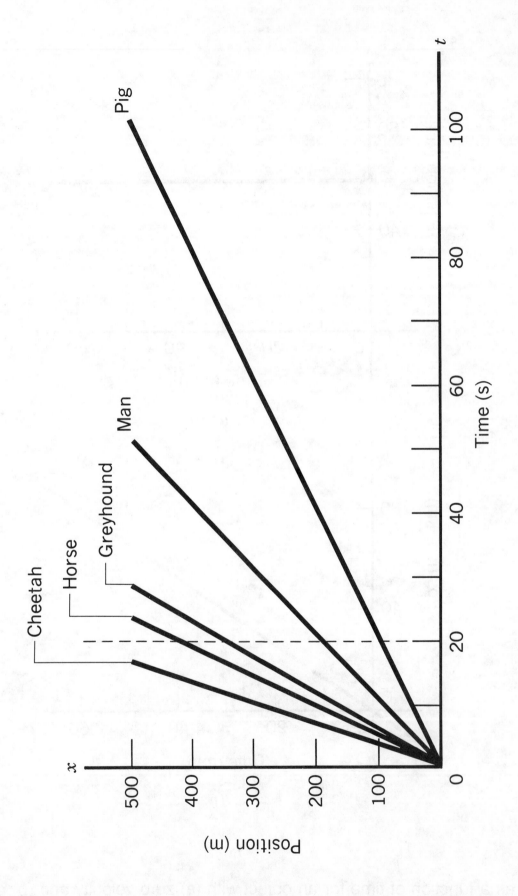

The velocity is the slope of the position as a function of time.

FIG. 2.2

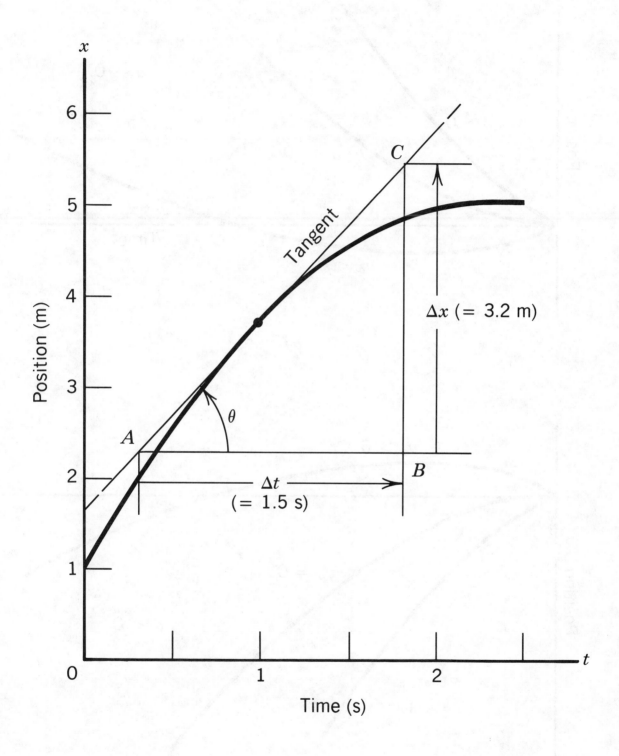

$$v = \lim_{\Delta t \to 0} \frac{\Delta x}{\Delta t}$$

FIG. 2.7

3

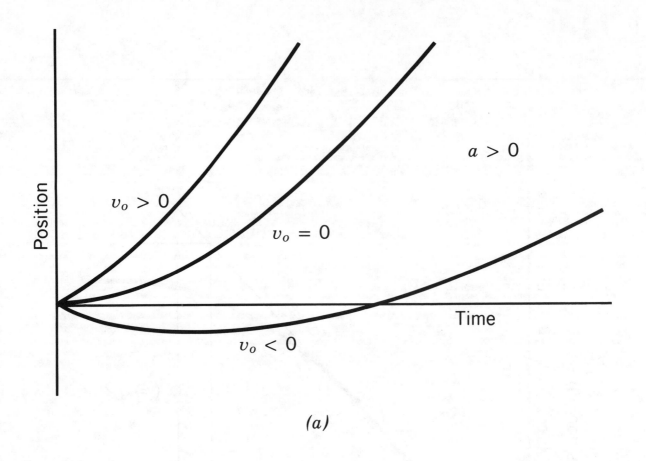

$v_o > 0$

$v_o = 0$

$a > 0$

Position

Time

$v_o < 0$

(a)

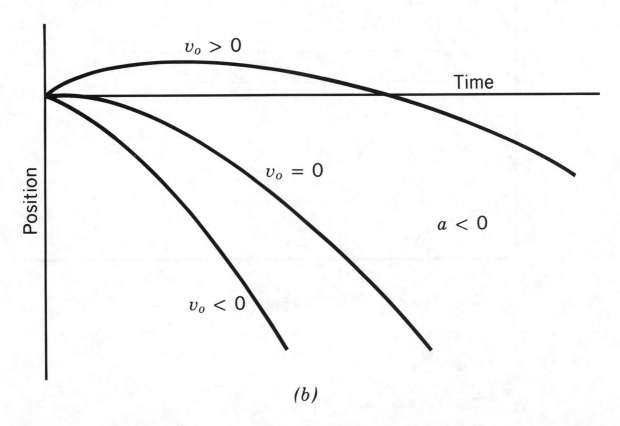

$v_o > 0$

Time

$v_o = 0$

$a < 0$

Position

$v_o < 0$

(b)

Position as a function of time for some motions with (a) positive acceleration and (b) negative acceleration.

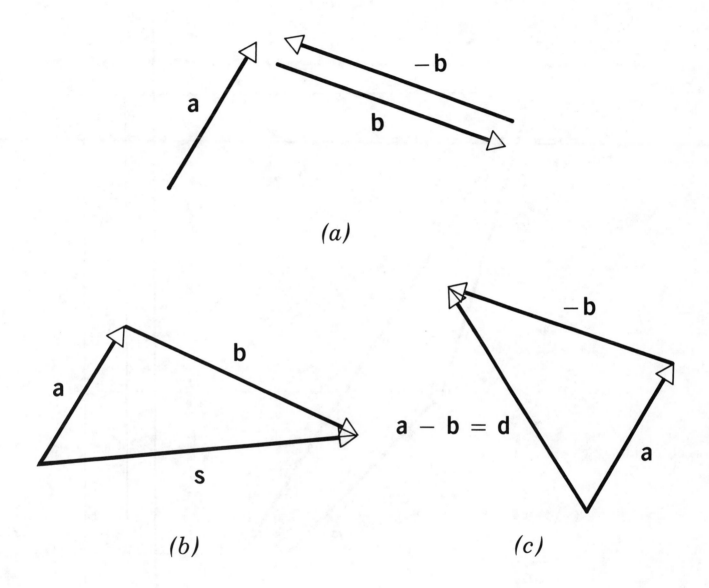

(a)

(b)

(c)

Graphical addition and subtraction of vectors.

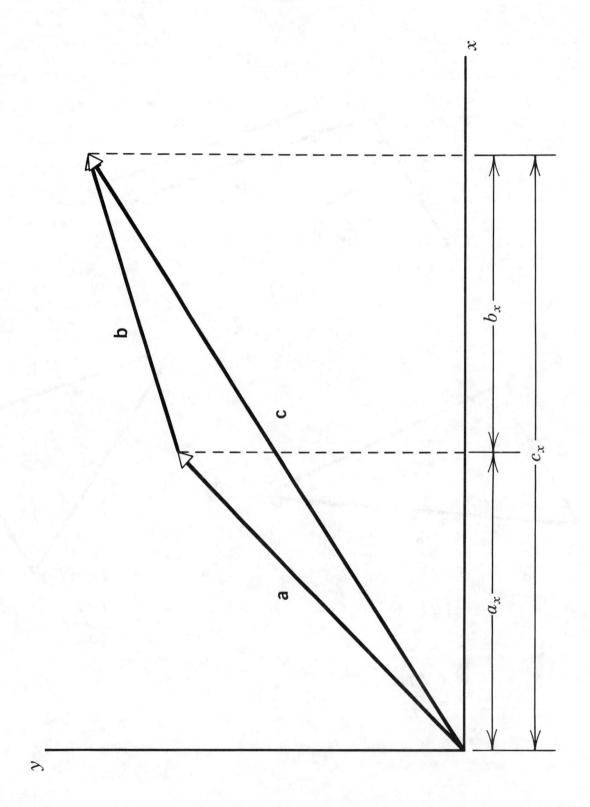

$$c_x = a_x + b_x$$
$$c_y = a_y + b_y$$

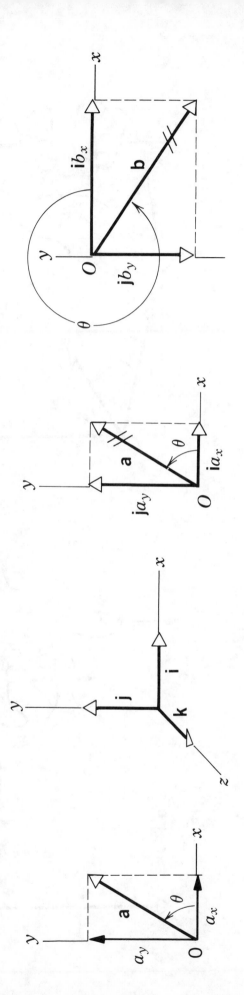

FIG. 3.8, 3.10 & 3.11

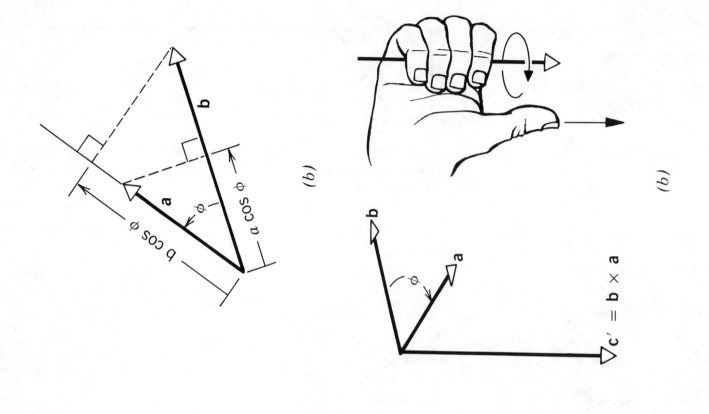

(a)

(b)

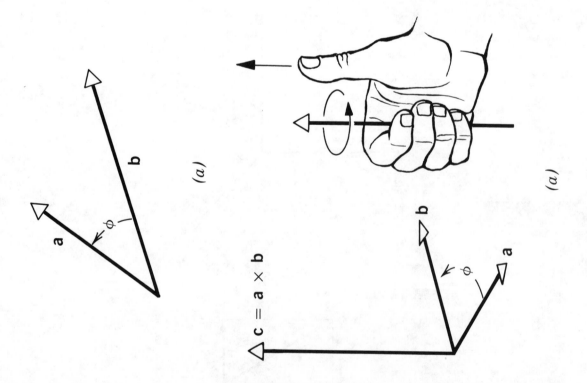

$c = a \times b$

(a)

$c' = b \times a$

(b)

Scalar product
Vector product

FIG. 3.18 & 3.19

8

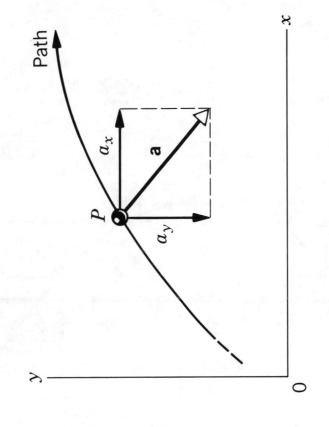

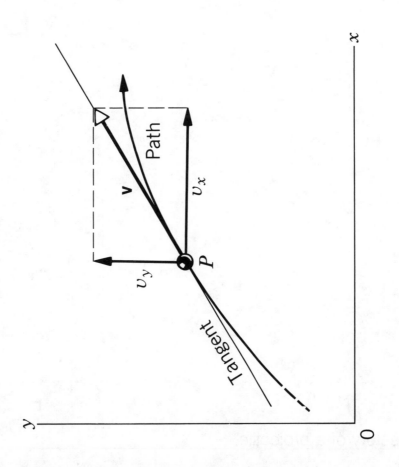

Velocity and acceleration for a body moving in two dimensions.

FIG. 4.2 & 4.5

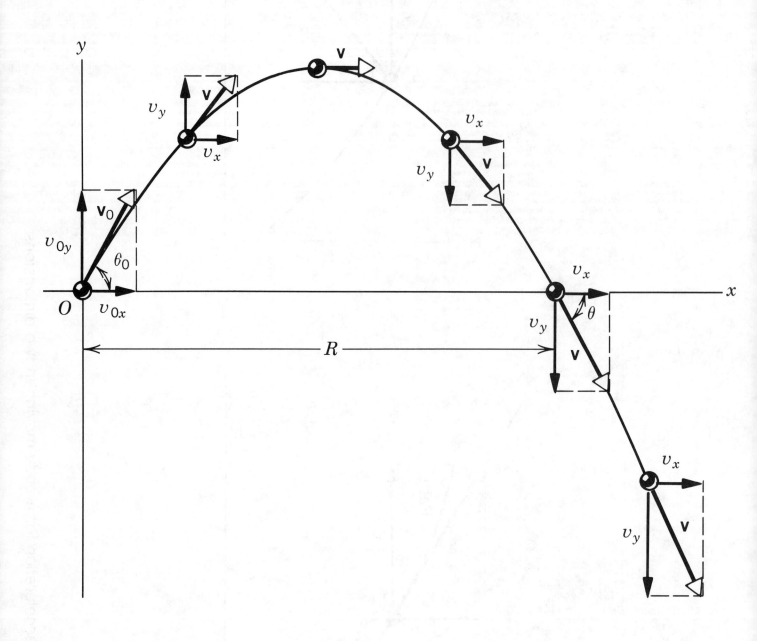

The trajectory of a projectile.

FIG. 4.9

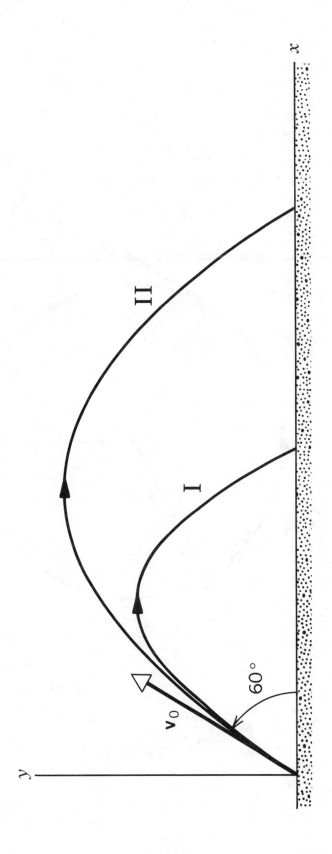

Projectile trajectories when air resistance is important.

FIG. 4.13

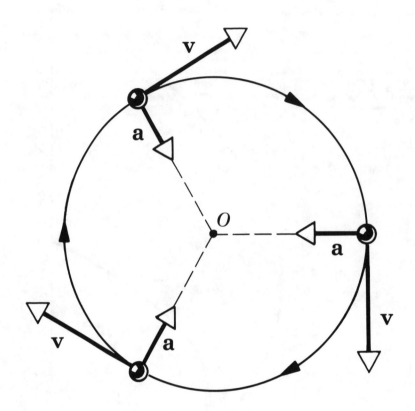

Uniform circular motion.

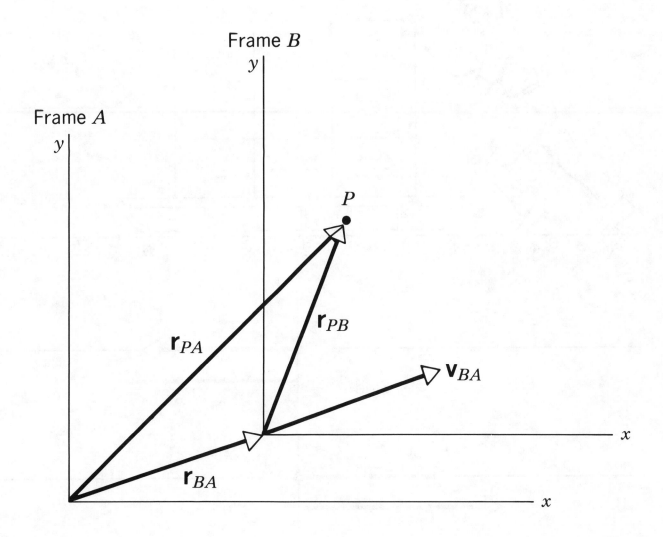

Frame B

y

Frame A

y

P

$\mathbf{r}_{PA}$

$\mathbf{r}_{PB}$

$\mathbf{v}_{BA}$

x

$\mathbf{r}_{BA}$

x

$$\mathbf{r}_{PA} = \mathbf{r}_{BA} + \mathbf{r}_{PB}$$
$$v_{PA} = v_{BA} + v_{PB}$$

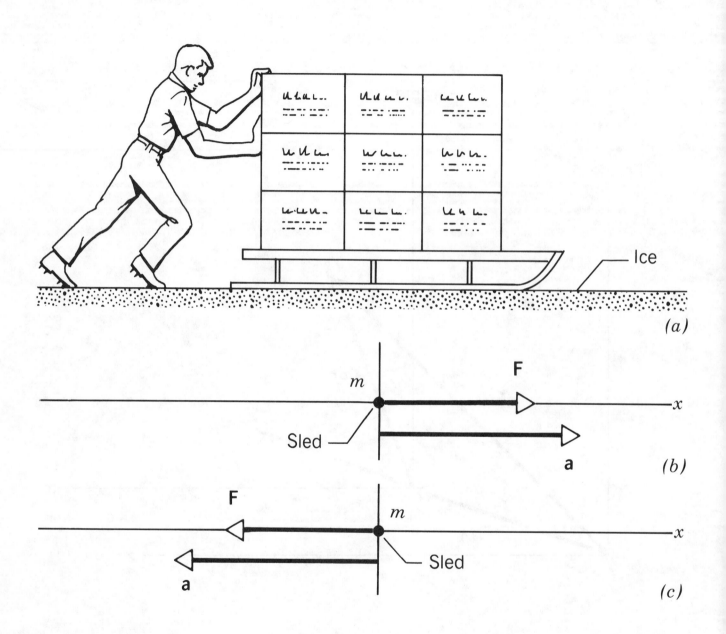

(a)

(b)

(c)

Sample problem 5-1.

FIG. 5.5

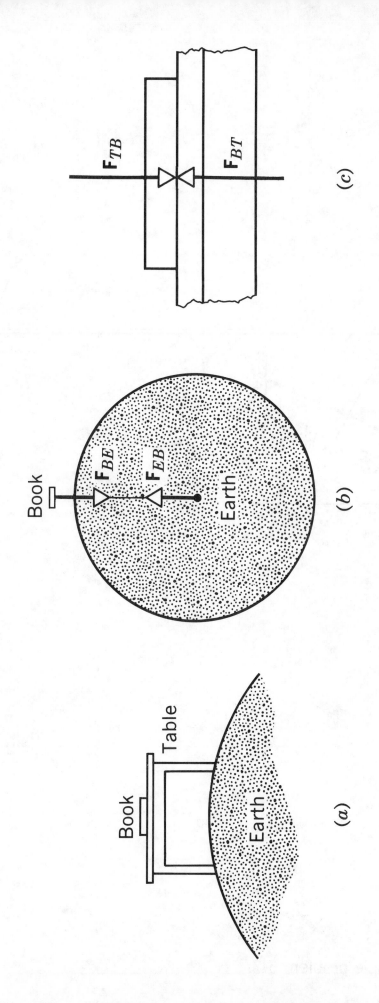

(c)

(b)

(a)

Newton's third law.

FIG. 5.10

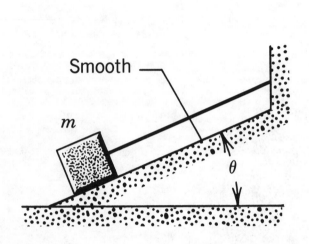

Smooth

m

θ

(a)

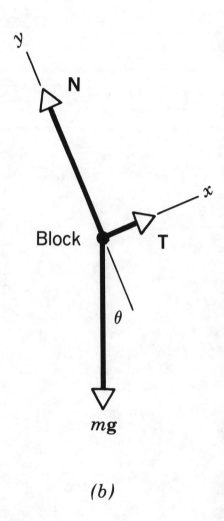

y

N

x

Block

T

θ

$m\mathbf{g}$

(b)

Sample problem 5-7.

16

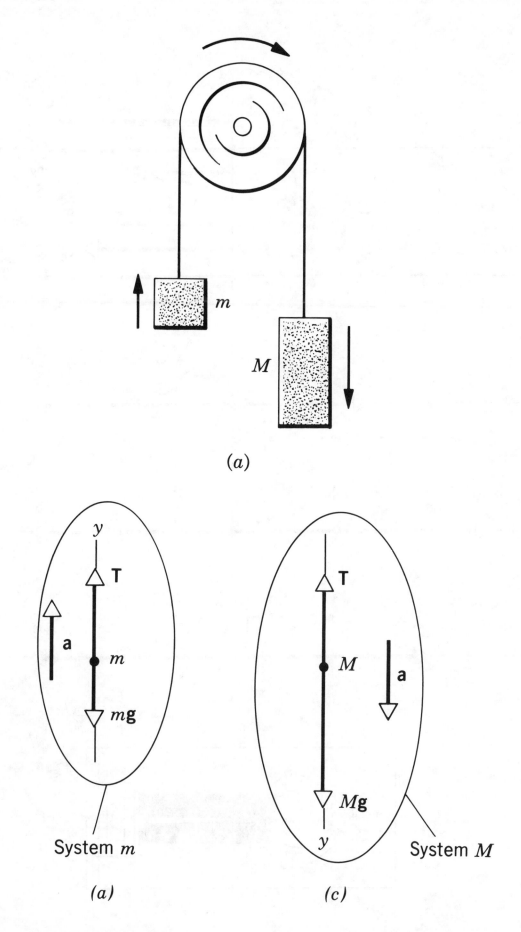

(a)

System m

(a)

System M

(c)

Sample problem 5-9.

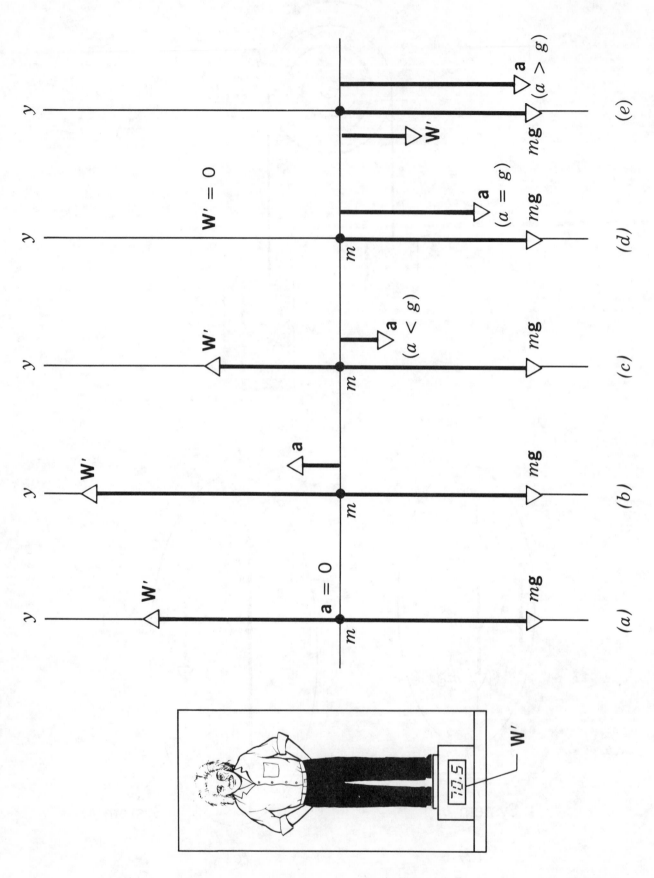

(a)

(b)

(c)

(d)

(e)

Sample problem 5-10.

FIG. 5.21

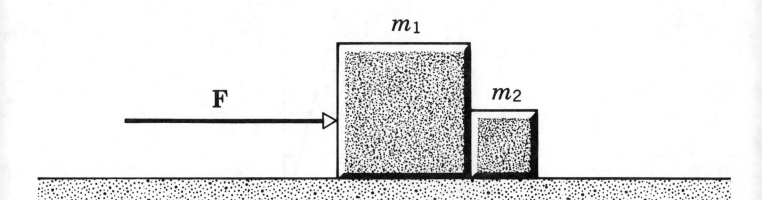

Problem 5-13.

FIG. 5.29

19

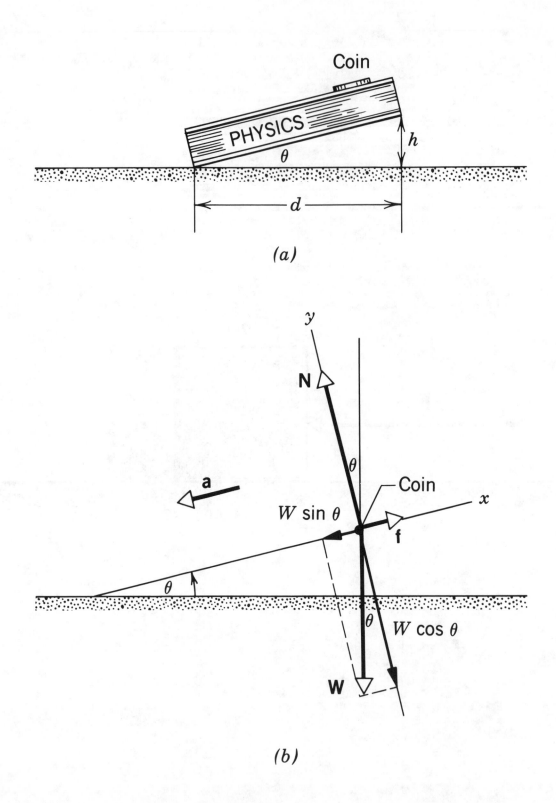

Coin

PHYSICS

θ

h

d

(a)

y

N

θ

a

$W \sin \theta$

Coin

x

f

θ

θ

$W \cos \theta$

W

(b)

Sample problem 6-1.

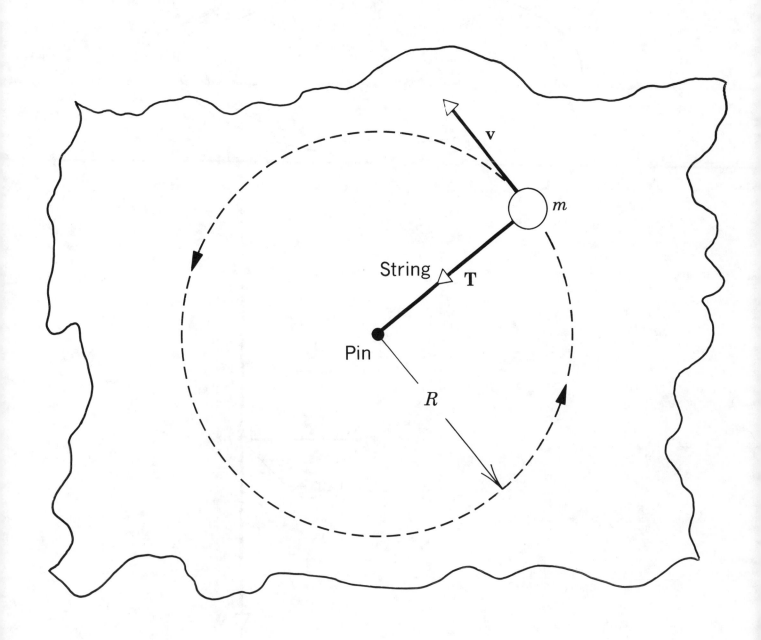

Force on a body in uniform circular motion.

FIG. 6.11

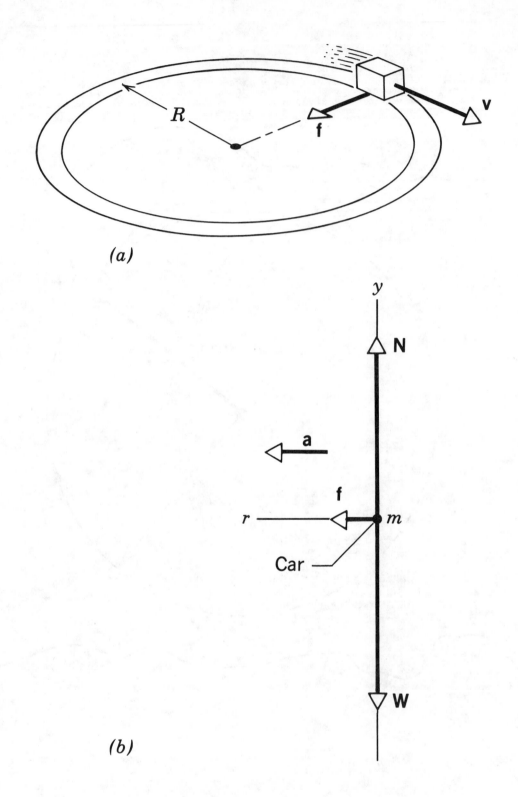

(a)

(b)

Sample problem 6-7.

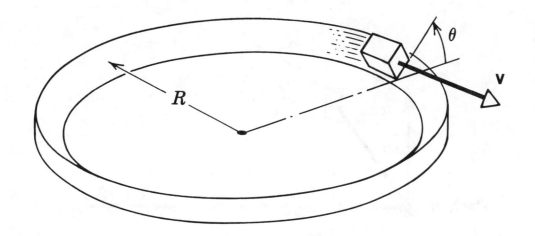

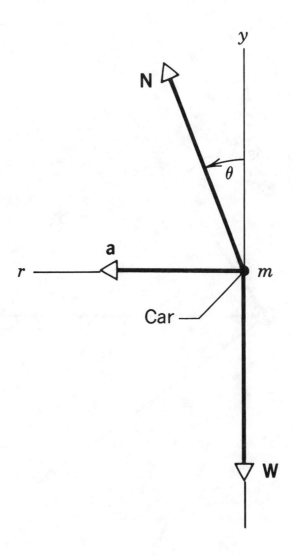

Sample problem 6-8.

FIG. 6.14

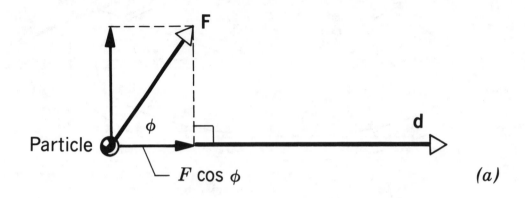

(a)

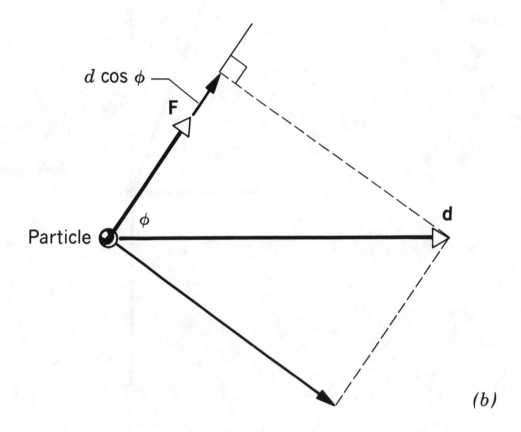

(b)

$$W = \mathbf{F} \cdot \mathbf{d}$$

FIG. 7.3

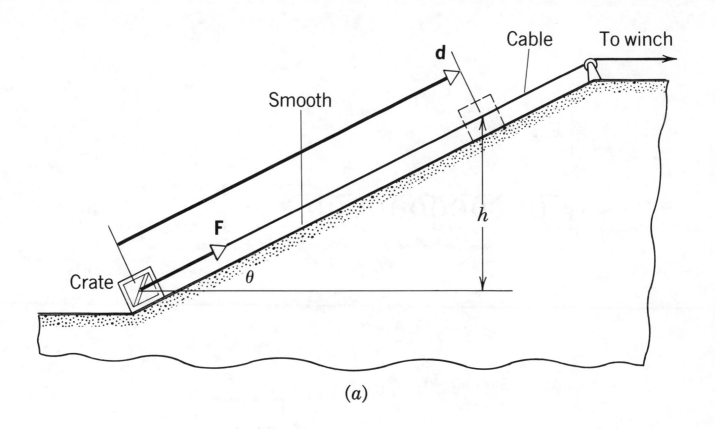

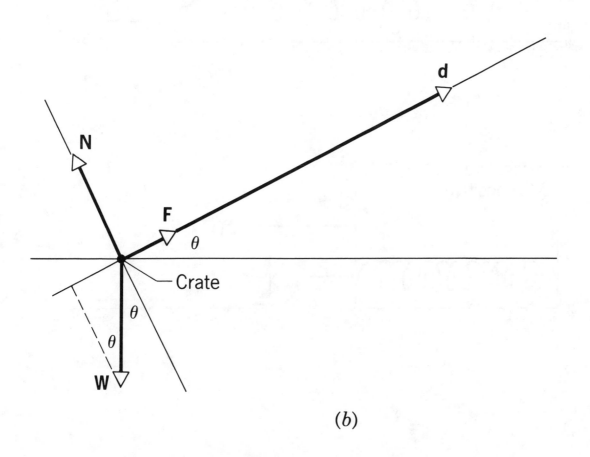

Sample problem 7-3.

FIG. 7.8

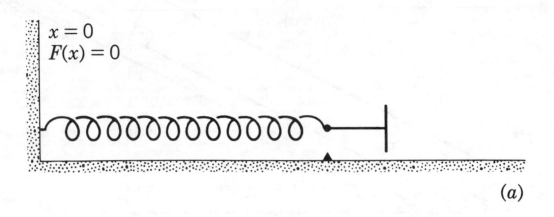

$x = 0$
$F(x) = 0$

(a)

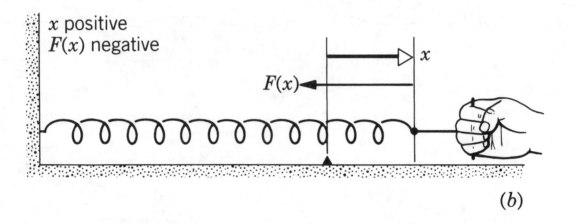

x positive
$F(x)$ negative

x

$F(x)$

(b)

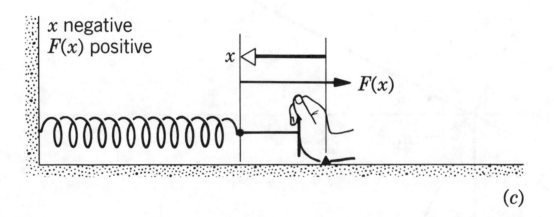

x negative
$F(x)$ positive

x

$F(x)$

(c)

Force of an ideal spring. $F(x) = -kx$

FIG. 7.10

26

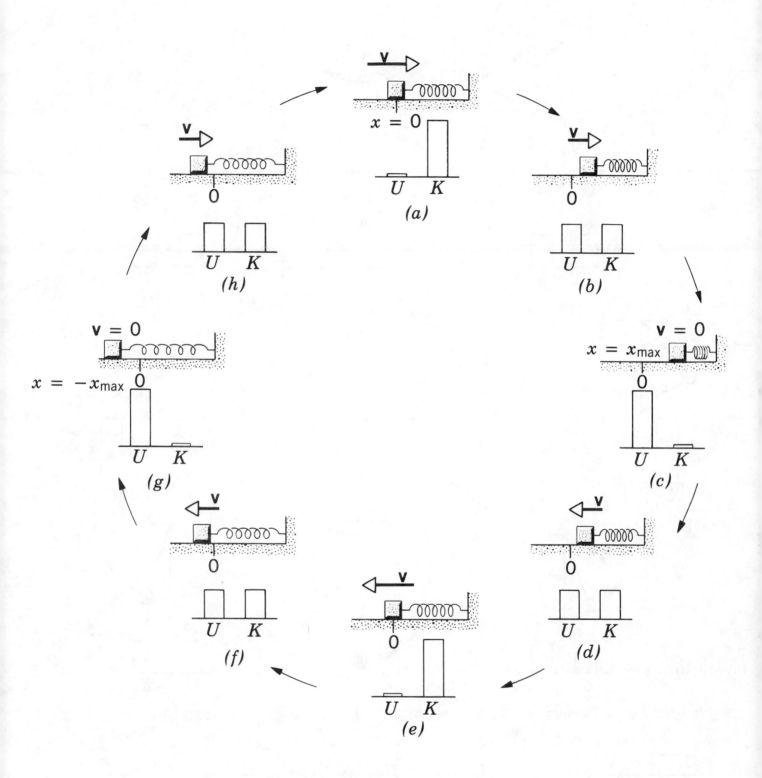

Potential and kinetic energies of a spring-mass system.

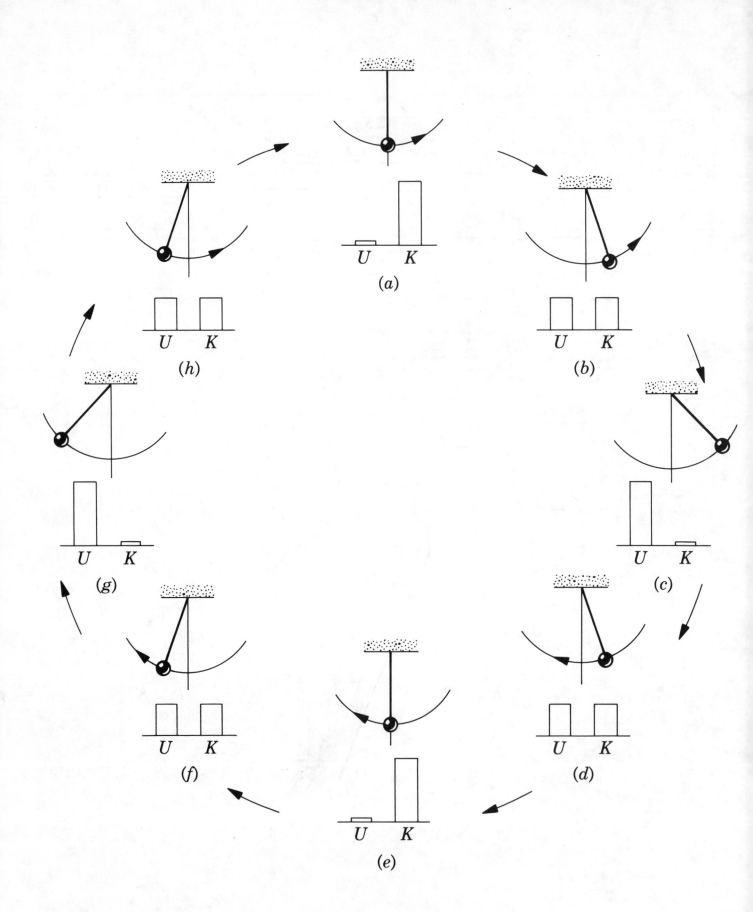

Potential and kinetic energies of a pendulum.

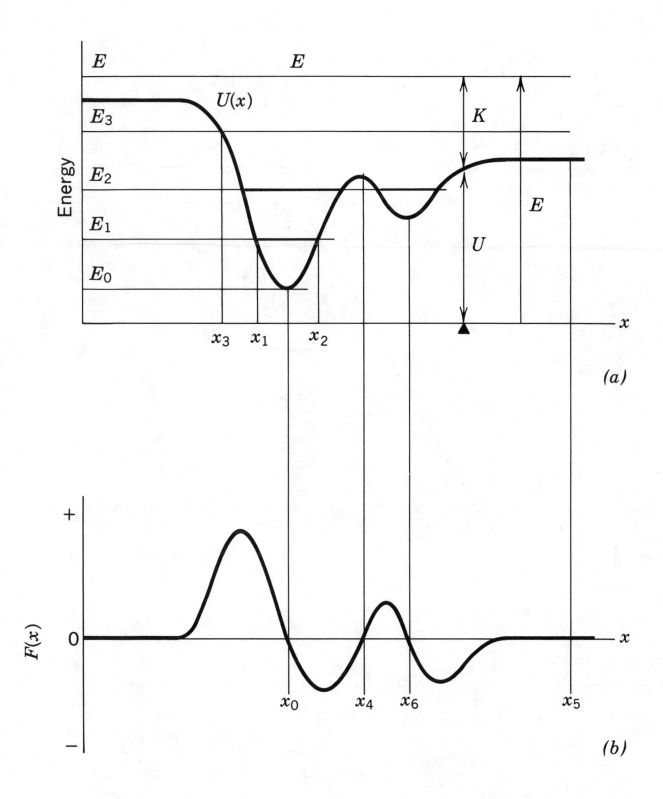

Potential energy of a particle acted on by a conservative force.

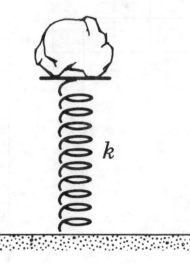

Problem 8-17.

FIG. 8.25

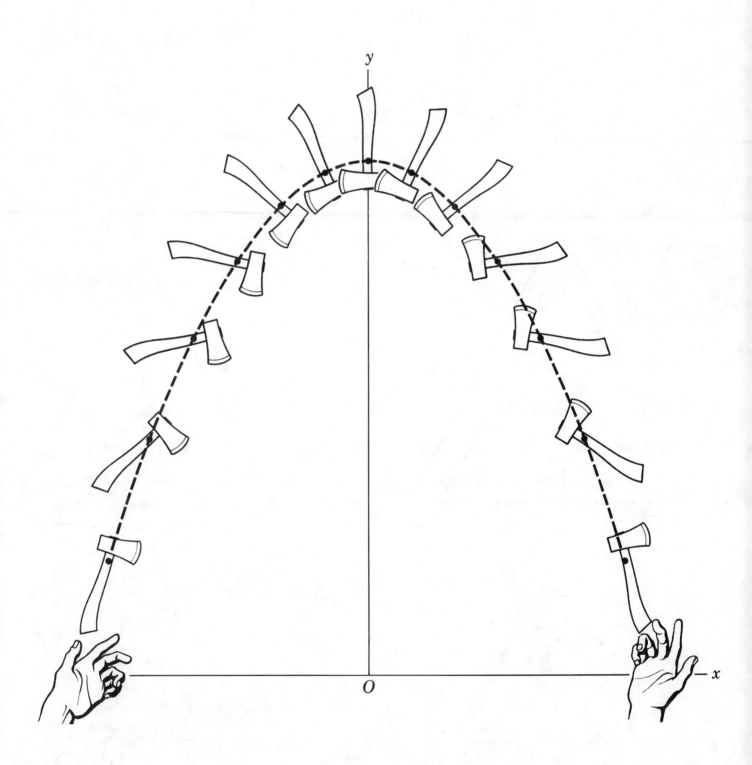

Center of mass of a projectile moves along a parabola while it rotates.

FIG. 9.1

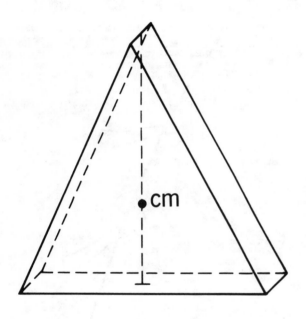

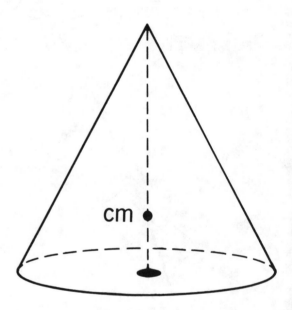

Question 9-2.

FIG. 9.18

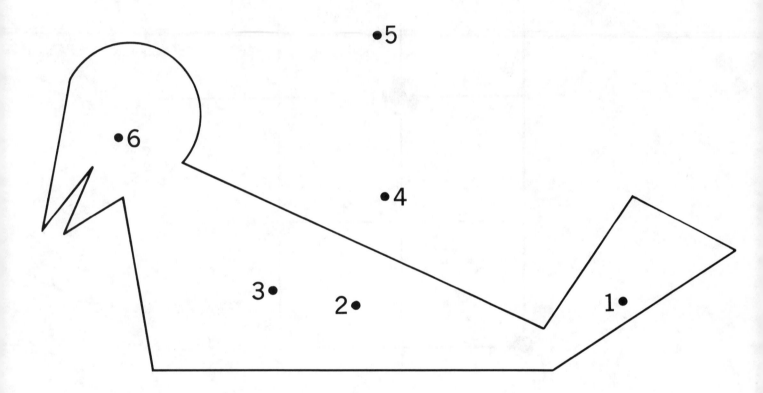

Question 9-5.

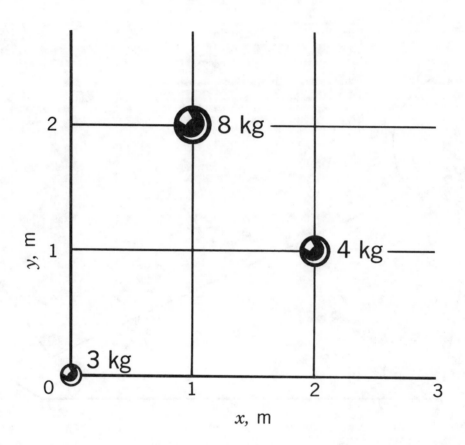

Problem 9-3.

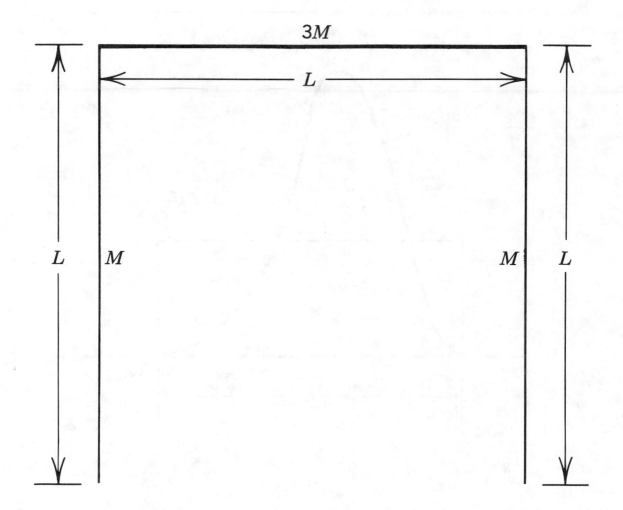

Problem 9-4.

FIG. 9.22   **35**

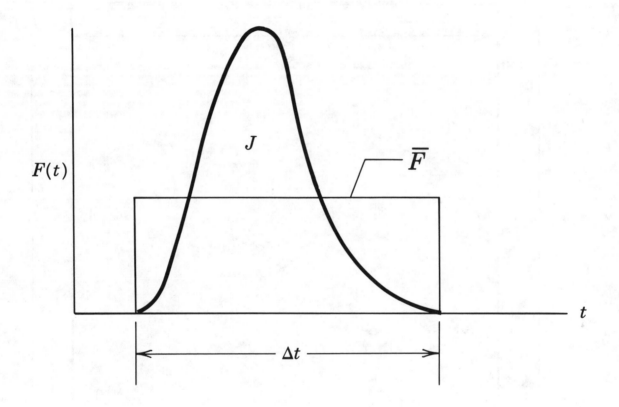

Impulse $\mathbf{J} = \int \mathbf{F}(t)\, dt = \bar{\mathbf{F}} \Delta t$

FIG. 10.7   **36**

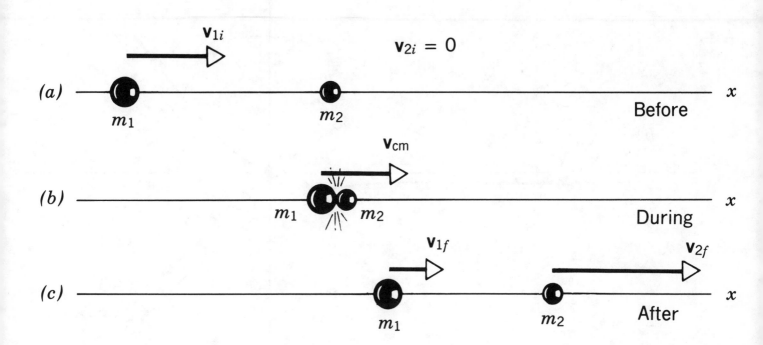

A collision in one dimension.

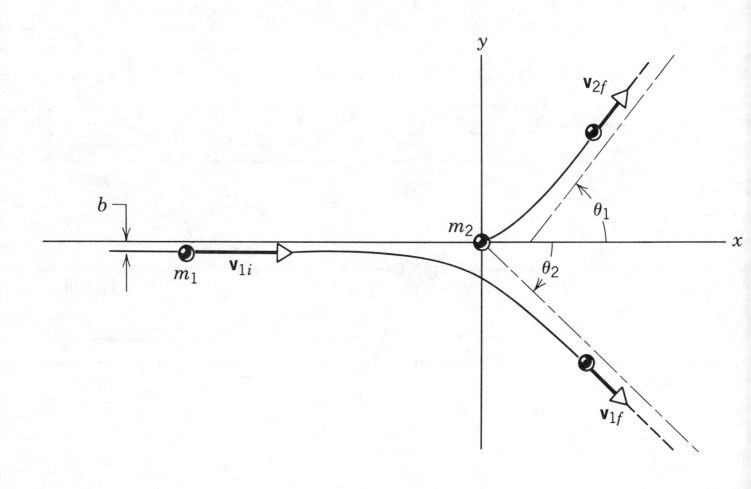

A collision in two dimensions.

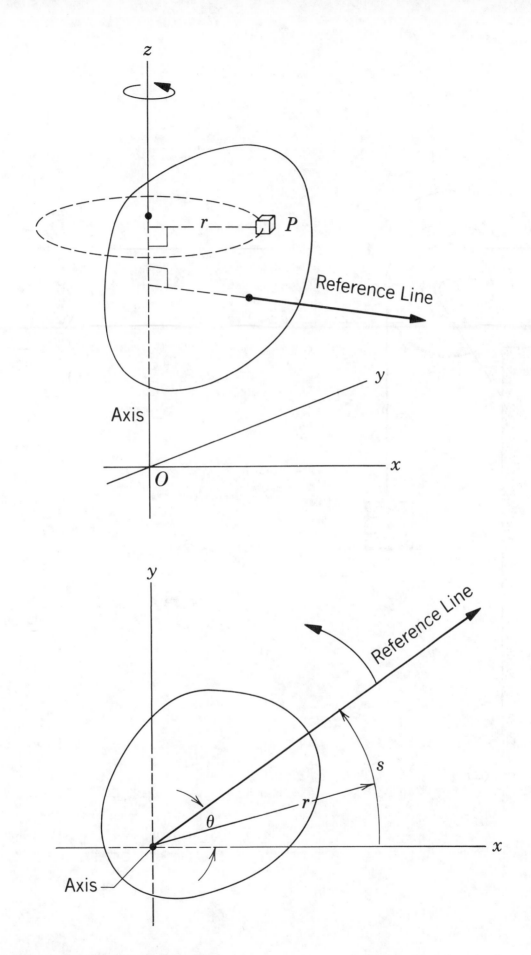

Rotation of a body about the z-axis.

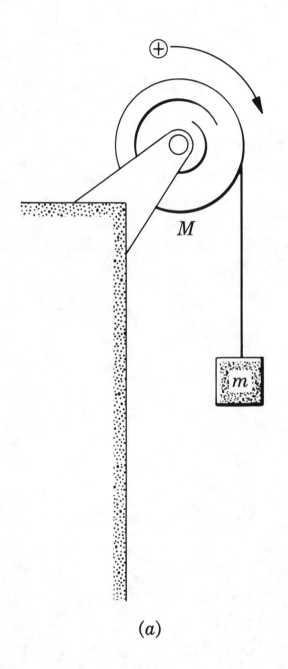

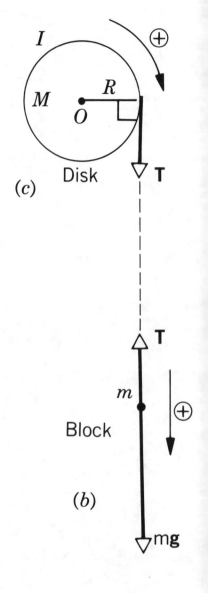

(a)

(b)

(c)

Disk

Block

Sample problem 11-10.

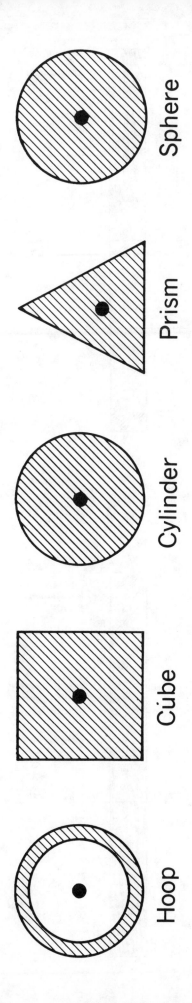

Sphere

Prism

Cylinder

Cube

Hoop

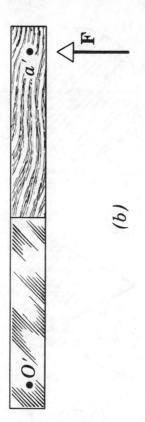

(b)

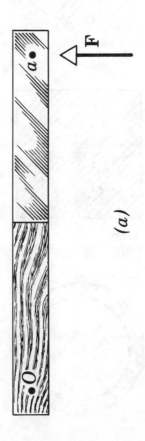

(a)

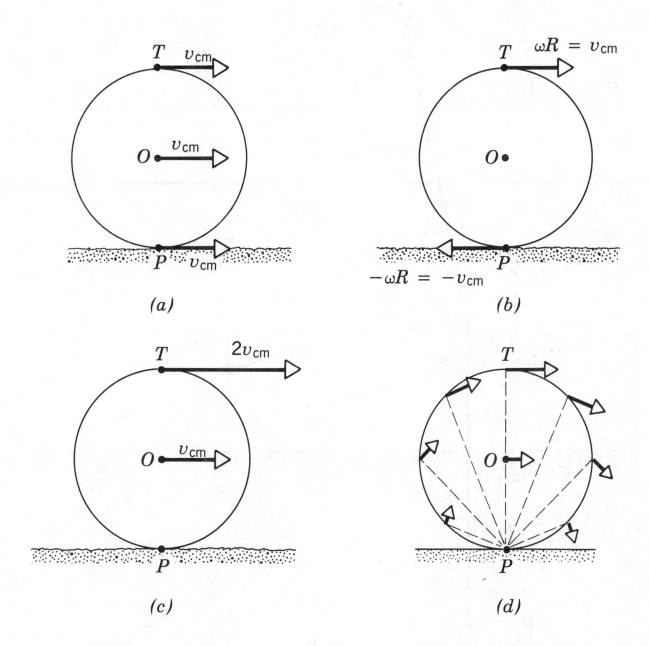

Velocities of points on a wheel rolling without slipping.
(*a*) Translational component
(*b*) Rotational component
(*c*) Combined
(*d*) Velocities of other points

FIG. 12.3 & 12.5 **43**

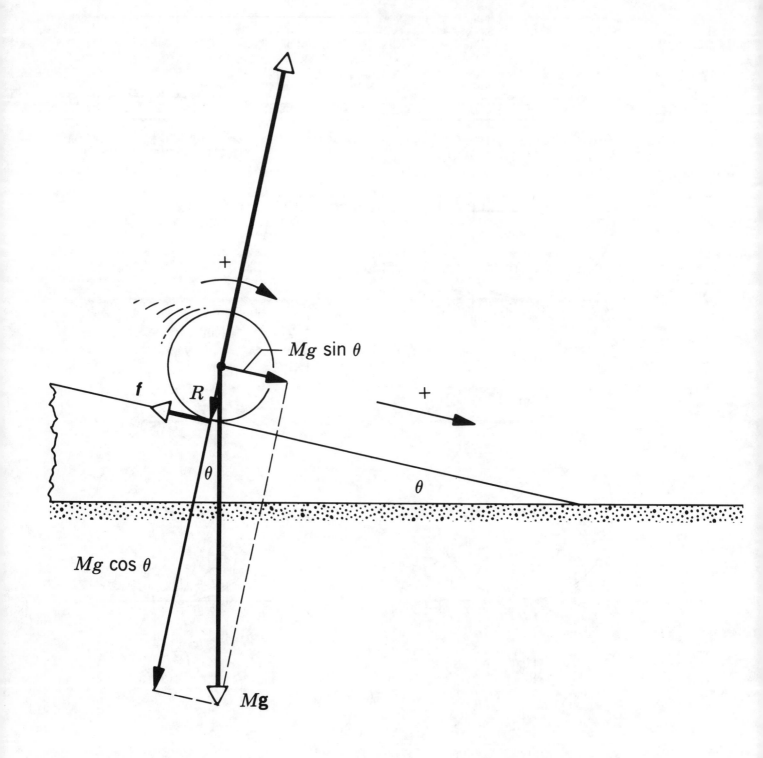

Sample problem 12-4.

FIG. 12.7 **44**

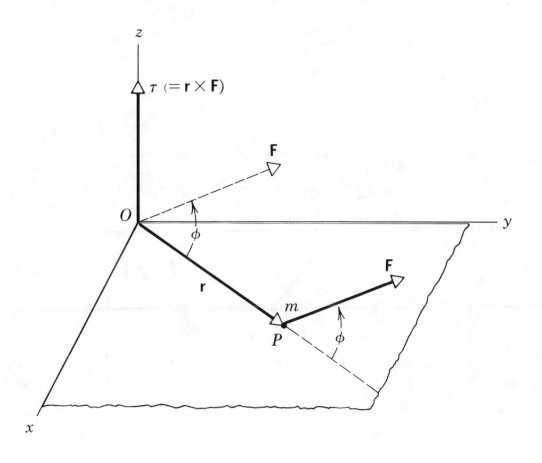

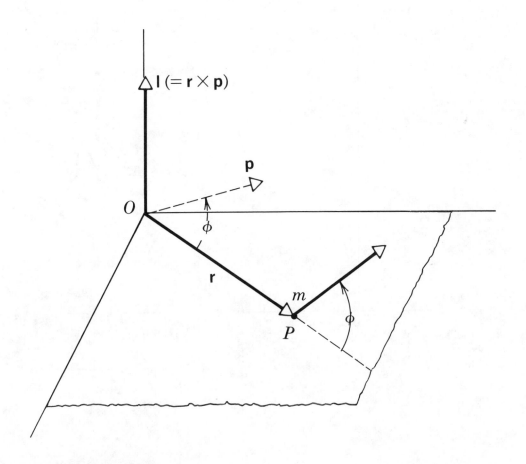

Torque and angular momentum.

FIG. 12.9 & 12.10

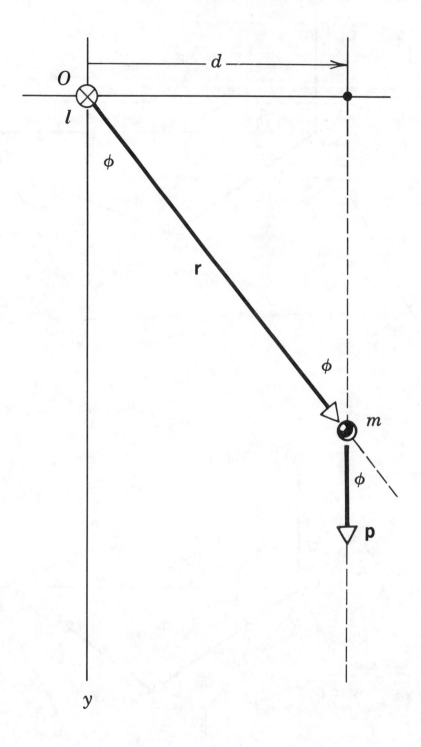

Angular momentum of an object moving in a straight line: $l = \mathbf{r} \times \mathbf{p}$

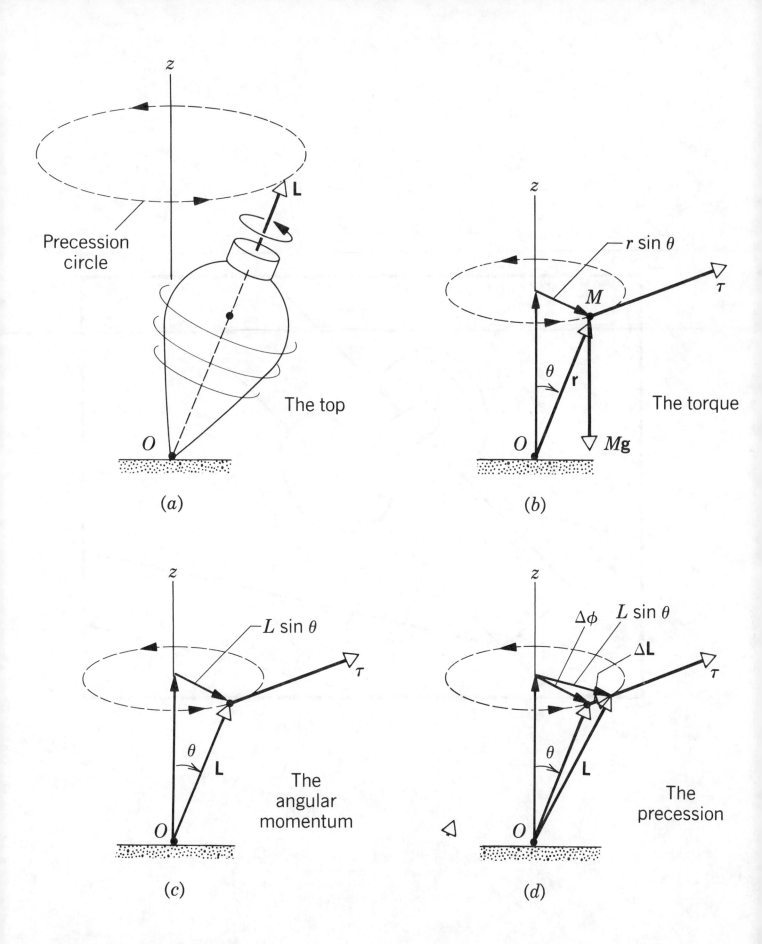

(a)

Precession circle

L

The top

O

(b)

z

$r \sin \theta$

M

θ

r

$M\mathbf{g}$

τ

The torque

O

(c)

z

$L \sin \theta$

τ

θ

L

The angular momentum

O

(d)

z

$\Delta\phi$

$L \sin \theta$

Δ**L**

τ

θ

L

The precession

O

Motion of a spinning top.

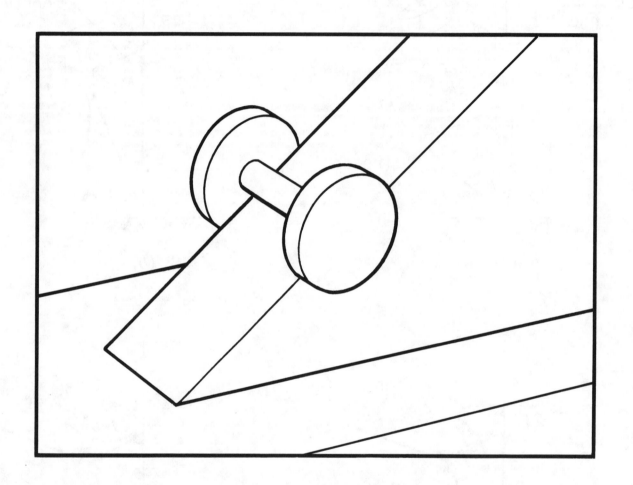

Question 12-8.

FIG. 12.24

48

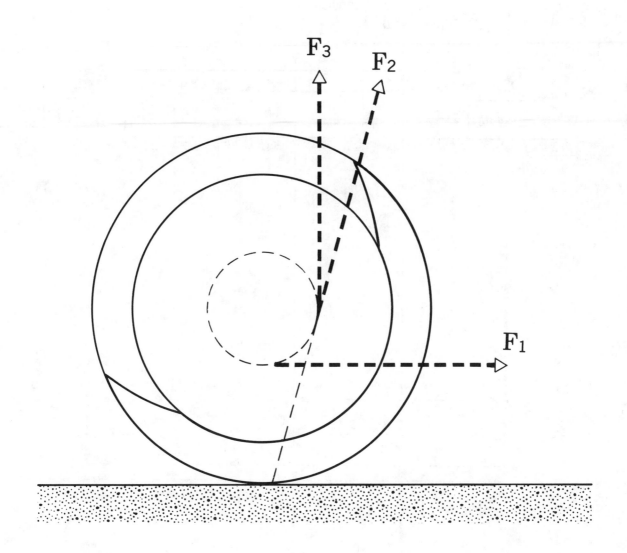

Question 12-10.

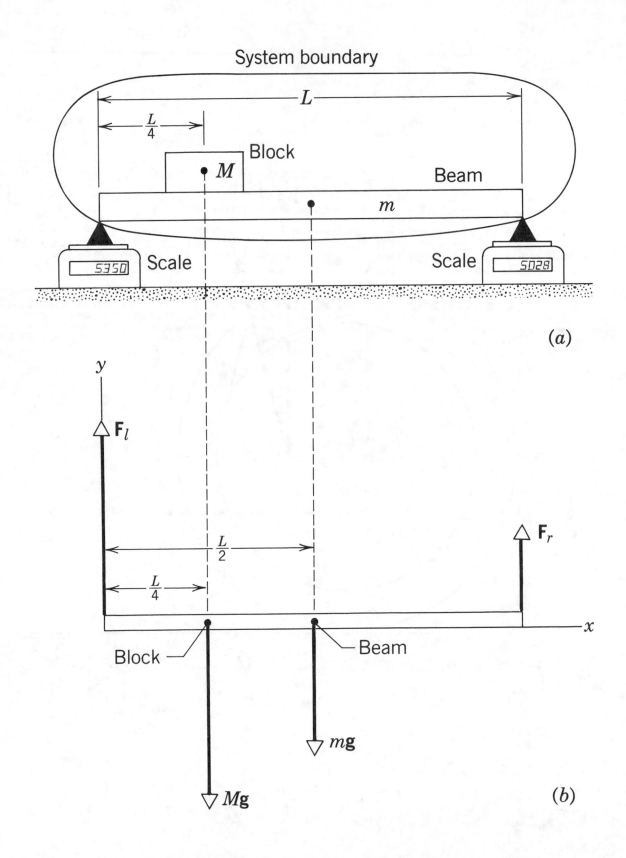

System boundary

L

$\dfrac{L}{4}$

Block

M

Beam

m

5350　Scale

Scale　5028

(a)

y

$\mathbf{F}_l$

$\dfrac{L}{2}$

$\dfrac{L}{4}$

$\mathbf{F}_r$

x

Block

Beam

$m\mathbf{g}$

$M\mathbf{g}$

(b)

Sample problem 13-1.

FIG. 13.5　**50**

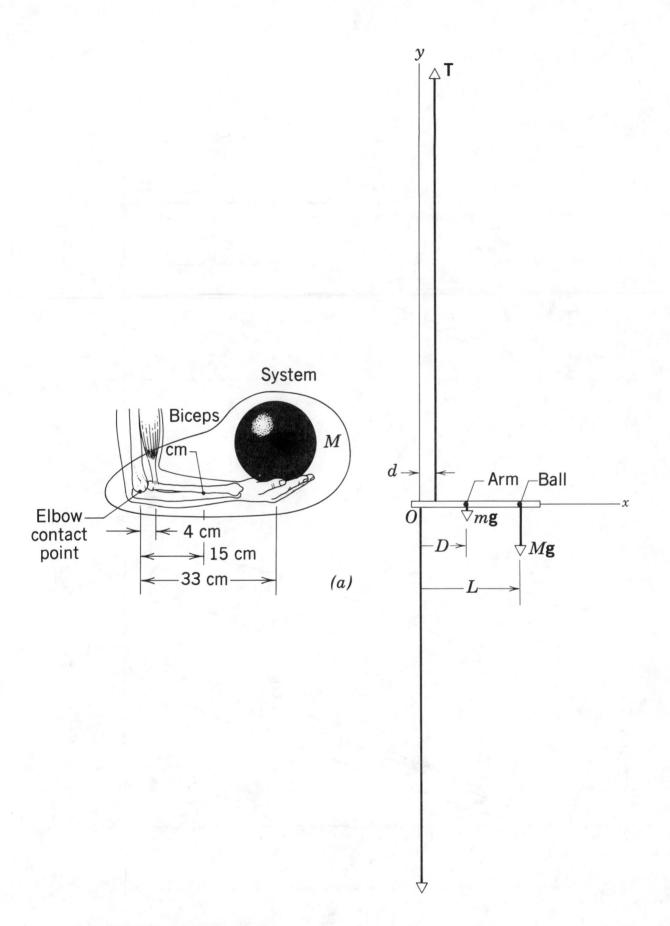

System

Biceps

cm

M

Elbow
contact
point

4 cm

15 cm

33 cm

(a)

y

T

d

Arm — Ball

O — *x*

▽*m***g**

D

▽*M***g**

L

Sample problem 13-2.

FIG. 13.6 **51**

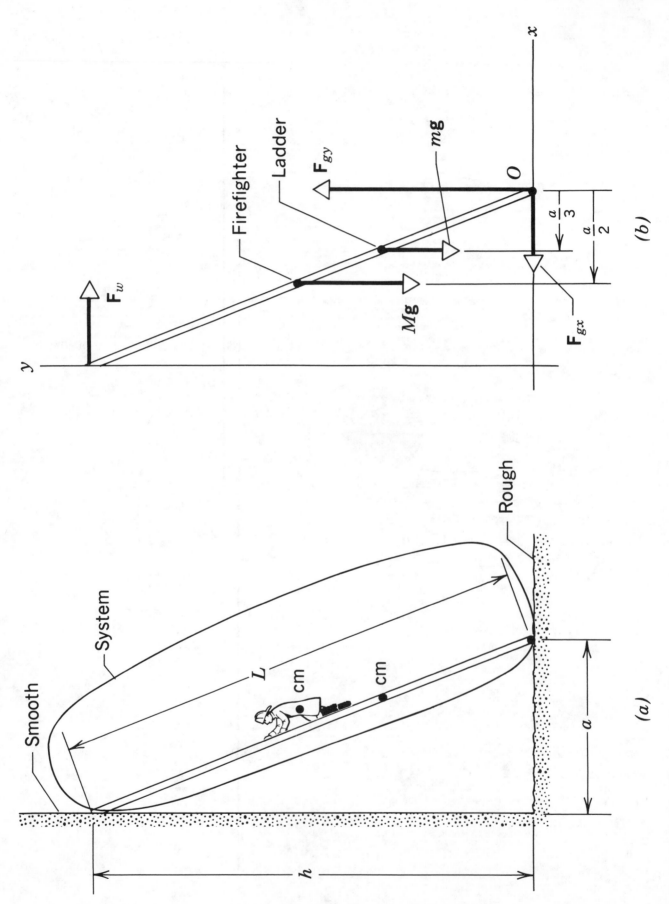

Firefighter

Ladder

$\mathbf{F}_{gy}$

mg

O

$\mathbf{F}_w$

Mg

$\mathbf{F}_{gx}$

$\dfrac{a}{3}$

$\dfrac{a}{2}$

x

y

(b)

System

Smooth

Rough

L

cm

cm

a

h

(a)

Sample problem 13-3.

FIG. 13.7

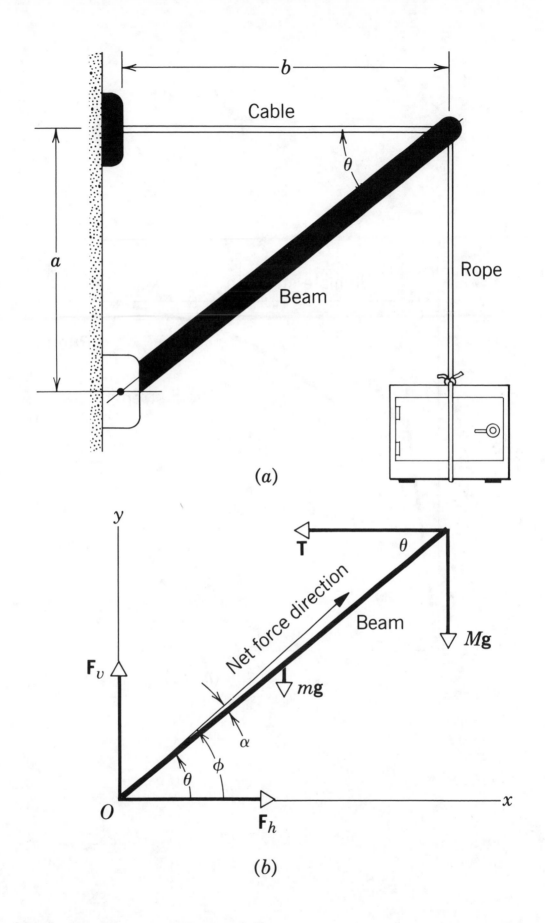

(a)

(b)

Sample problem 13-5.

FIG. 13.8

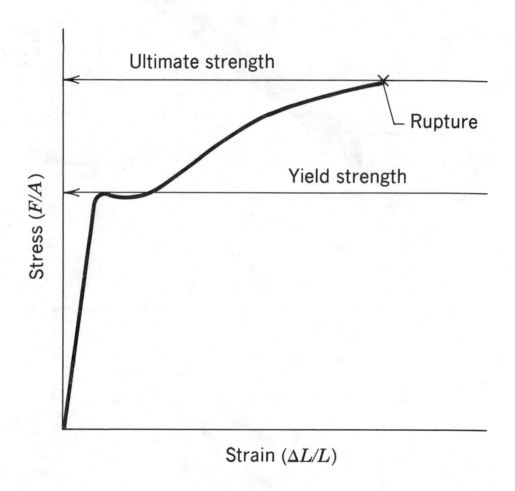

Stress as a function of strain for a steel cylinder.

54

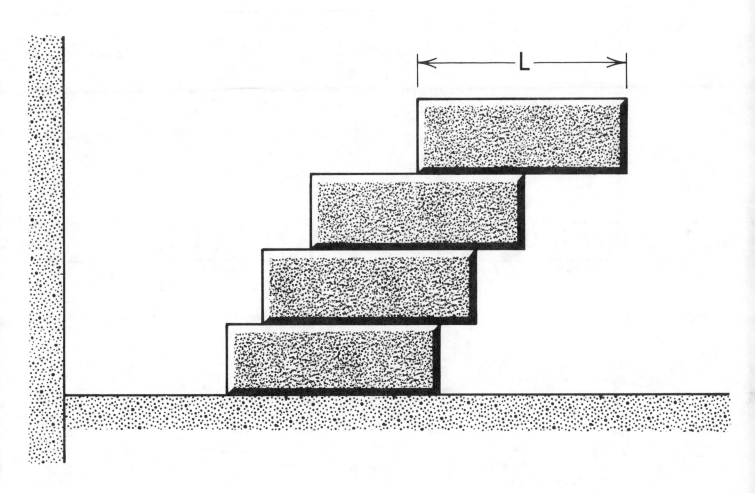

Problem 13-29.

FIG. 13.33

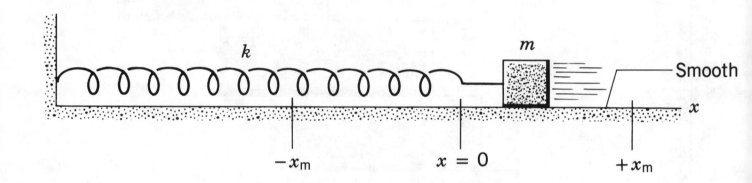

Mass on an ideal spring.

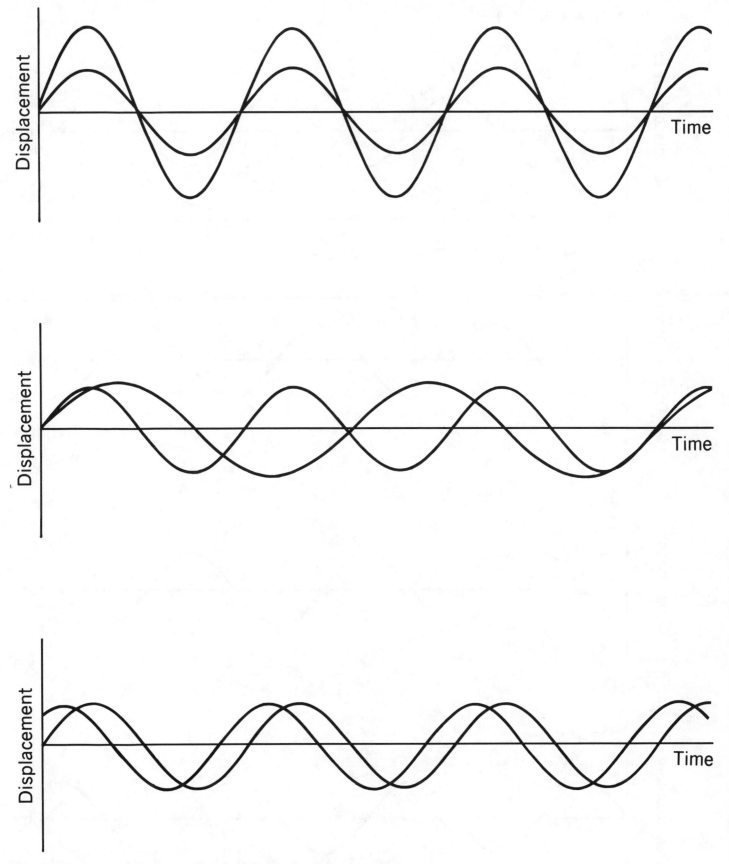

Same period, phase constant; different amplitude:

Same amplitude, phase constant; different period:

Same amplitude, period; different phase constant:

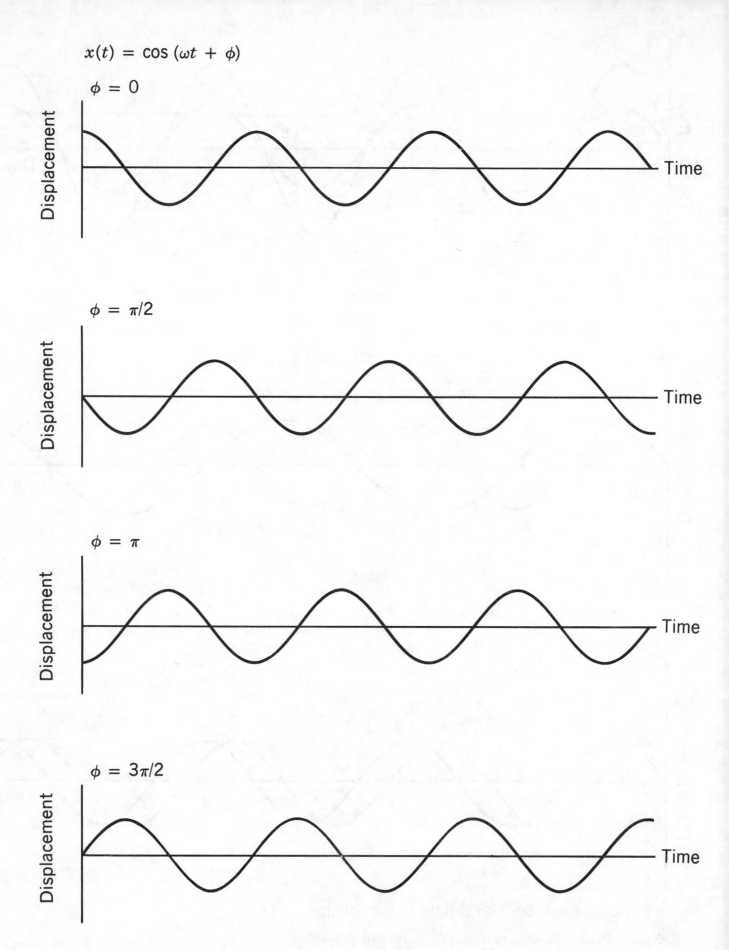

$$x(t) = \cos(\omega t + \phi)$$

$\phi = 0$

$\phi = \pi/2$

$\phi = \pi$

$\phi = 3\pi/2$

Dependence of simple harmonic motion on the phase constant.

From Chapter 14

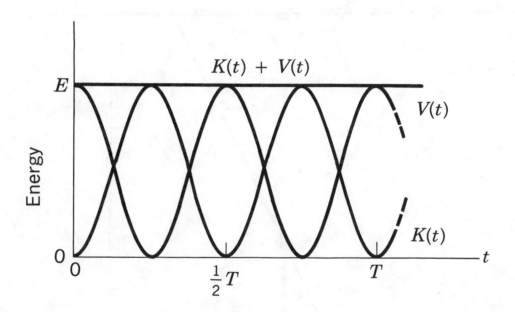

Potential and kinetic energies of an oscillating spring-mass system.
$E = U(t) + K(t)$ = constant.

FIG. 14.6   **59**

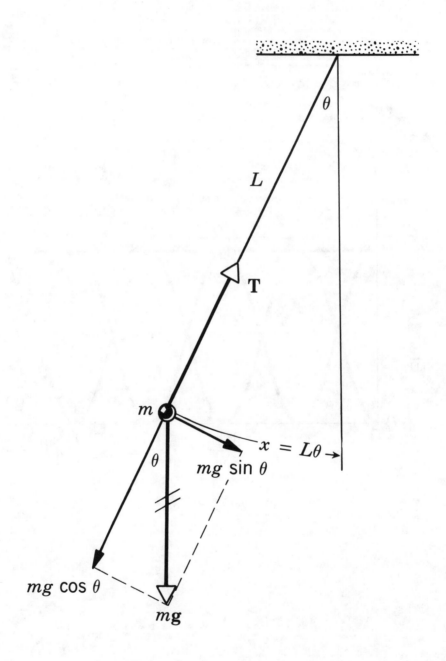

A simple pendulum.

FIG. 14.9 **60**

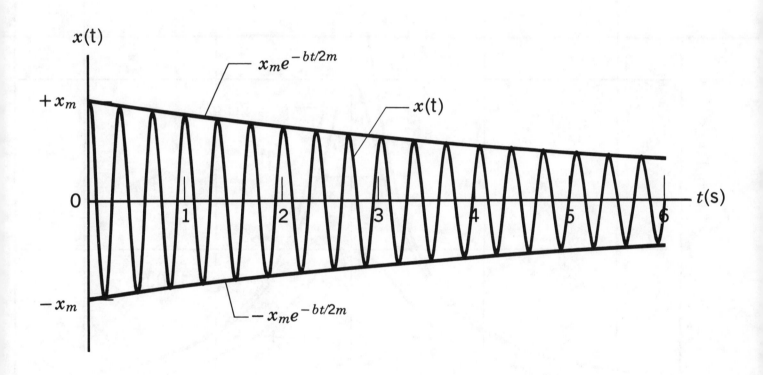

Displacement as a function of time for a damped oscillator.

FIG. 14.18   **61**

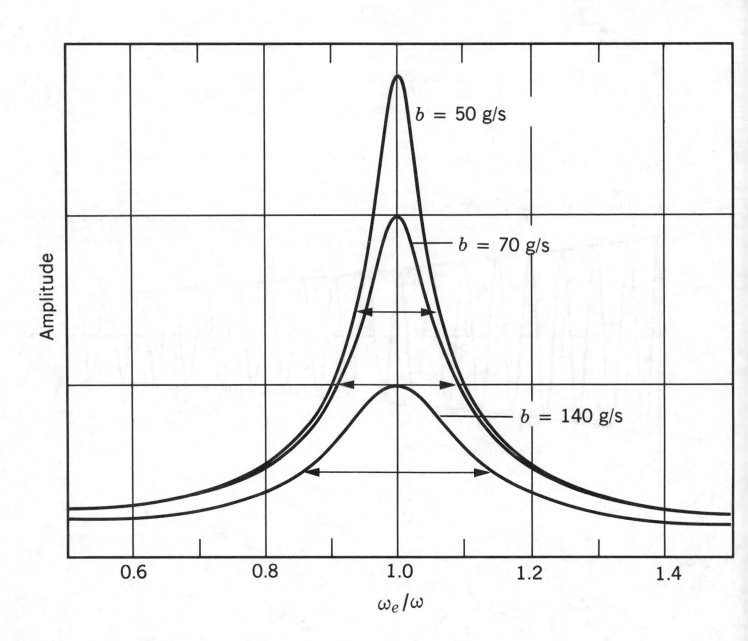

Amplitude as a function of forcing frequency for a damped forced oscillator.

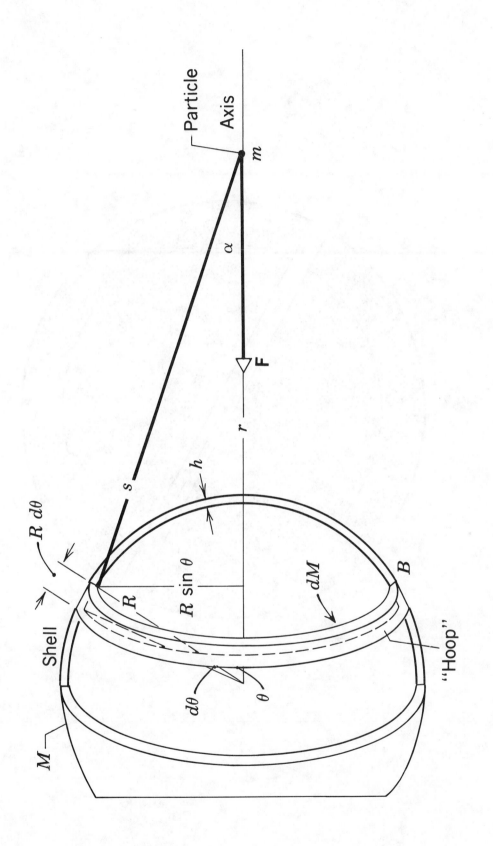

Proof of the gravitational shell theorem.

FIG. 15.7

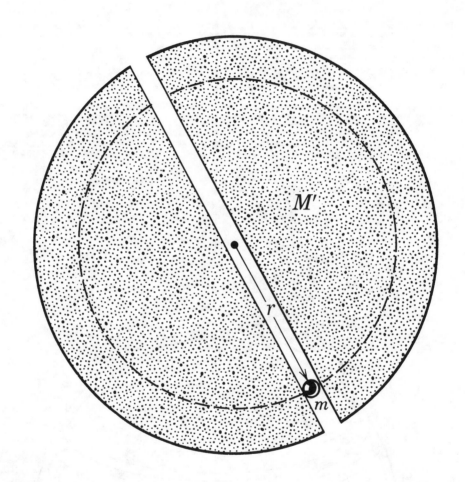

Sample problem 15-3.

FIG. 15.8

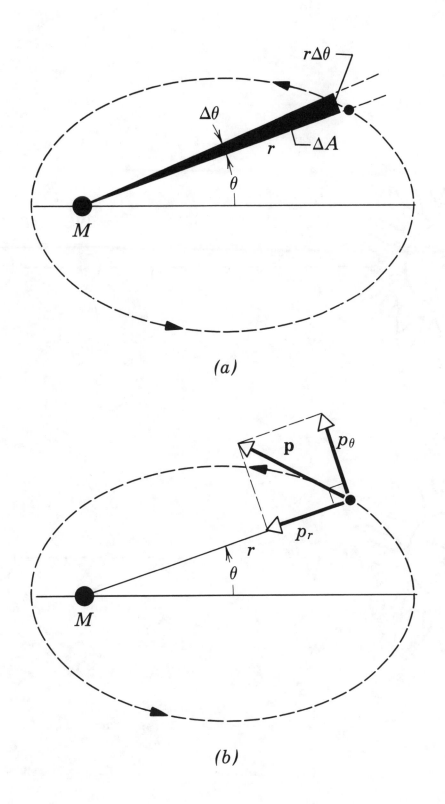

(a)

(b)

Planets sweep out equal areas in equal times.

FIG. 15.20

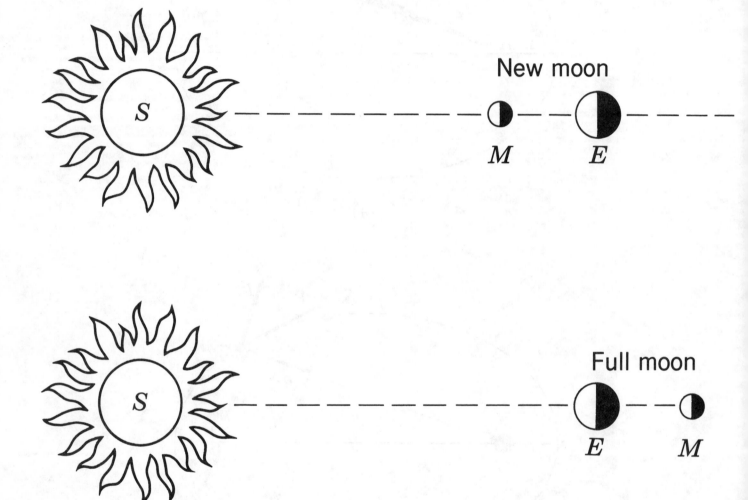

New moon

M   *E*

Full moon

E   *M*

Question 15-18.

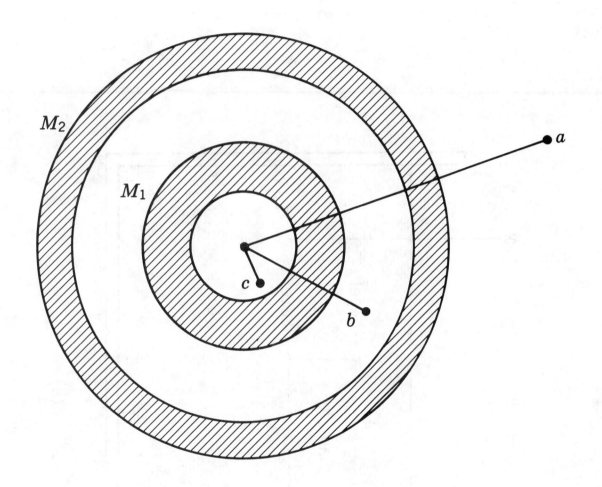

Problem 15-9.

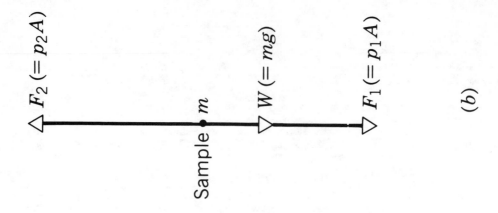

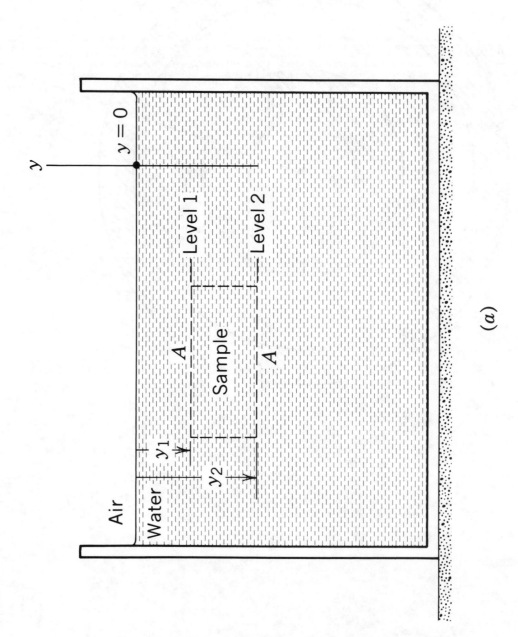

For a static fluid.
$$p_2 = p_1 + \rho g(y_1 - y_2).$$

FIG. 16.2

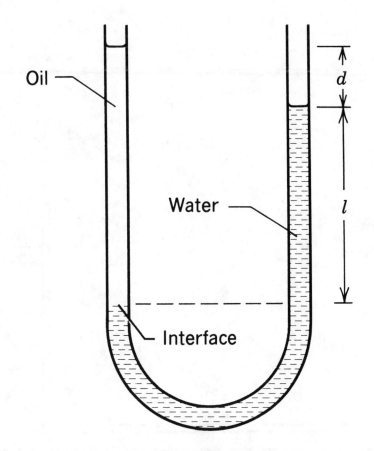

A U-tube containing oil and water.

FIG. 16.6 **69**

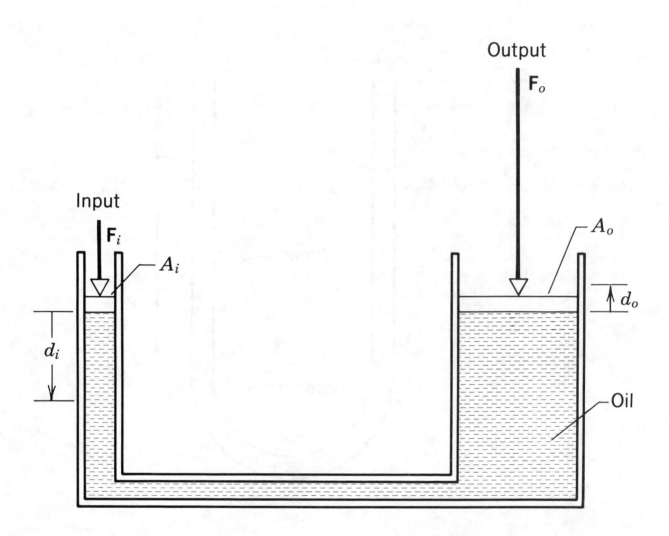

A hydraulic lever.

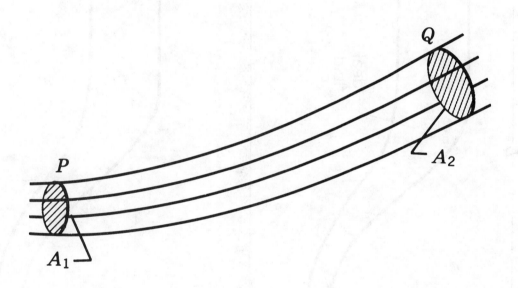

A tube of fluid flow $\rho_1 A_1 v_1 = \rho_2 A_2 v_2$ for a steady flow.

FIG. 16.18 **71**

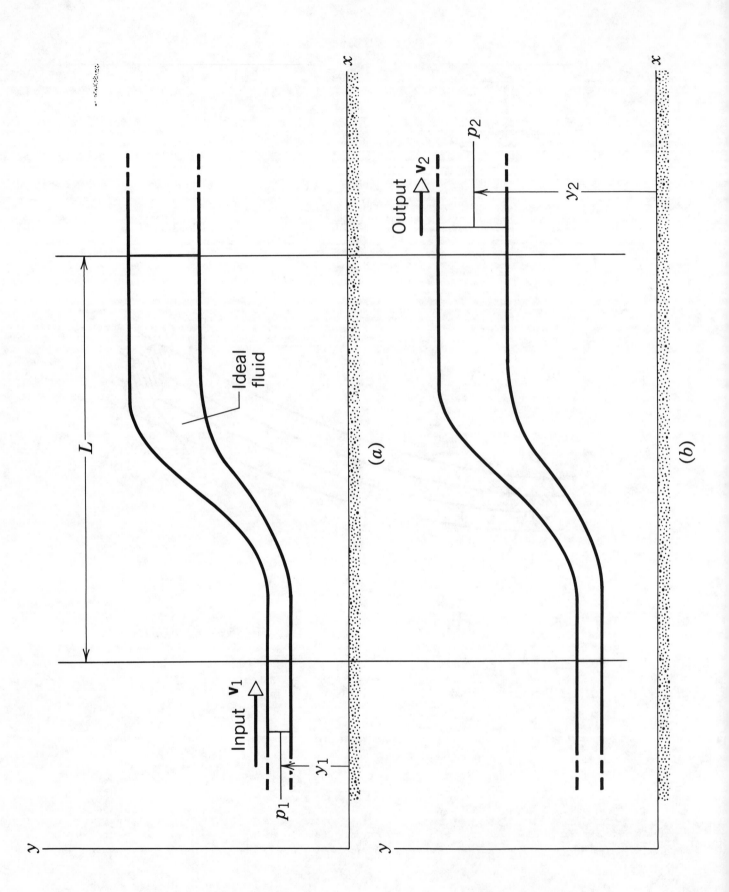

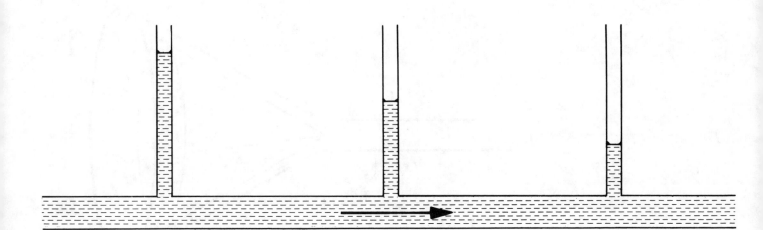

Question 16-33.

FIG. 16.29

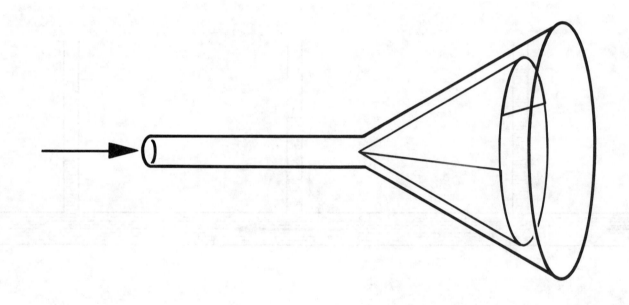

Question 16-46.

FIG. 16.30 **74**

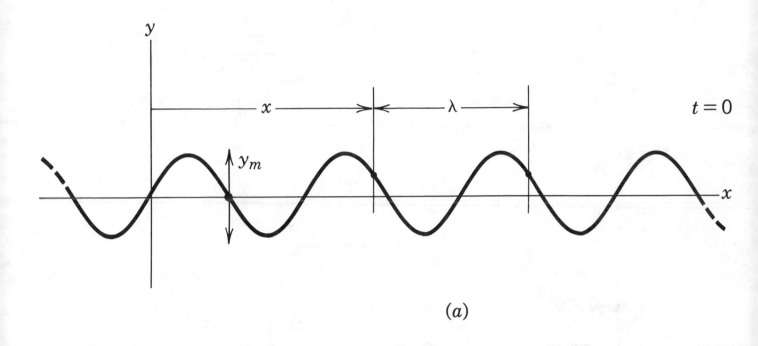

(a)

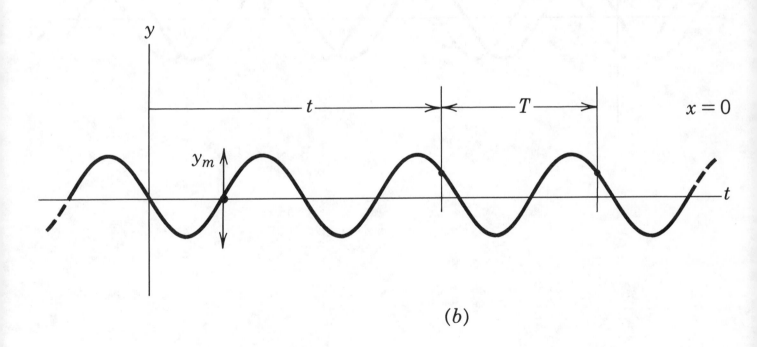

(b)

Displacement as a function of position (for given time) and as a function of time (for a given position) of a sinusoidal wave on a string.

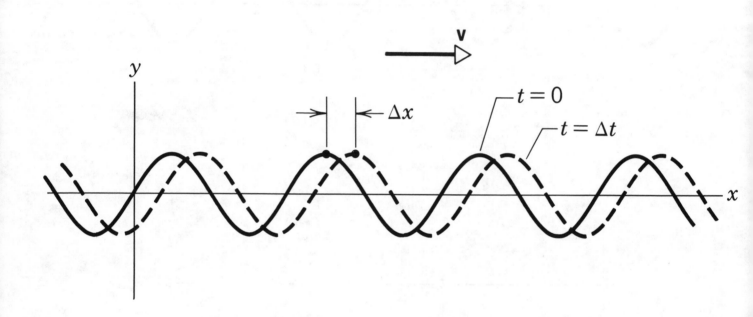

The string shape moves distance Δx in time Δt. The wave velocity is $\Delta x/\Delta t$.

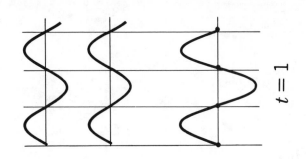

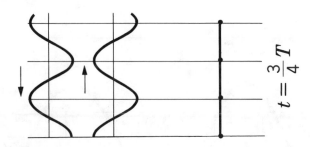

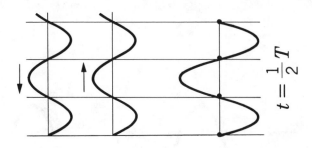

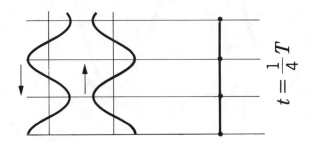

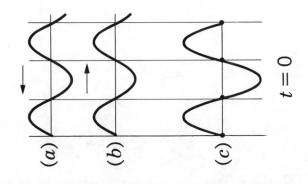

$t = 1$

$t = \frac{3}{4}T$

$t = \frac{1}{2}T$

$t = \frac{1}{4}T$

$t = 0$

(a)

(b)

(c)

A standing wave can be constructed from two traveling waves.

FIG. 17.16

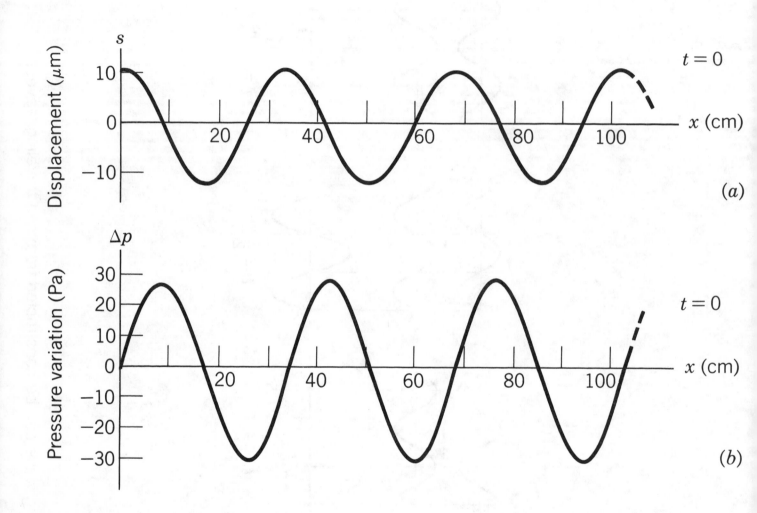

Sinusoidal sound wave.

FIG. 18.5   **78**

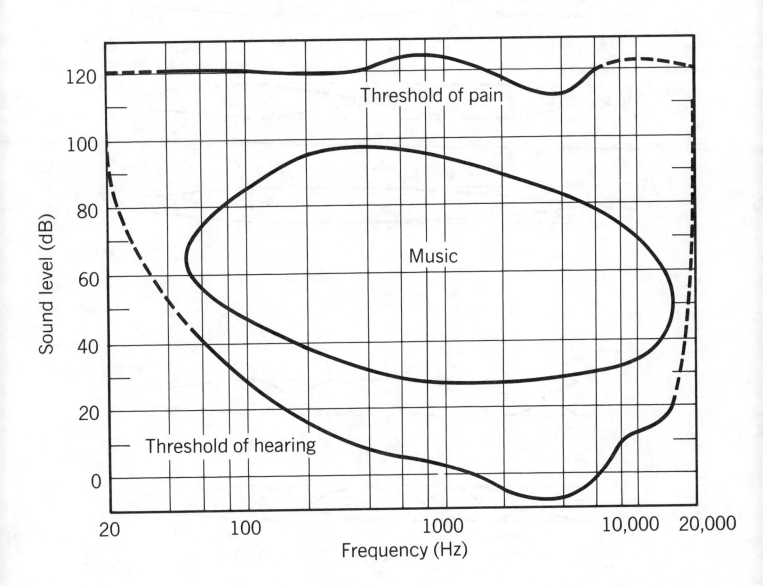

Average range of sound intensities for human hearing.

FIG. 18.6

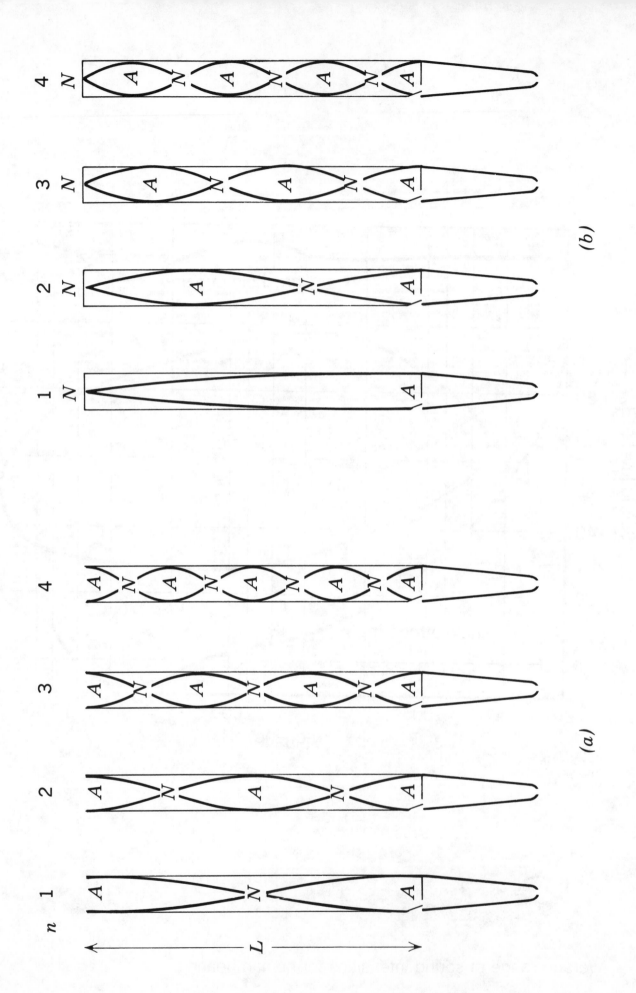

(a)

(b)

Standing waves in pipes.

FIG. 18.9

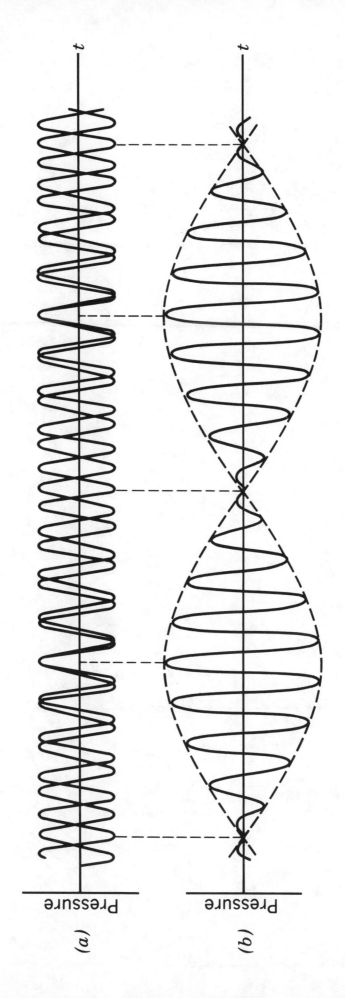

Beats.

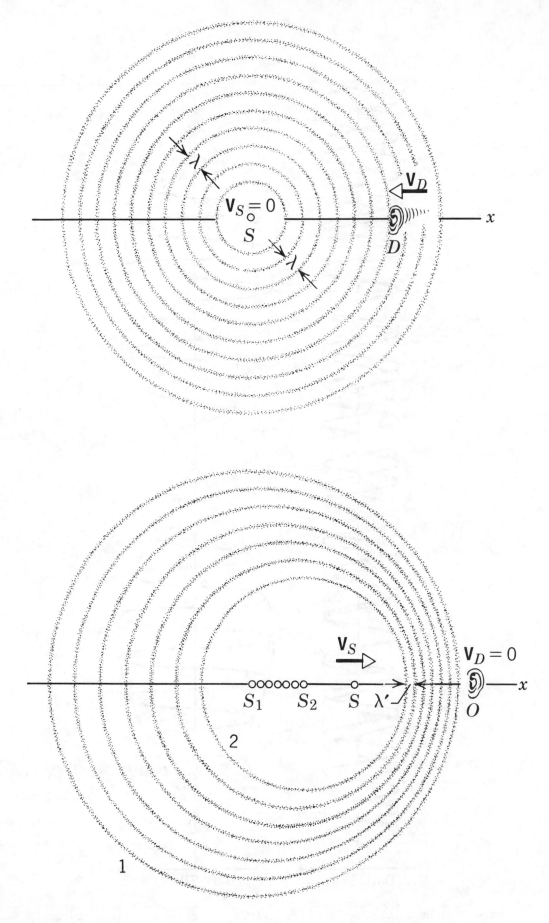

Doppler effect for sound.

FIG. 18.15 & 18.16 **82**

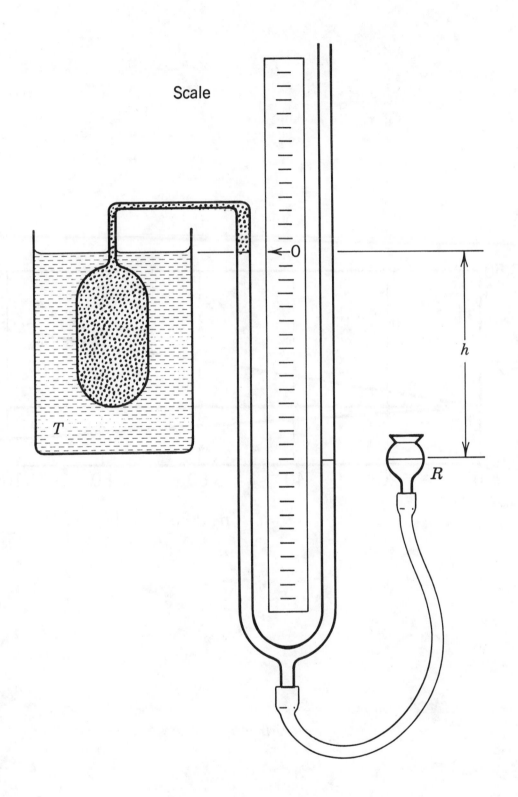

Scale

h

R

T

A constant-volume gas thermometer.

FIG. 19.5   **83**

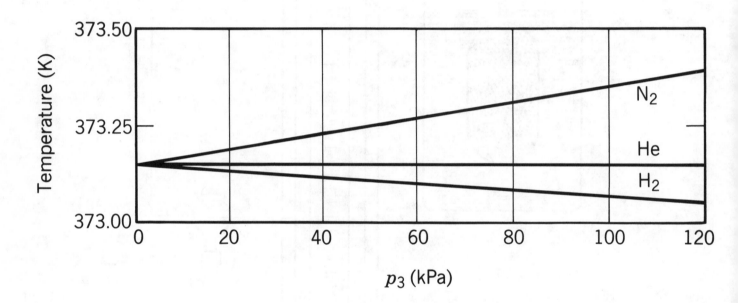

As the amount of gas in a thermometer is decreased, all gases tend to indicate the same temperature.

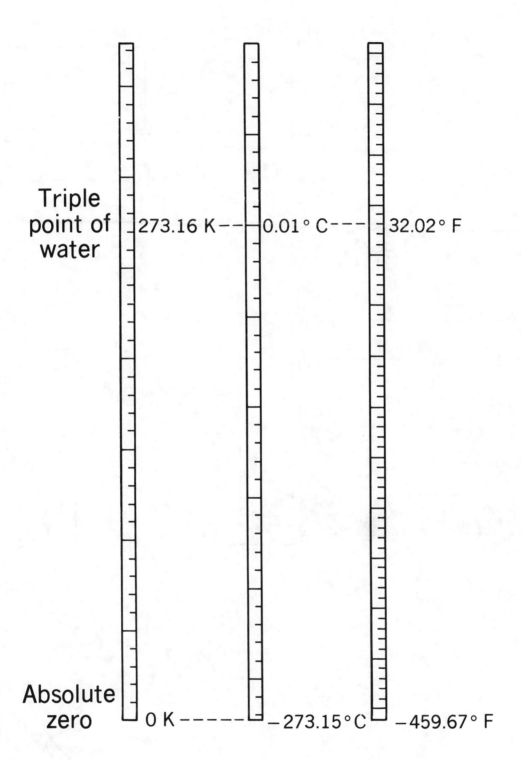

Triple point of water — 273.16 K — — 0.01° C — — — 32.02° F

Absolute zero — 0 K — — — — — −273.15°C — −459.67° F

Kelvin, Celsius and Fahrenheit temperature scales.

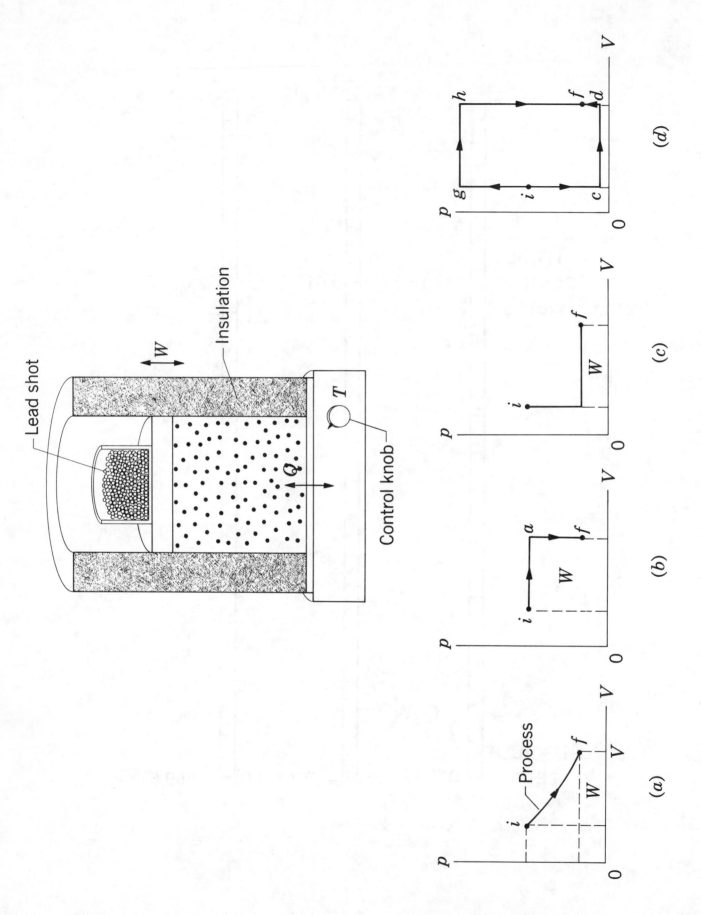

Representation on a pV diagram of the compression of a gas.

FIG. 20.3 & 20.4

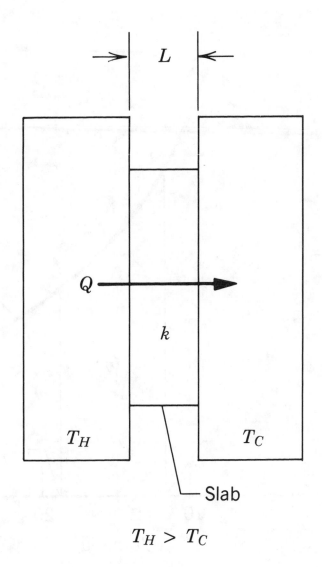

$T_H > T_C$

Thermal conduction.

FIG. 20.8

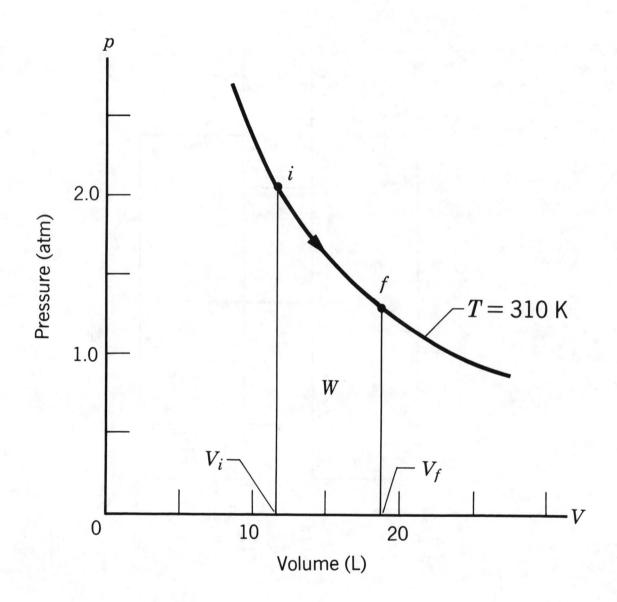

A pV diagram.

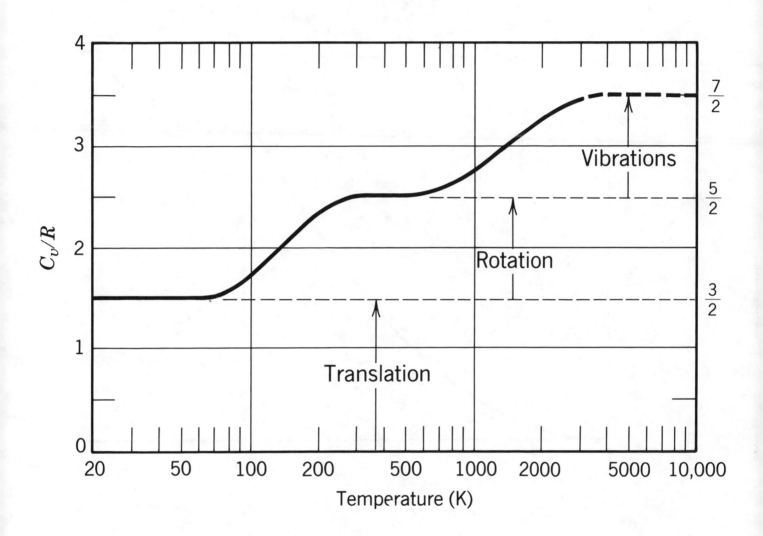

Heat capacity of hydrogen gas as a function of temperature.

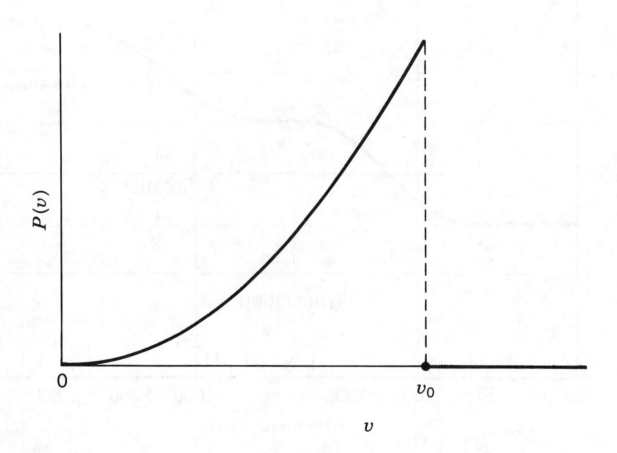

Problem 21-56.

FIG. 21.18

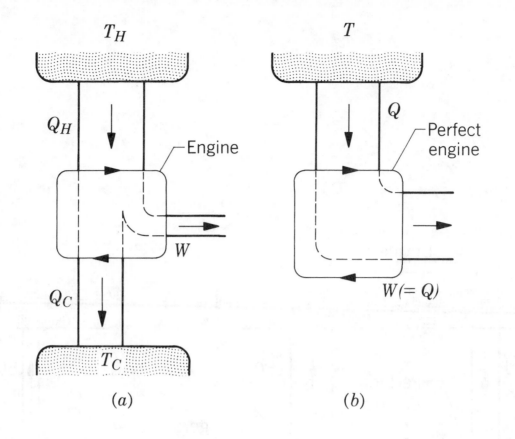

(a) (b)

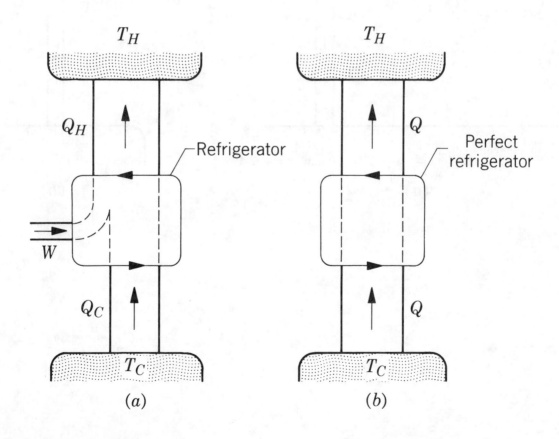

(a) (b)

Engines and refrigerators.

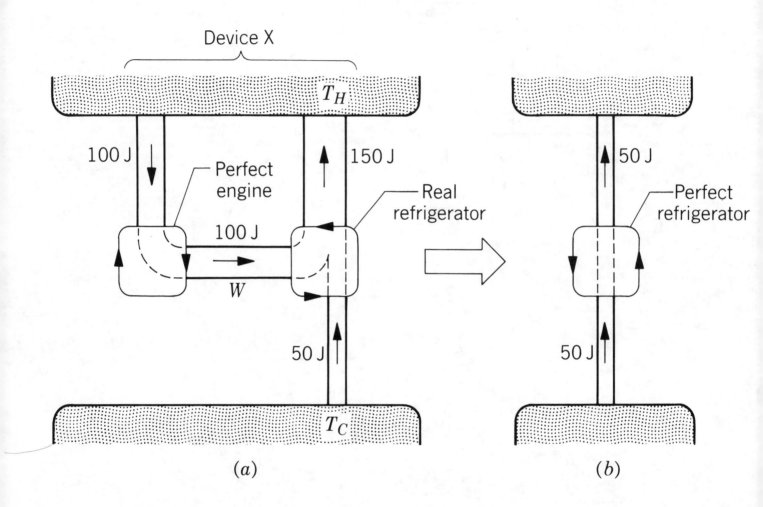

(a) (b)

A "perfect" refrigerator constructed from a "perfect" engine and a "real" refrigerator.

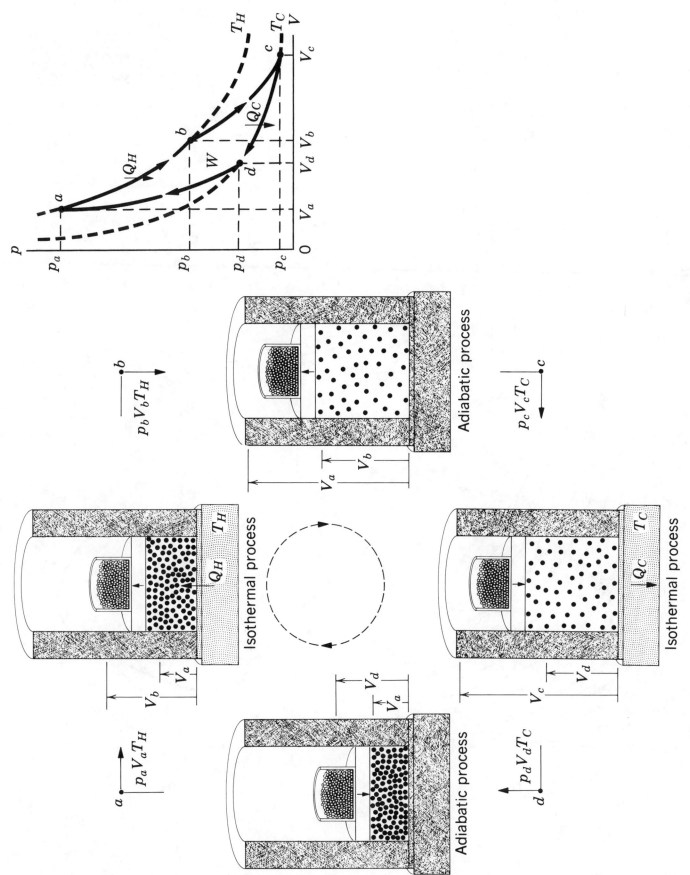

Carnot cycle.

FIG. 22.6 & 22.7 **93**

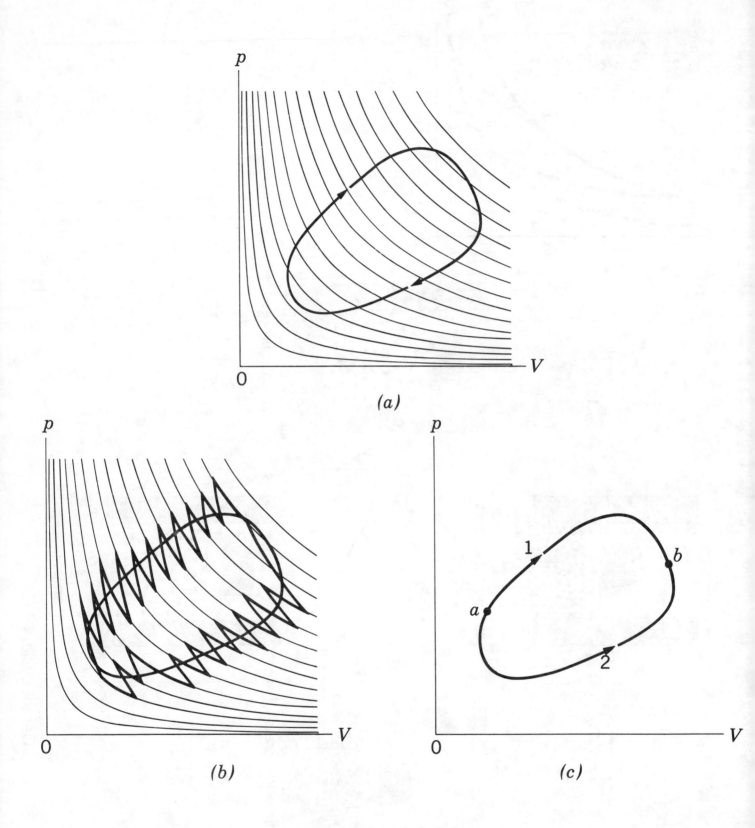

(a)

(b)

(c)

A cyclic process as a series of Carnot cycles.

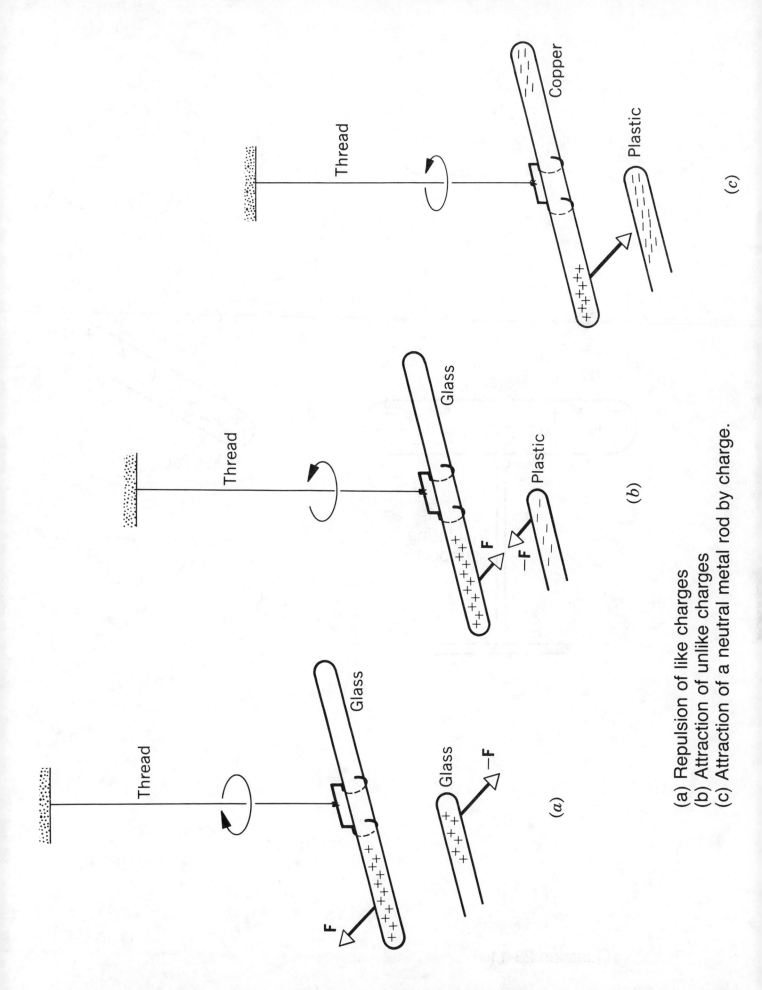

(a) Repulsion of like charges
(b) Attraction of unlike charges
(c) Attraction of a neutral metal rod by charge.

FIG. 23.1 & 23.4

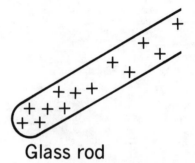

Glass rod

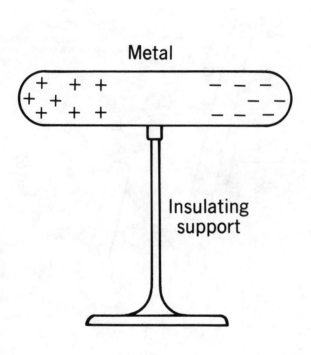

Metal

Insulating
support

Question 23-11.

FIG. 23.8

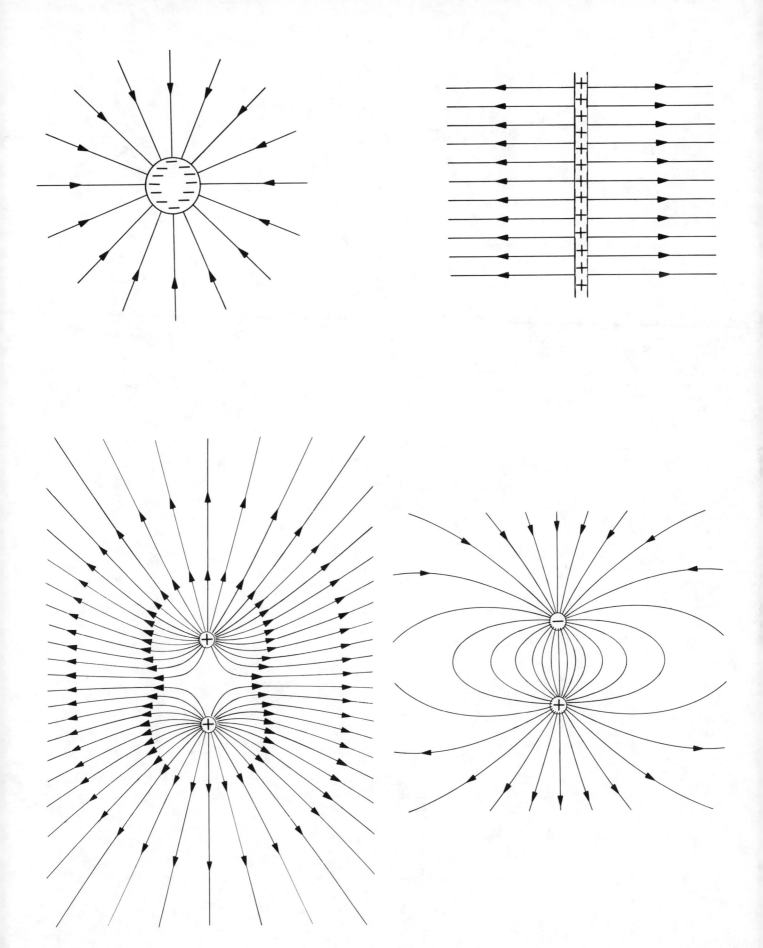

Electric field lines.

FIG. 24.2, 24.3, 24.4 & 24.5   **97**

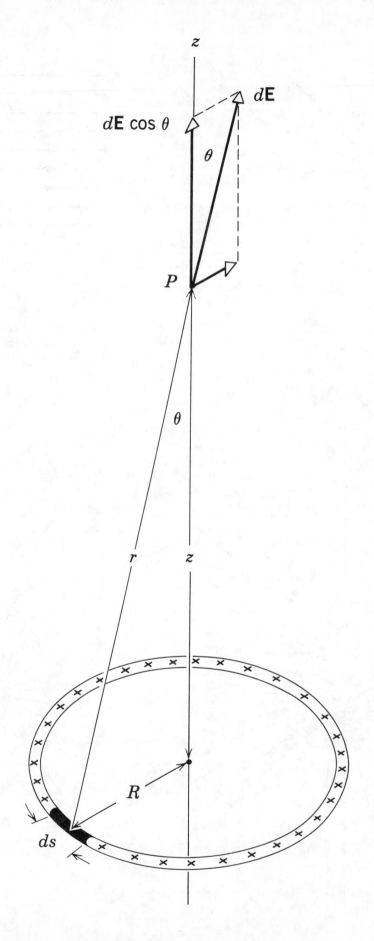

Calculation of the electric field of a charged ring.

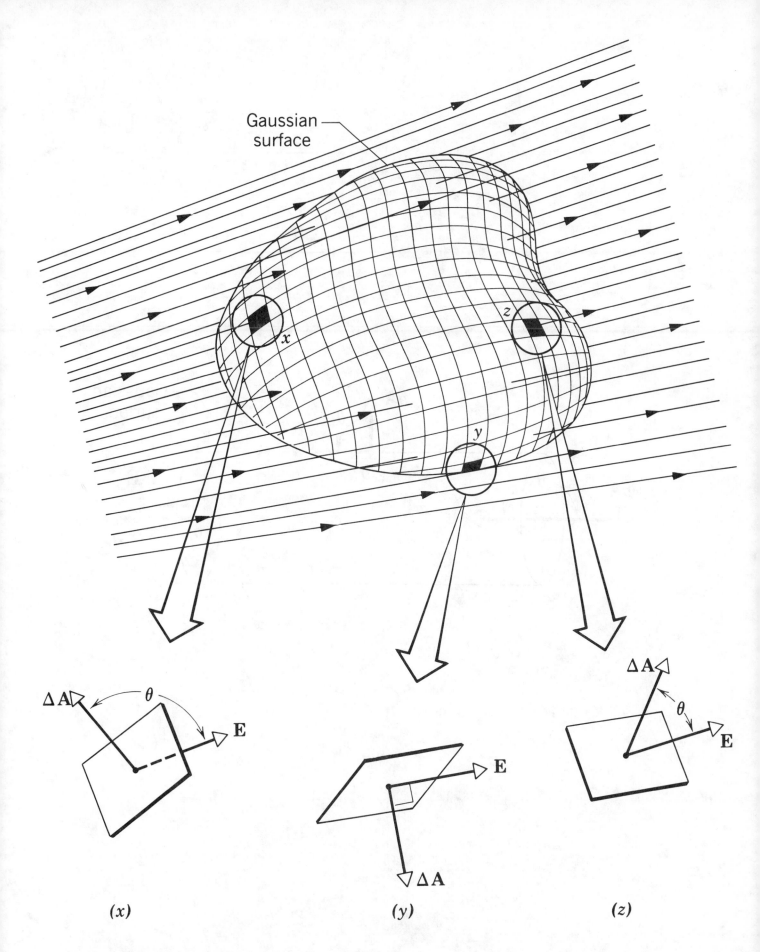

Gaussian surface

x
z
y

ΔA θ E

(x)

E ΔA

(y)

ΔA θ E

(z)

Calculation of electric flux.

FIG. 25.3

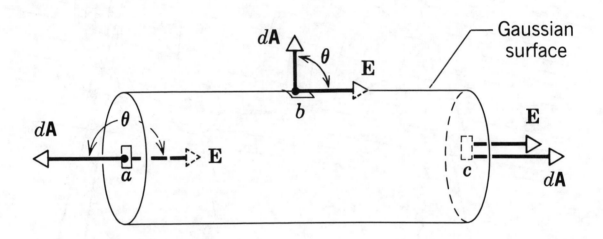

Sample problem 25-1.

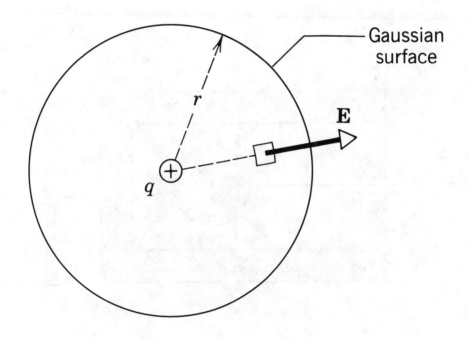

Gauss's law used to calculate the electric field of a point charge.

FIG. 25.7

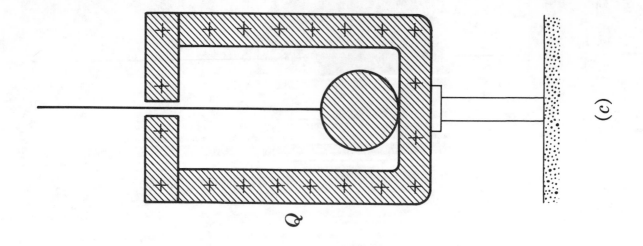

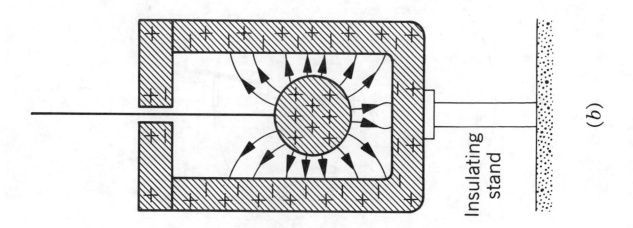

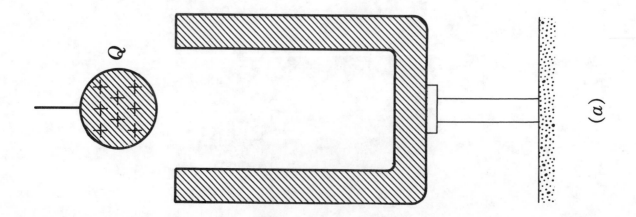

(c)

(b)

Insulating
stand

(a)

Charge on a conductor is on its surface.

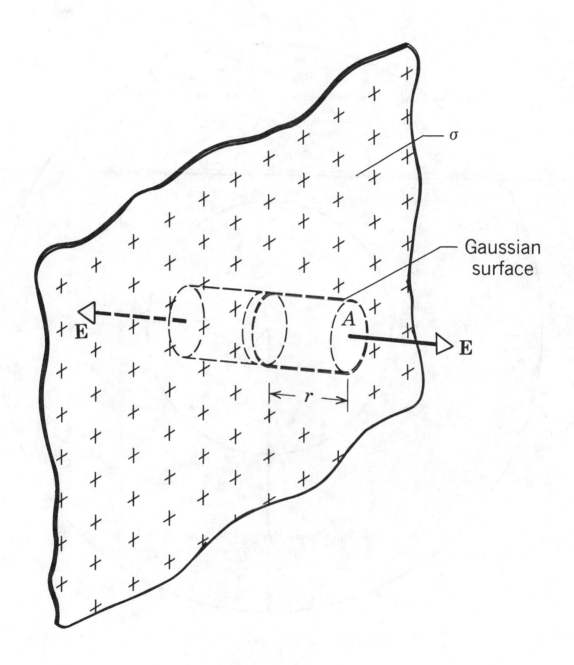

Gaussian surface used to find the electric field of an infinate sheet of charge.

FIG. 25.12 **103**

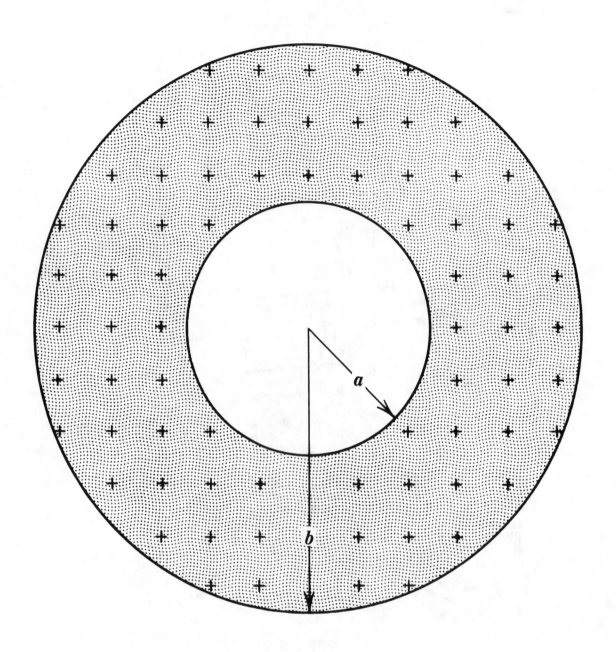

Problem 25-48.

FIG. 25.34

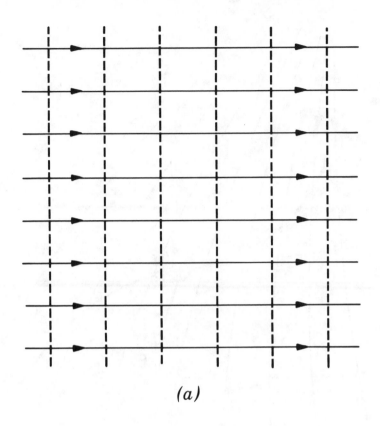

(a)

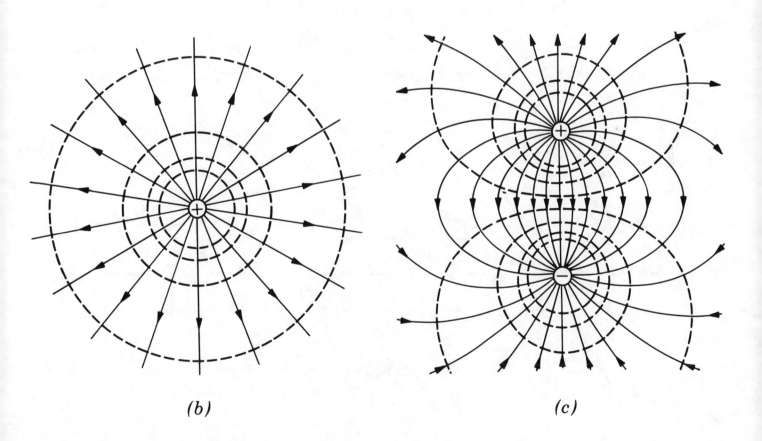

(b)

(c)

Electric field lines and equipotential surfaces.

FIG. 26.3

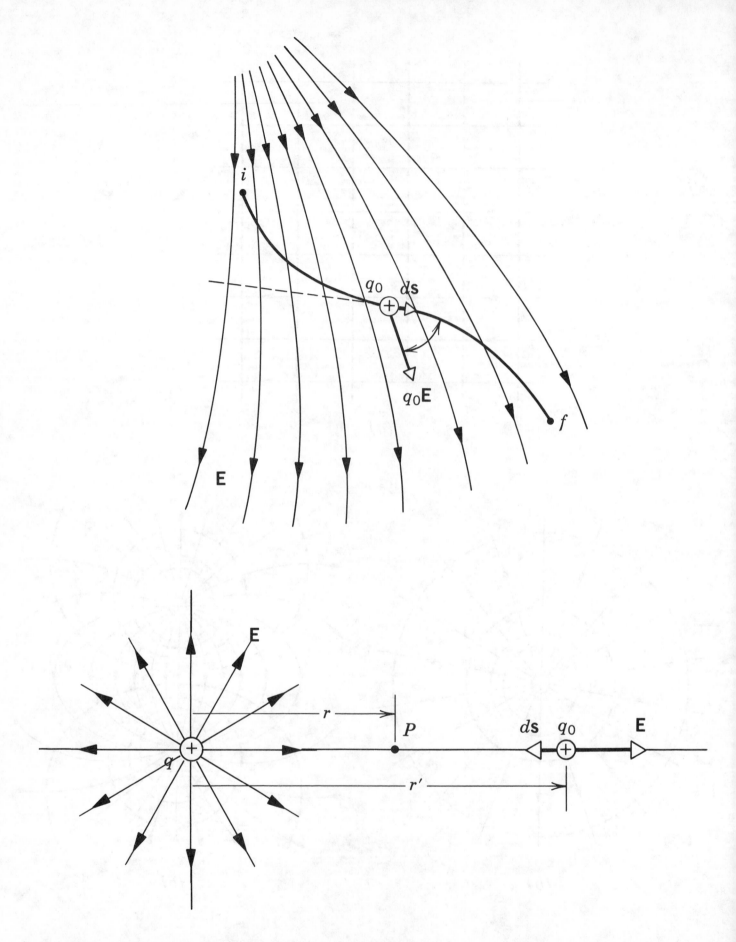

Calculation of the electric potential.

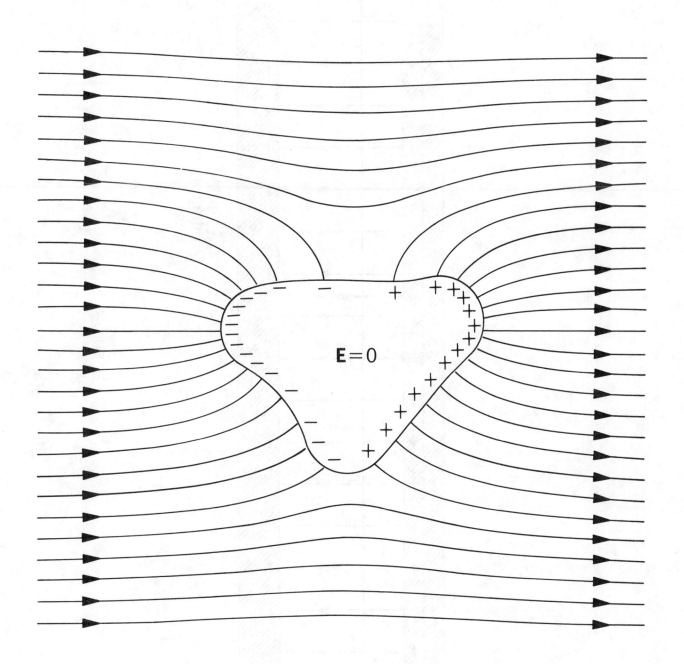

Electric field lines near a conductor in an external field.

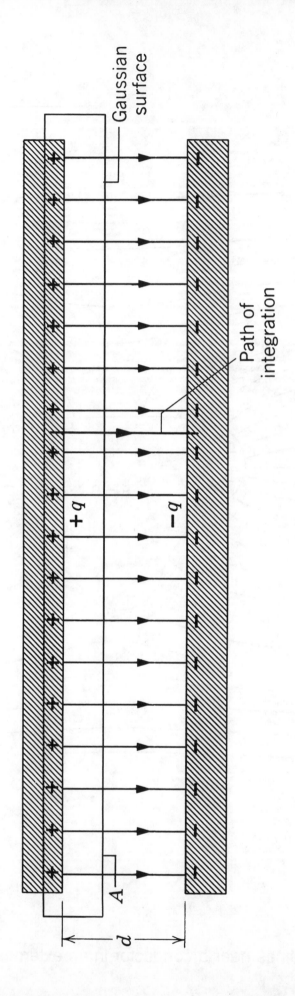

A parallel-plate capacitor.

FIG. 27.4

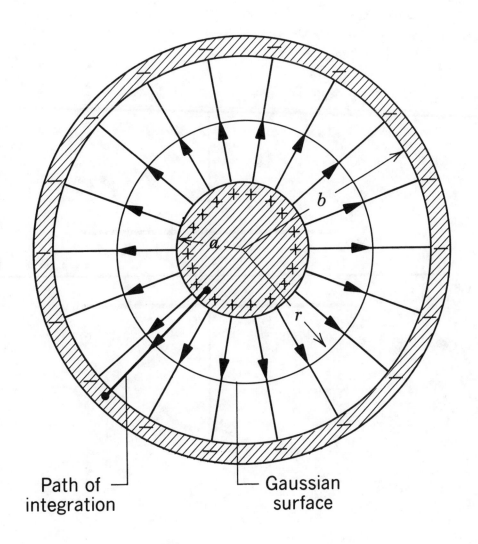

Path of integration

Gaussian surface

A cylindrical capacitor.

FIG. 27.5

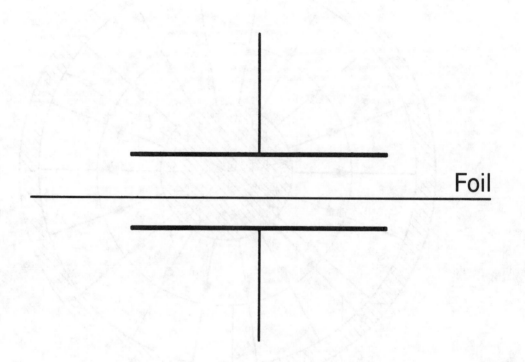

Foil

Question 27-3.

FIG. 27.17

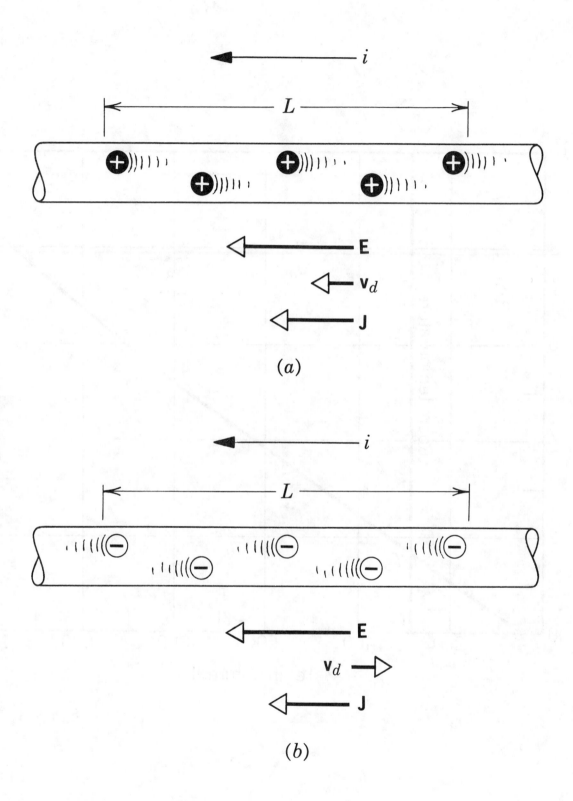

(a)

(b)

Positive and negative charge current, both to the left.

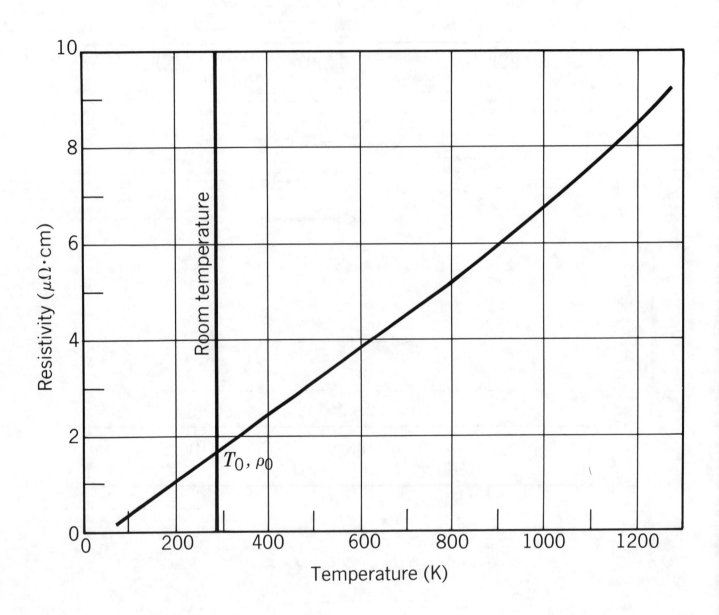

Resistivity of copper as a function of temperature.

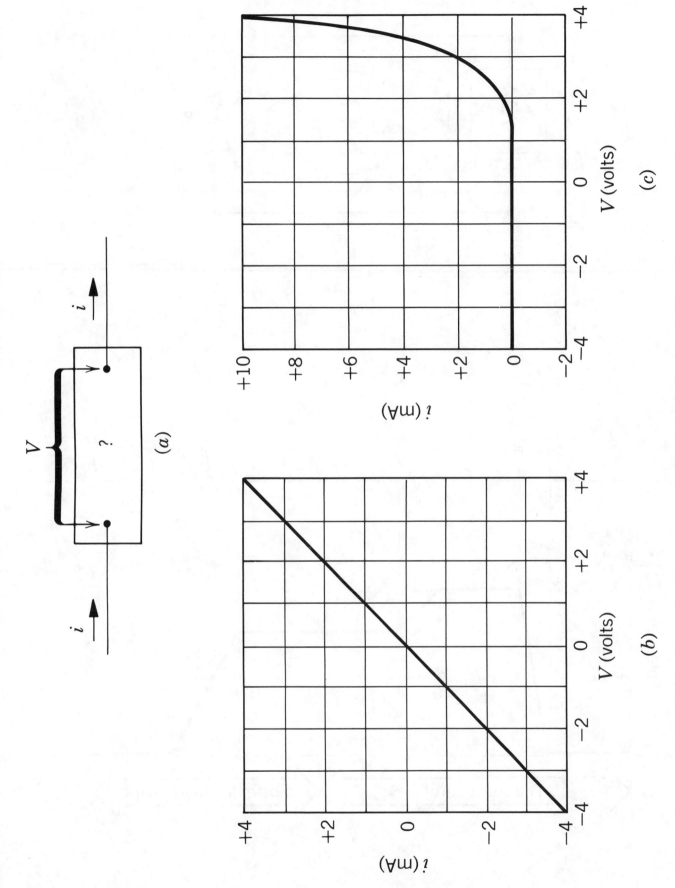

Current as a function of potential difference for an ohmic and a non-ohmic substance.

FIG. 28.11   **113**

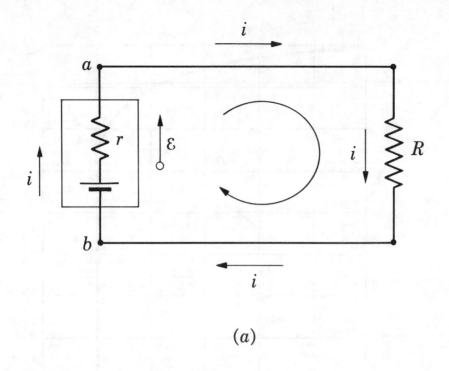

(a)

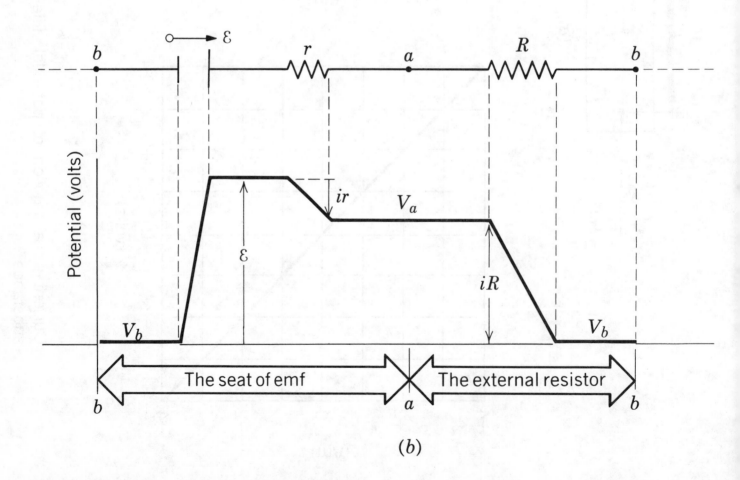

(b)

Potential in a single-loop circuit.

FIG. 29.5

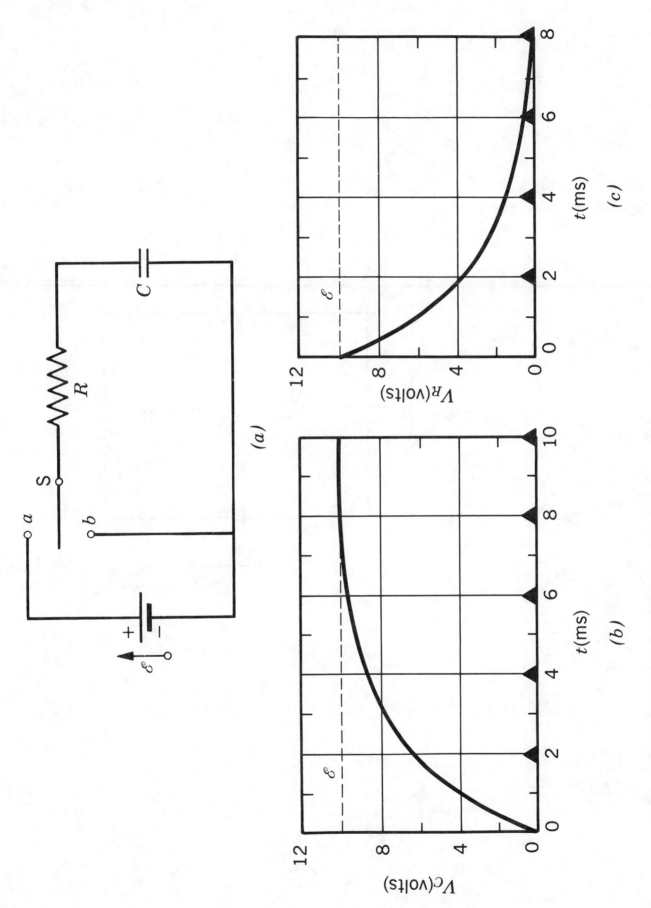

Charging and discharging a capacitor.

FIG. 29.15 & 29.16

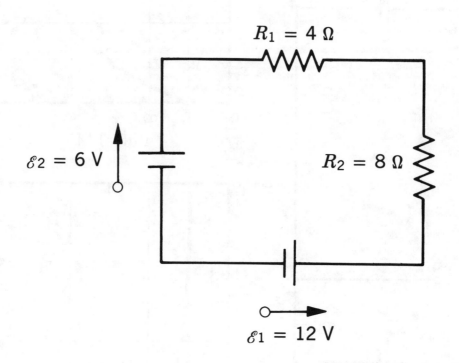

$R_1 = 4\ \Omega$

$\mathcal{E}_2 = 6$ V

$R_2 = 8\ \Omega$

$\mathcal{E}_1 = 12$ V

Problem 29-8.

FIG. 29.19   **116**

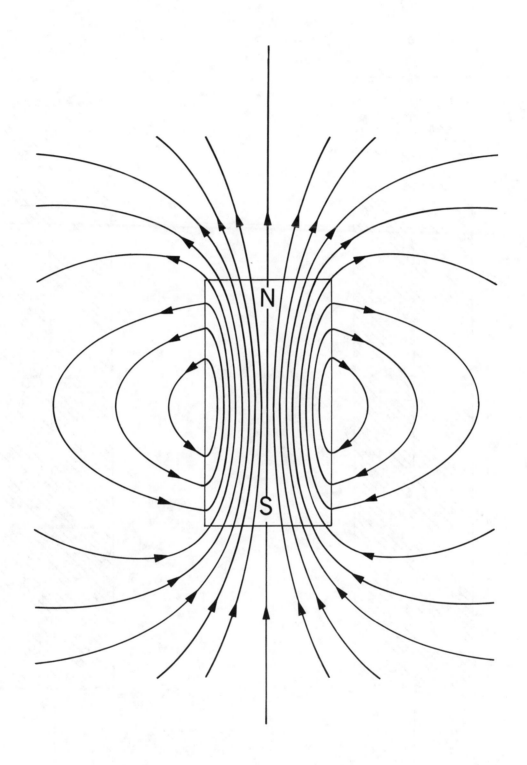

Magnetic field lines of a bar magnet.

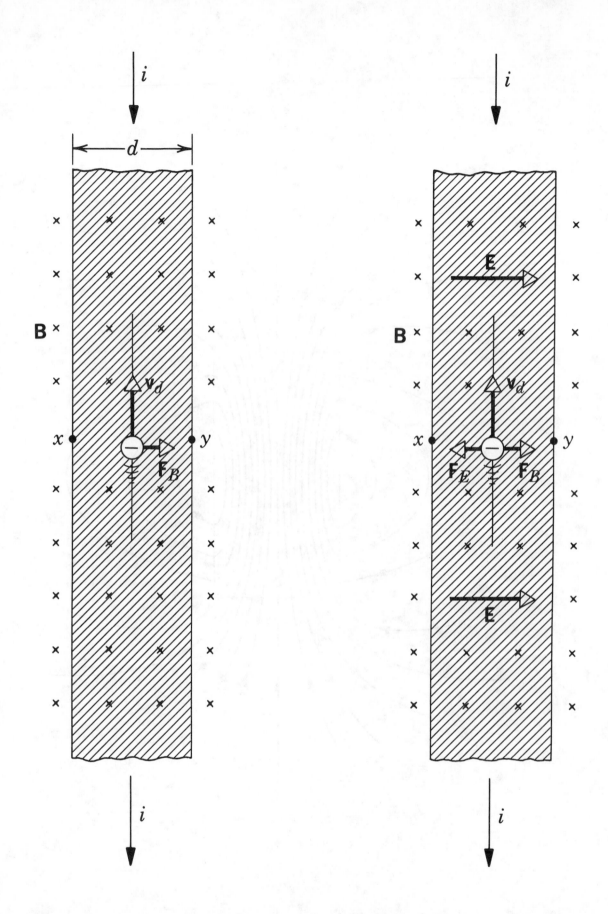

The Hall effect.

FIG. 30.10

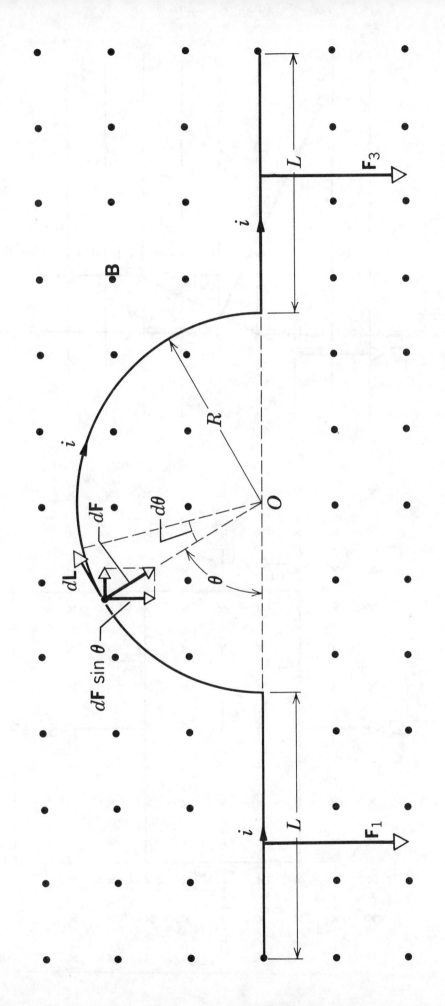

Sample problem 30-7.

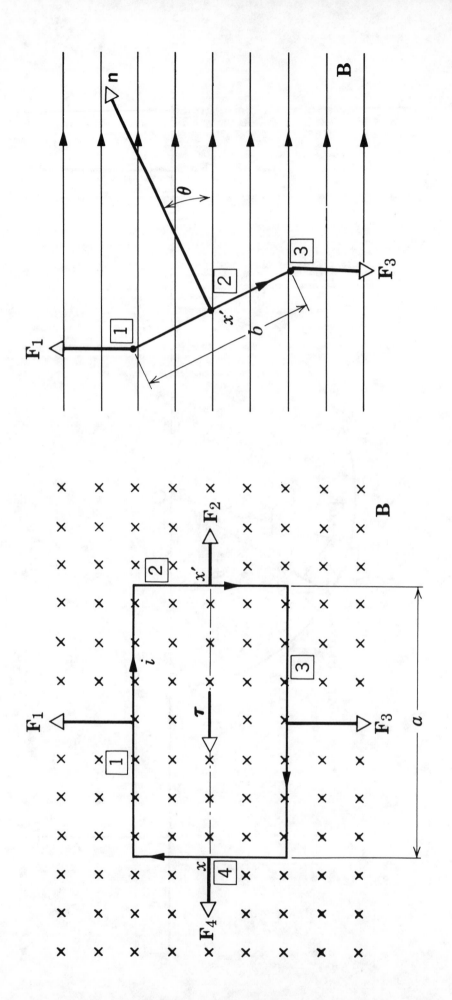

Calculation of magnetic torque on a current loop.

FIG. 30.26

(a) The electric field of a charge element (b) The magnetic field of a current element.

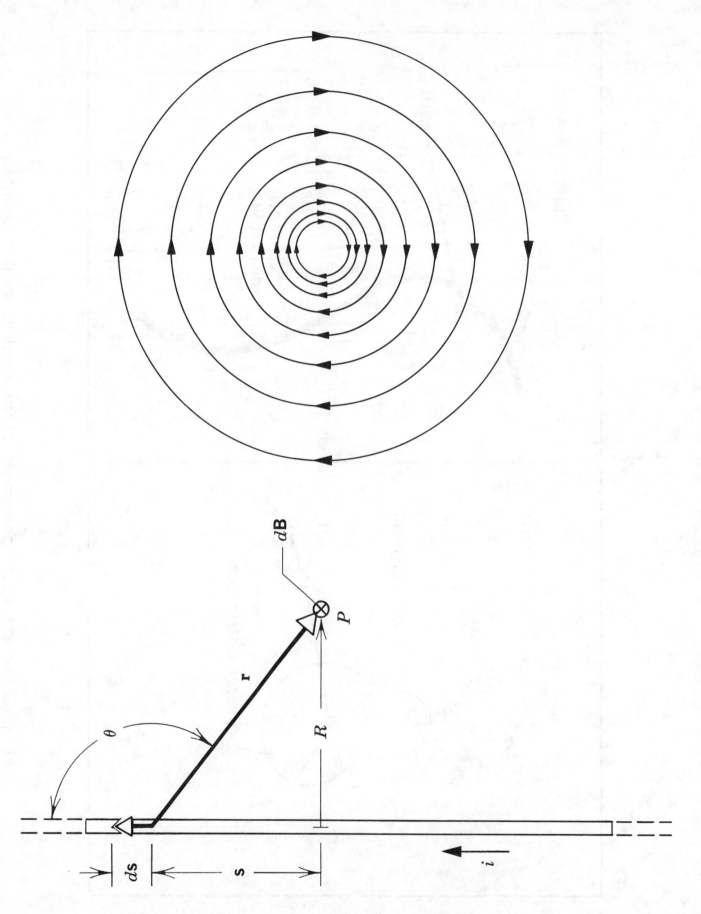

The magnetic field of a long straight wire.

FIG. 31.2 & 31.3 **122**

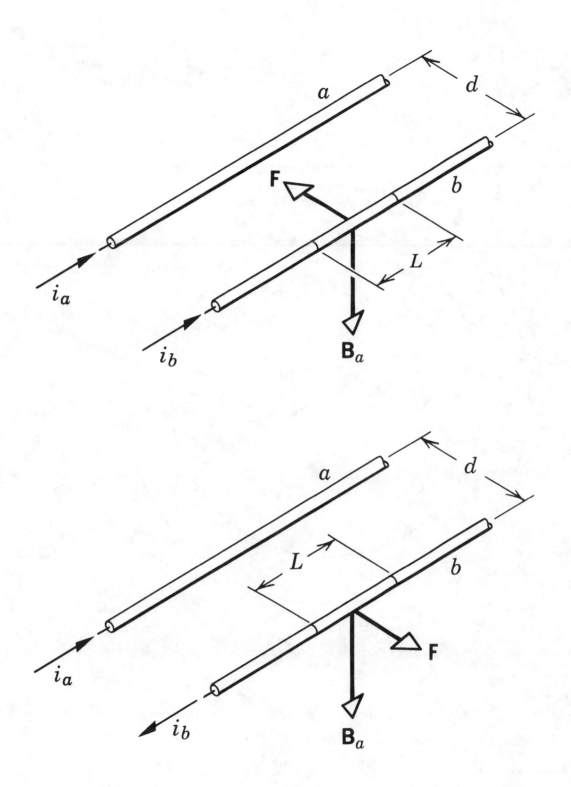

Force of one long straight wire on another.

FIG. 31.5   **123**

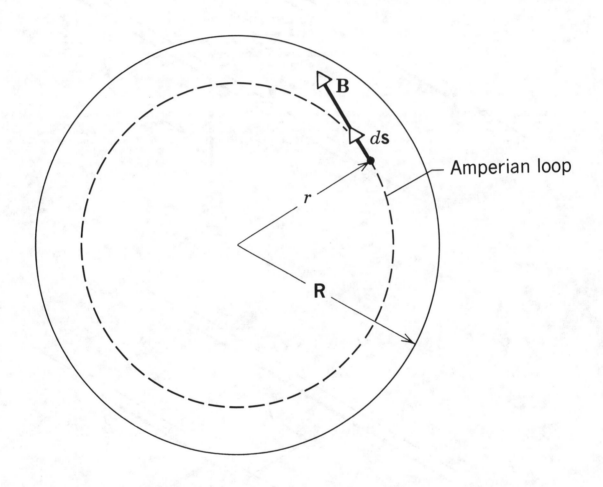

Amperian loop used to find the magnetic field inside a long straight wire.

FIG. 31.10 **124**

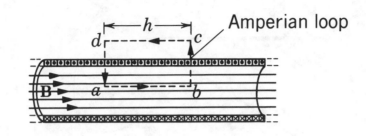

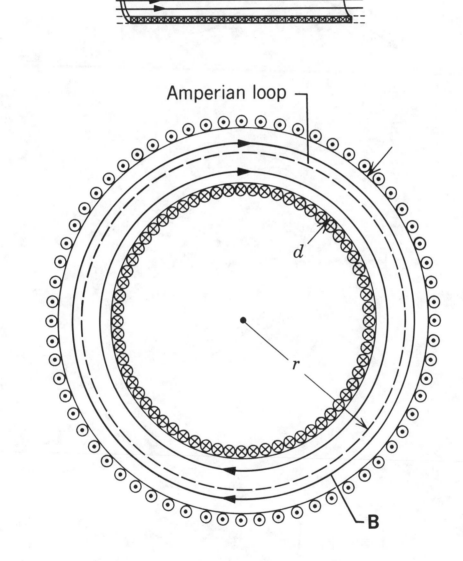

Amperian loops used to find the magnetic fields inside a solenoid and inside a torroid.

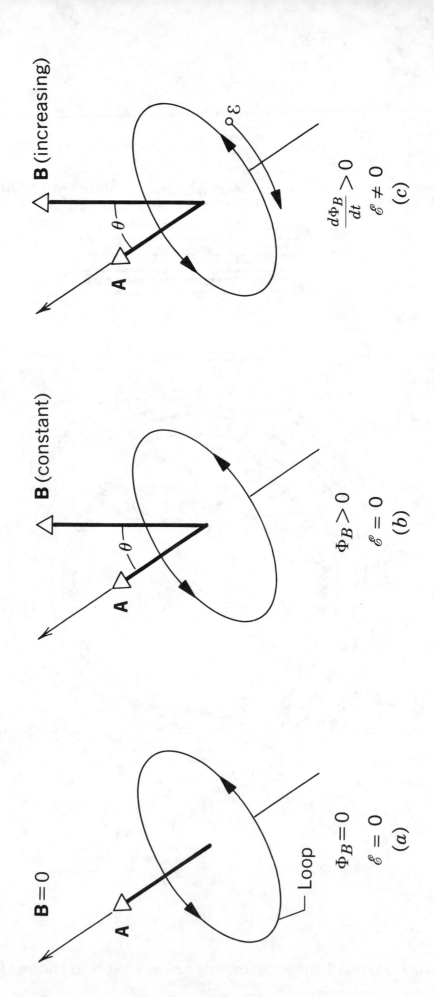

Relation between magnetic flux and induced emf.

FIG. 32.3

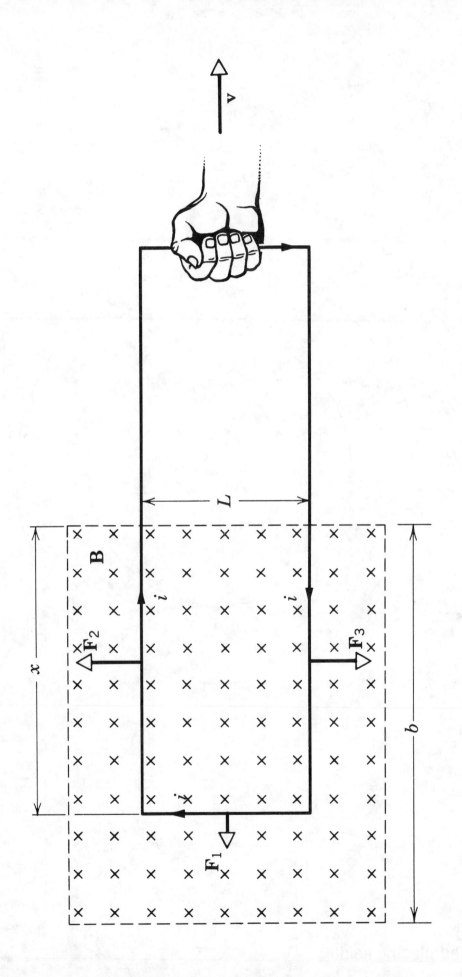

Emf induced by motion.

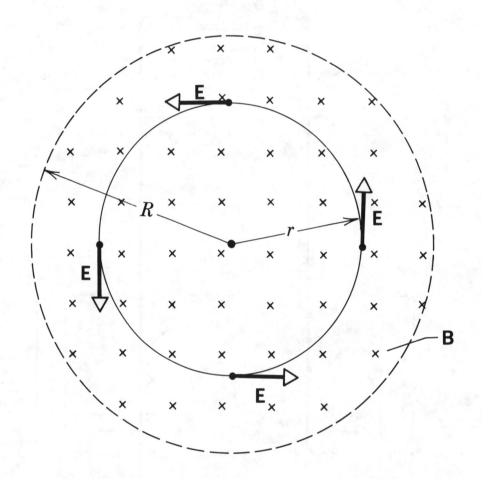

An induced electric field.

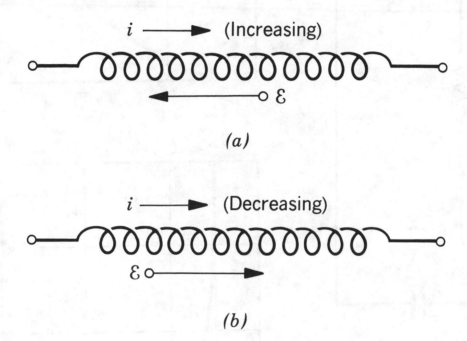

$i \longrightarrow$ (Increasing)

$\mathcal{E}$

(a)

$i \longrightarrow$ (Decreasing)

$\mathcal{E}$

(b)

Emf induced in an inductor.

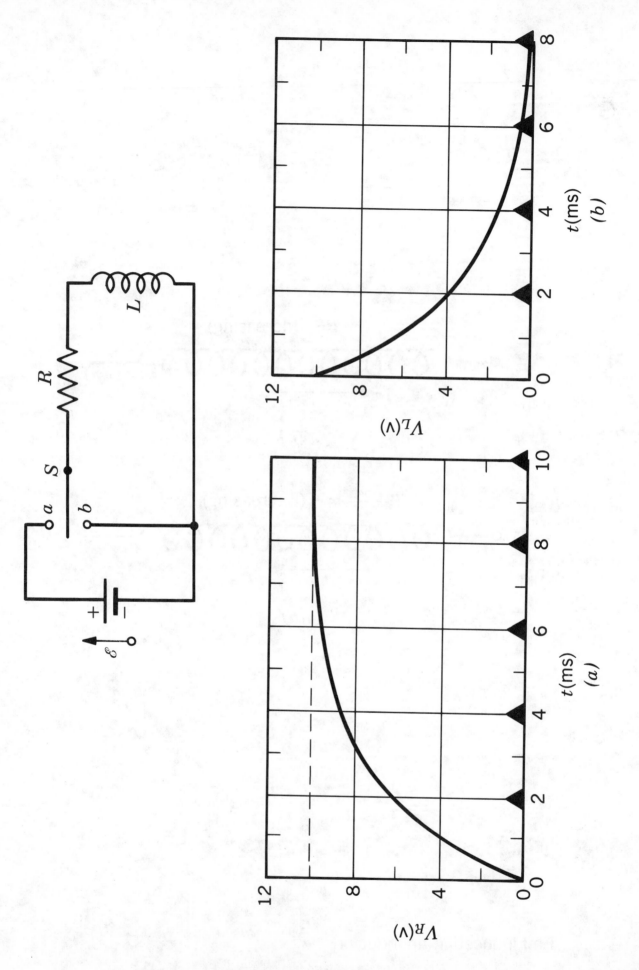

An LR circuit.

FIG. 33.5 & 33.7

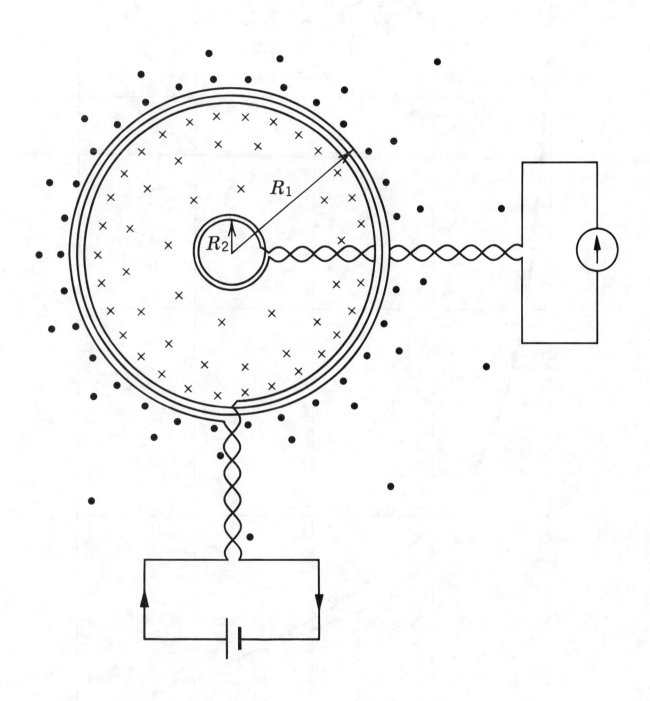

Mutual inductance.

FIG. 33.10

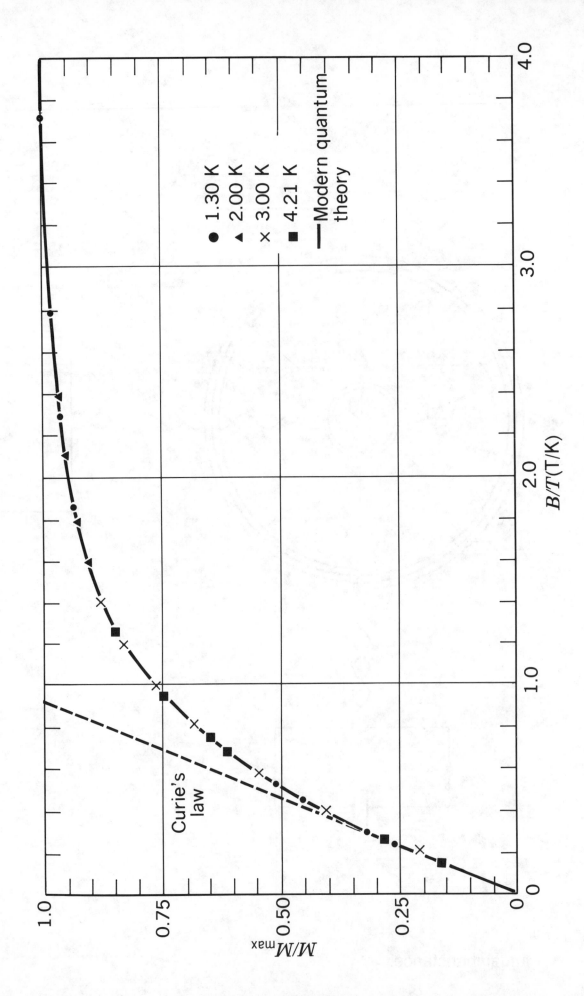

Magnetization for a paramagnetic salt (From W.E. Henry).

FIG. 34.11

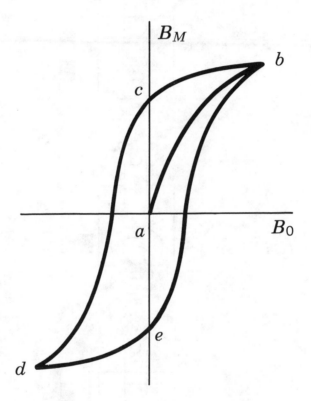

Hysteresis curve for a ferromagnet.

FIG. 34.15

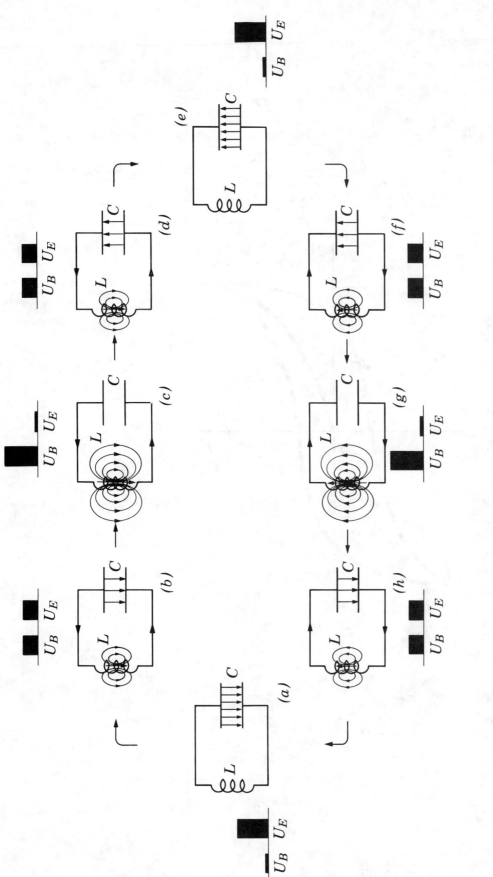
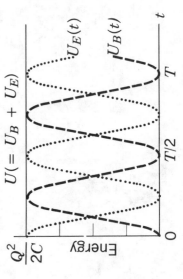

Energy in an oscillating LC circuit.

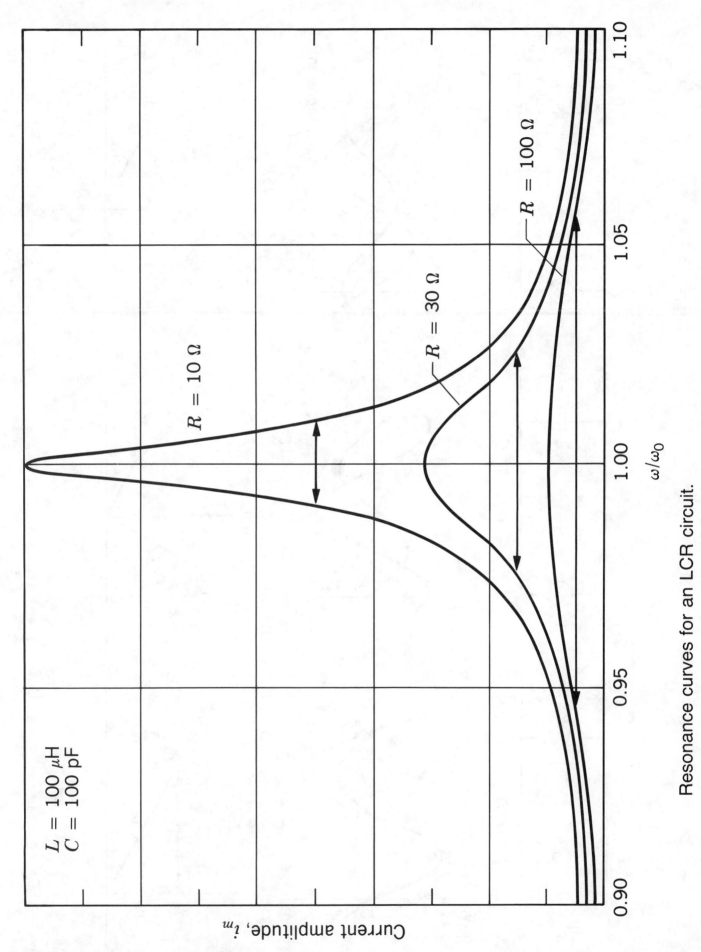

Resonance curves for an LCR circuit.

FIG. 35.6

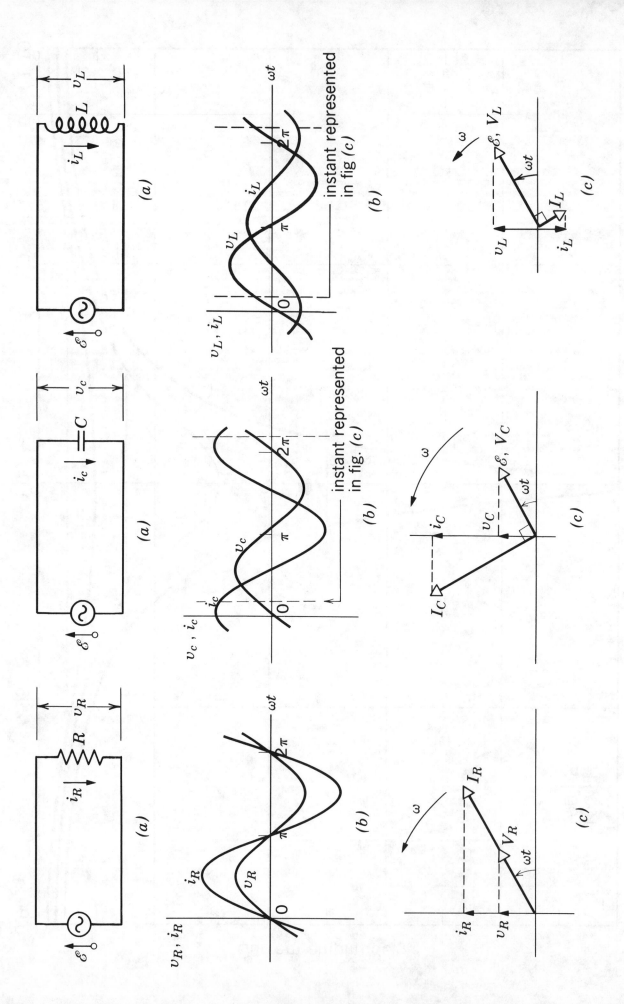

Current-voltage phase relations for resistance, capacitance and inductance.

FIG. 36.3, 36.4 & 36.5 **136**

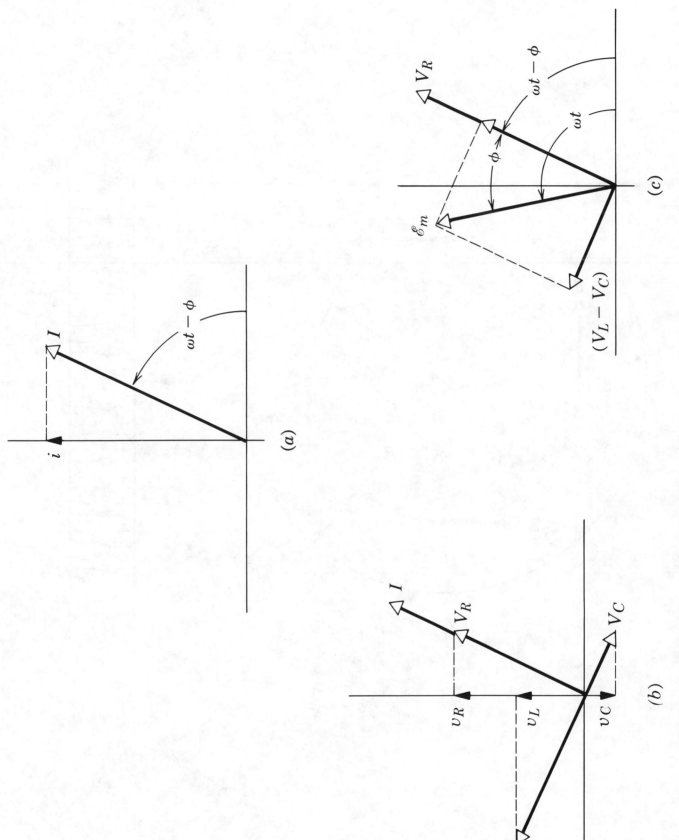

Phase relations for an LCR series circuit.

FIG. 36.6

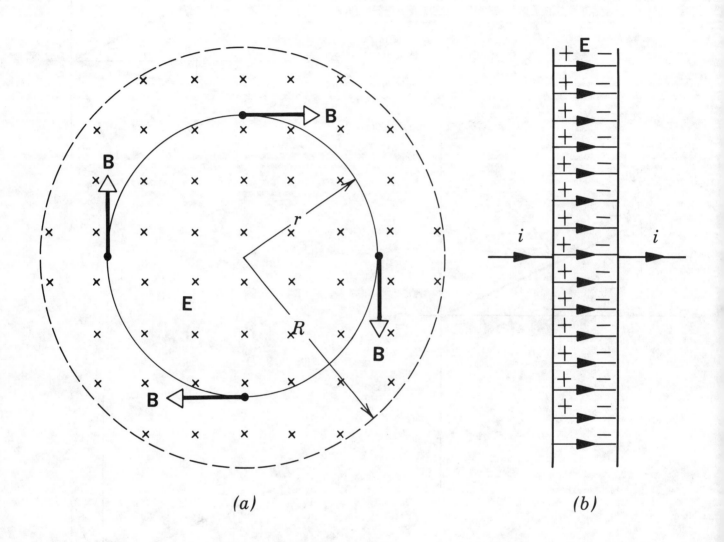

(a)

(b)

Magnetic field in a charging parallel-plate capacitor.

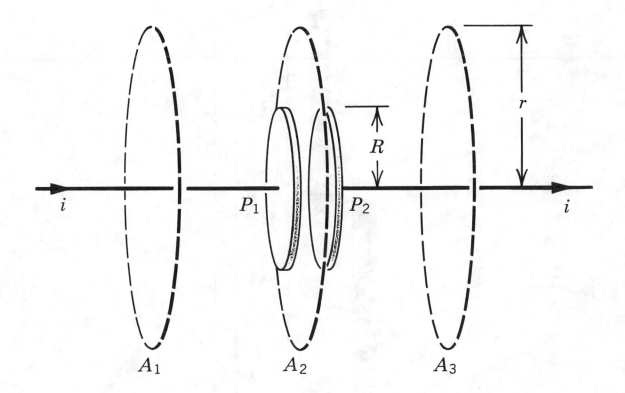

Problem 37-9.

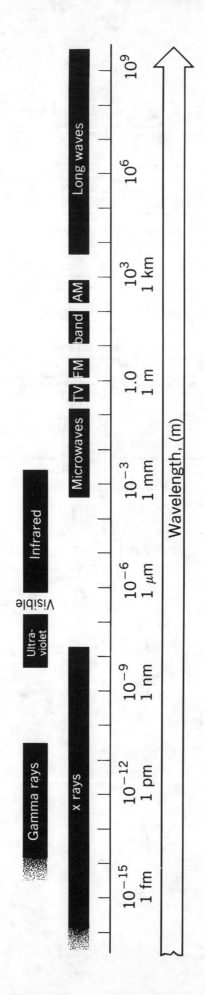

The electromagnetic spectrum.

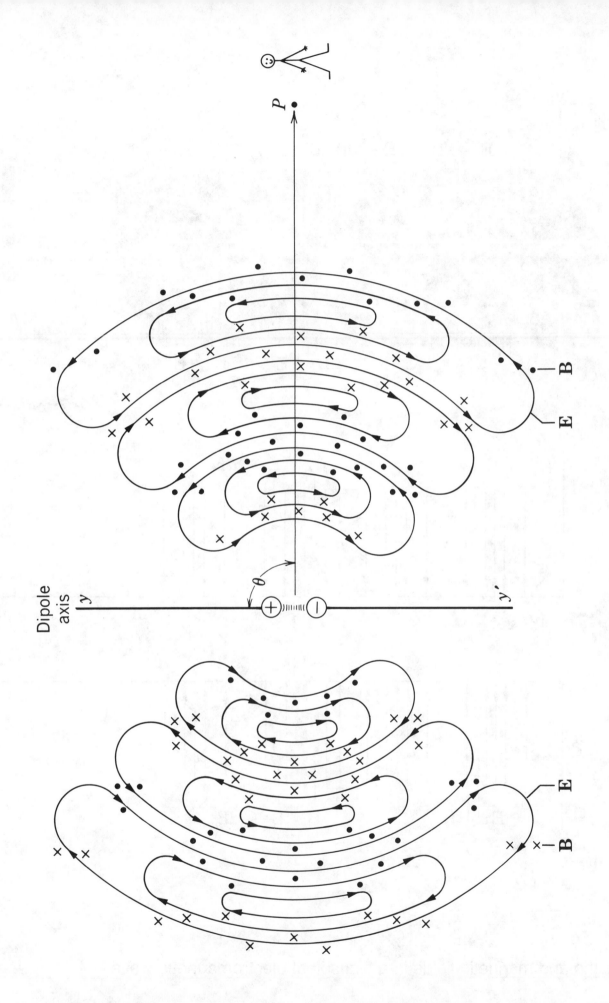

Dipole axis

y

y'

θ

P

B

E

E

B

Electric and magnetic fields around an oscillating electric dipole.

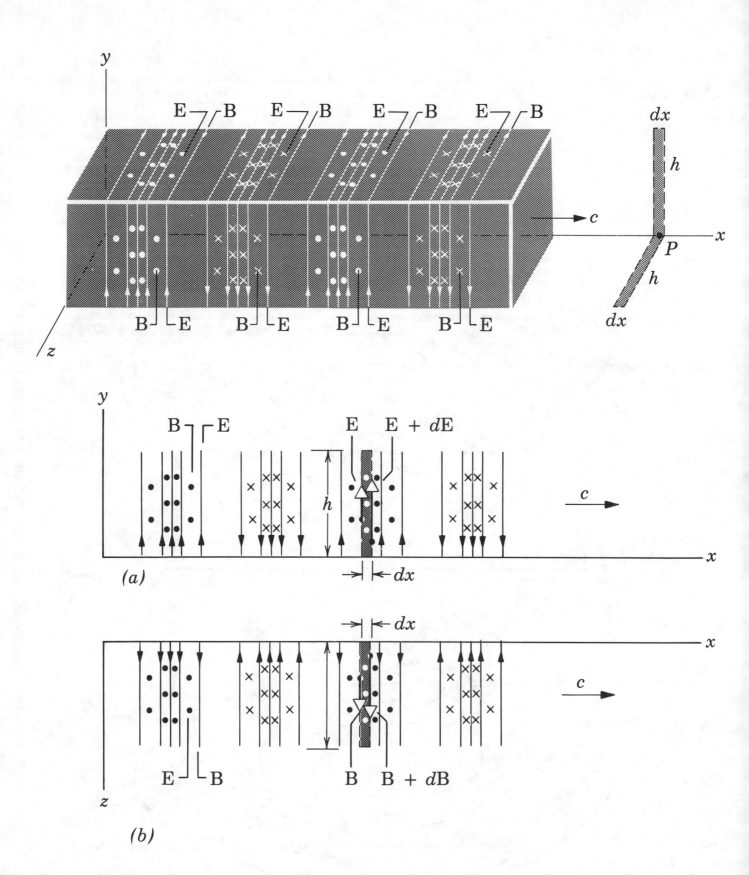

Electric and magnetic fields in a sinusoidal electromagnetic wave.

FIG. 38.6 & 38.7

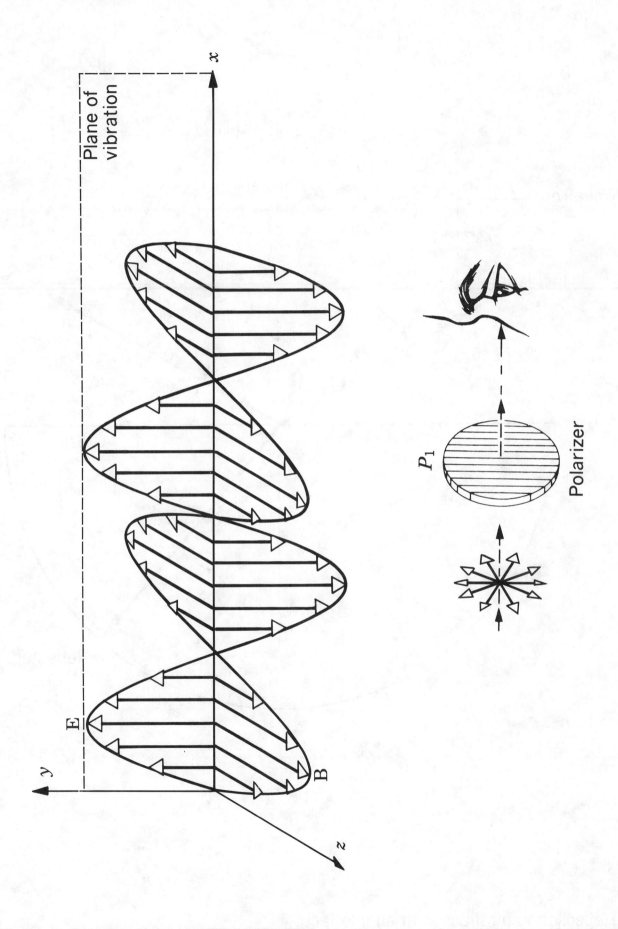

Plane of vibration

x

y

z

E

B

P_1

Polarizer

FIG. 38.10 & 38.13   **143**

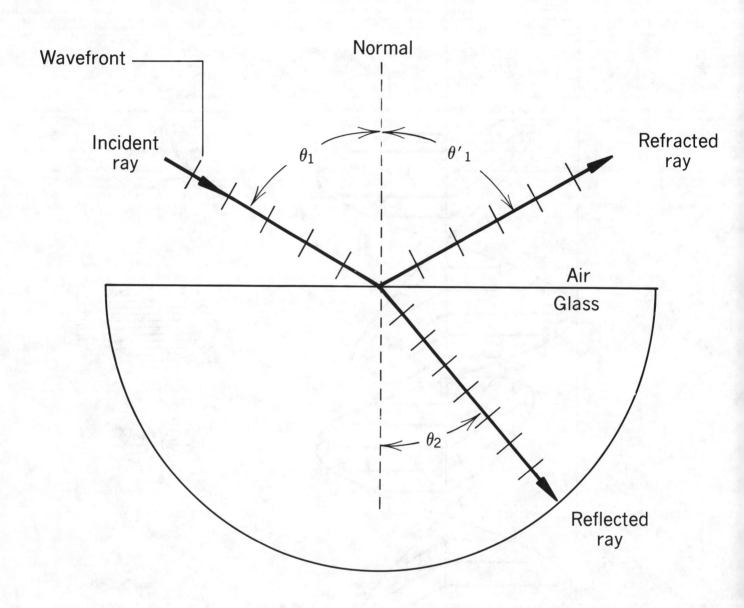

Reflection and refraction at an interface.

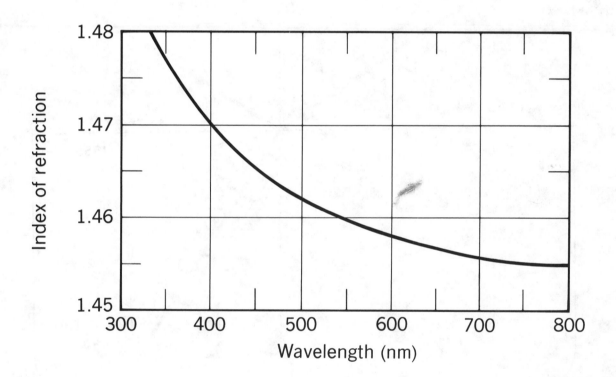

Index of refraction as a function of wavelength for fused quartz.

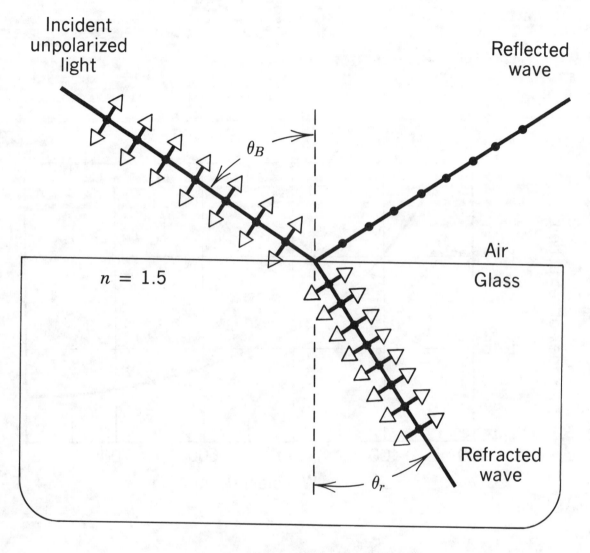

Incident
unpolarized
light

Reflected
wave

θ_B

$n = 1.5$

Air

Glass

θ_r

Refracted
wave

● Perpendicular component

◁——▷ Parallel component

Reflection at the Brewster angle.

FIG. 39.9   **146**

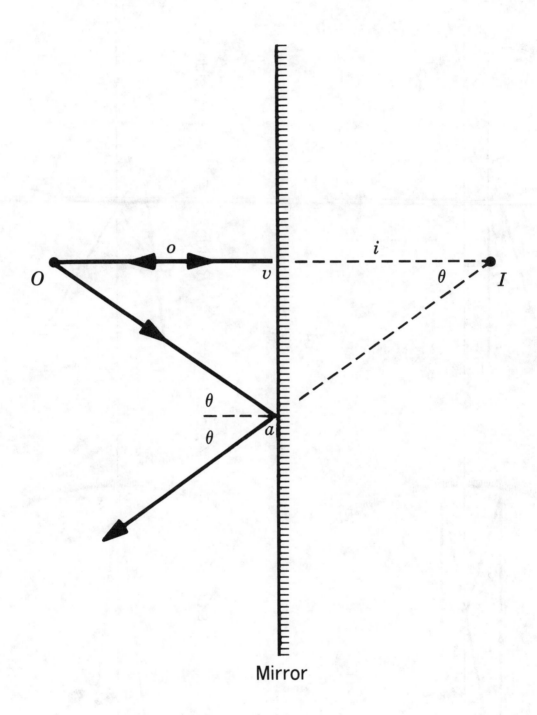

Reflection at a plane mirror.

FIG. 39.11

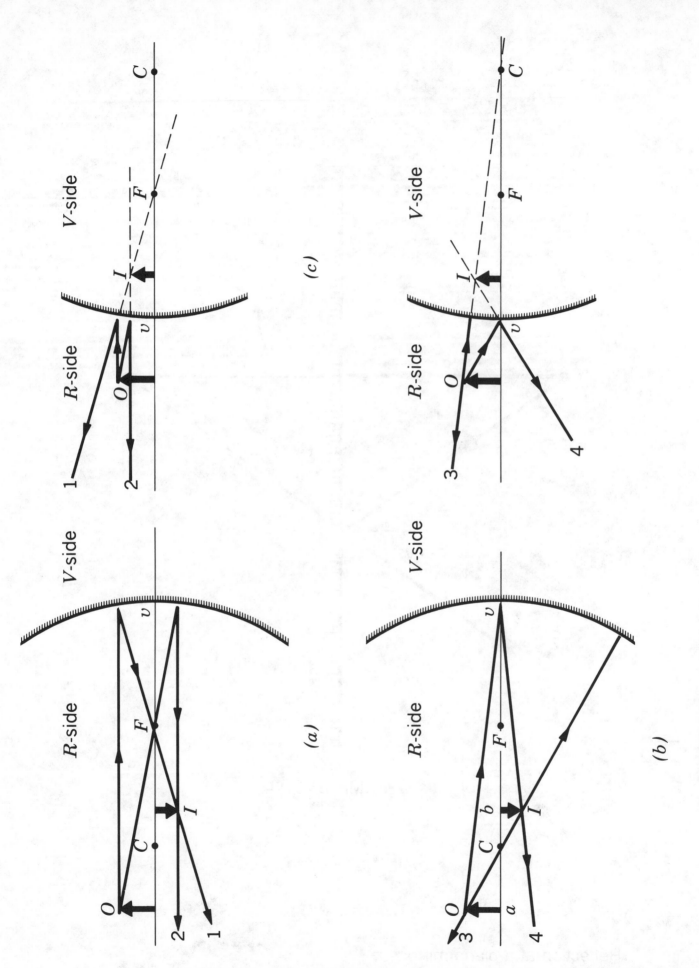

Reflection at a spherical mirror.

FIG. 39.18

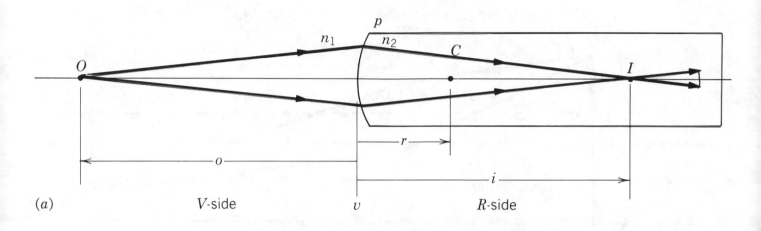

(a)

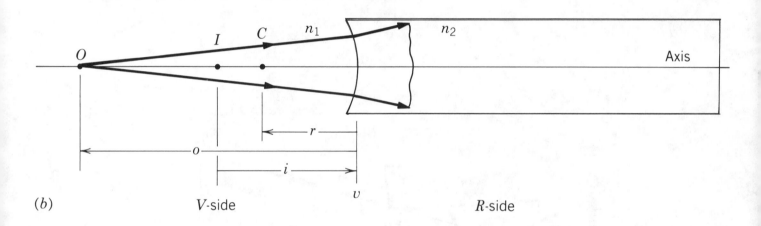

(b)

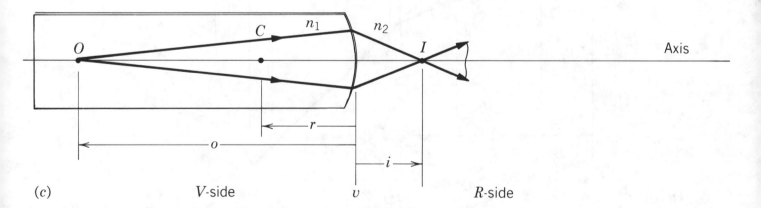

(c)

Refraction by a spherical surface.

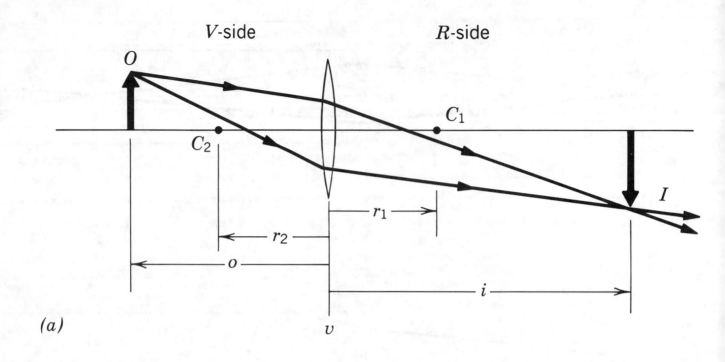

(a)

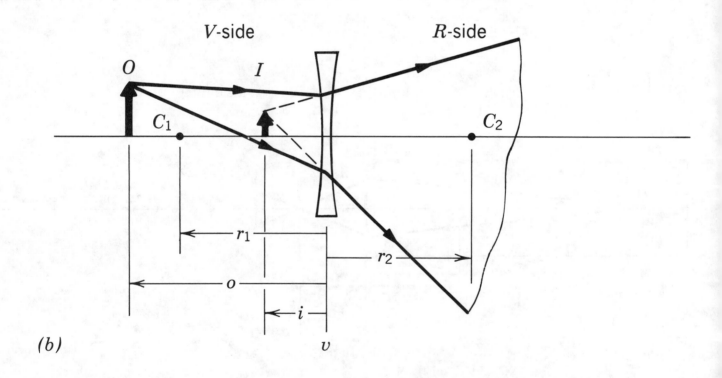

(b)

Ray diagrams for thin lenses.

FIG. 39.21

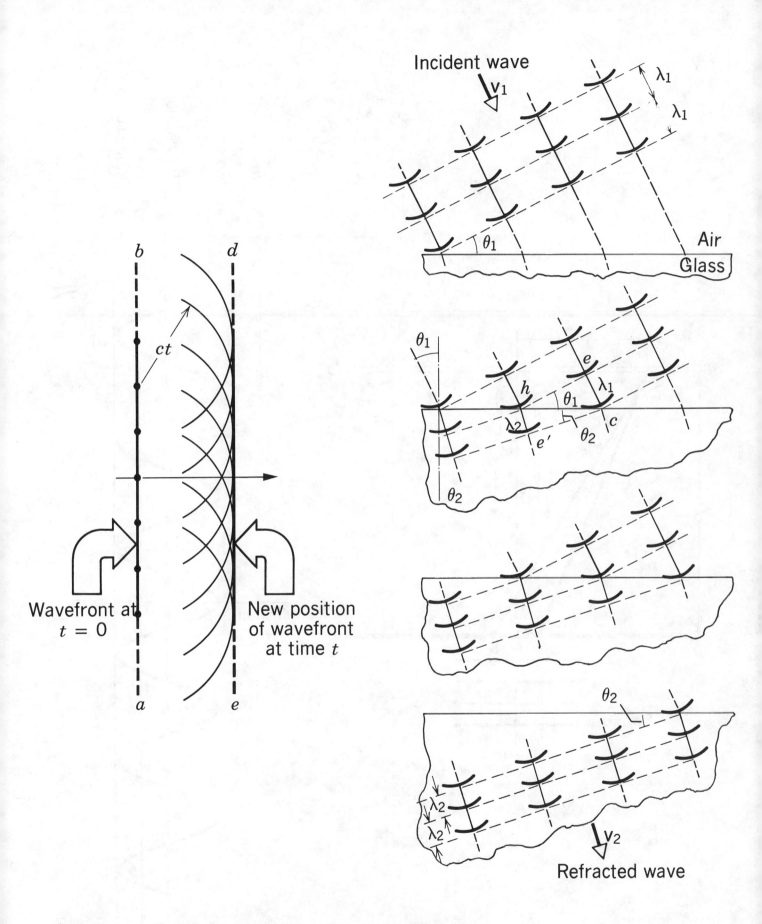

Incident wave

v_1

λ_1

λ_1

θ_1

Air

Glass

b

d

ct

Wavefront at
$t = 0$

New position
of wavefront
at time t

a

e

θ_1

e

h

θ_1

λ_1

λ_2

e'

θ_2

c

θ_2

θ_2

λ_2

λ_2

v_2

Refracted wave

Construction of wavefronts from Huygens' wavelets.

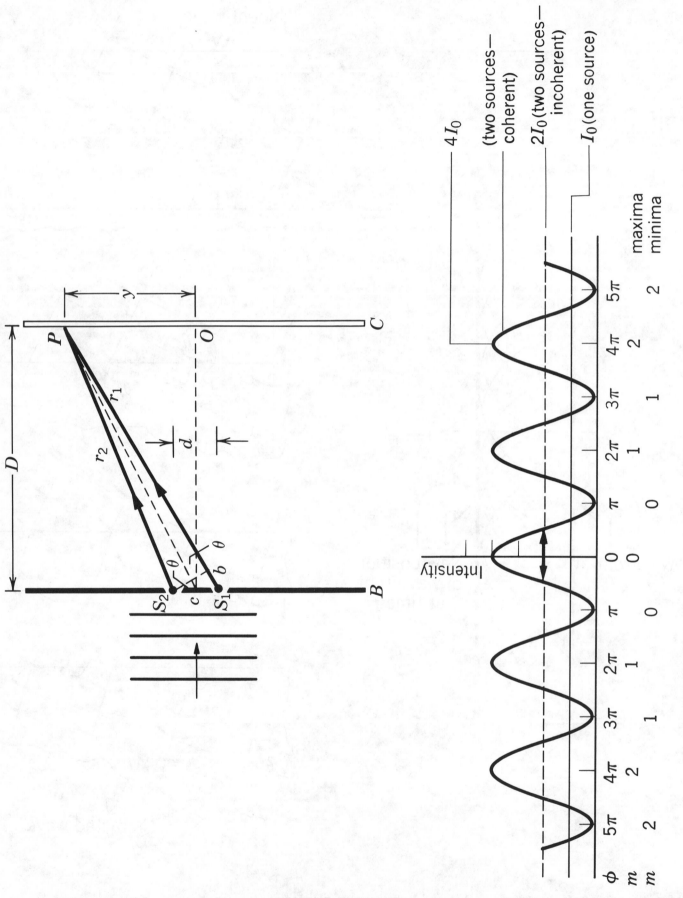

Double-slit interference.

FIG. 40.8 & 40.10

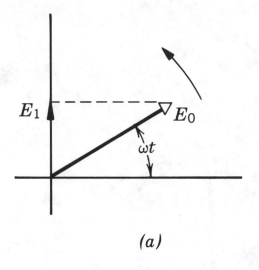

(a)

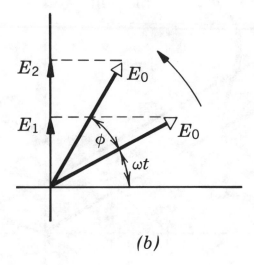

(b)

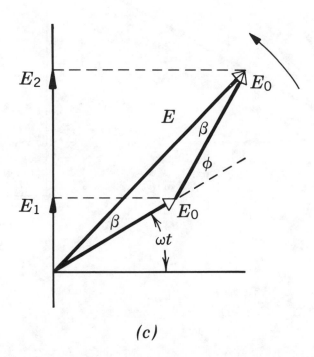

(c)

Phasors for double-slit interference.

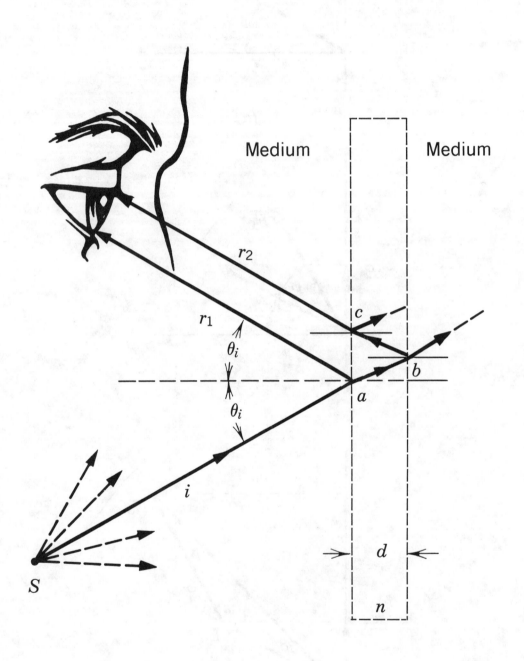

Medium

Medium

Thin-film interference.

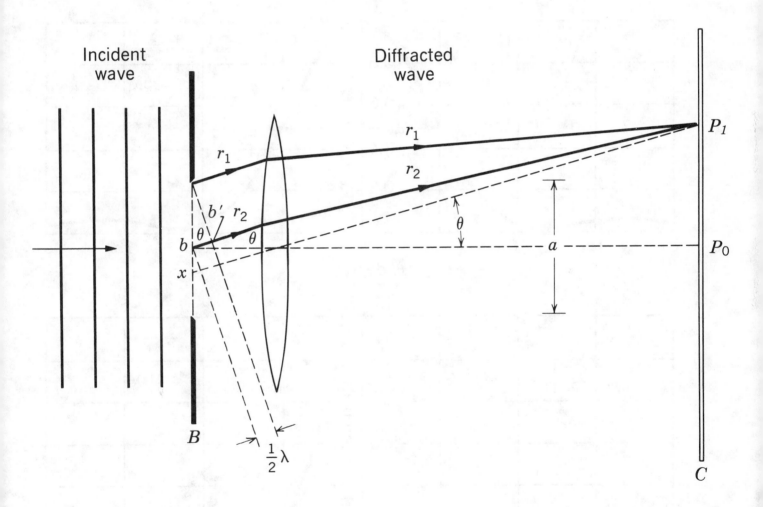

Single-slit diffraction.

FIG. 41.4 & 41.5

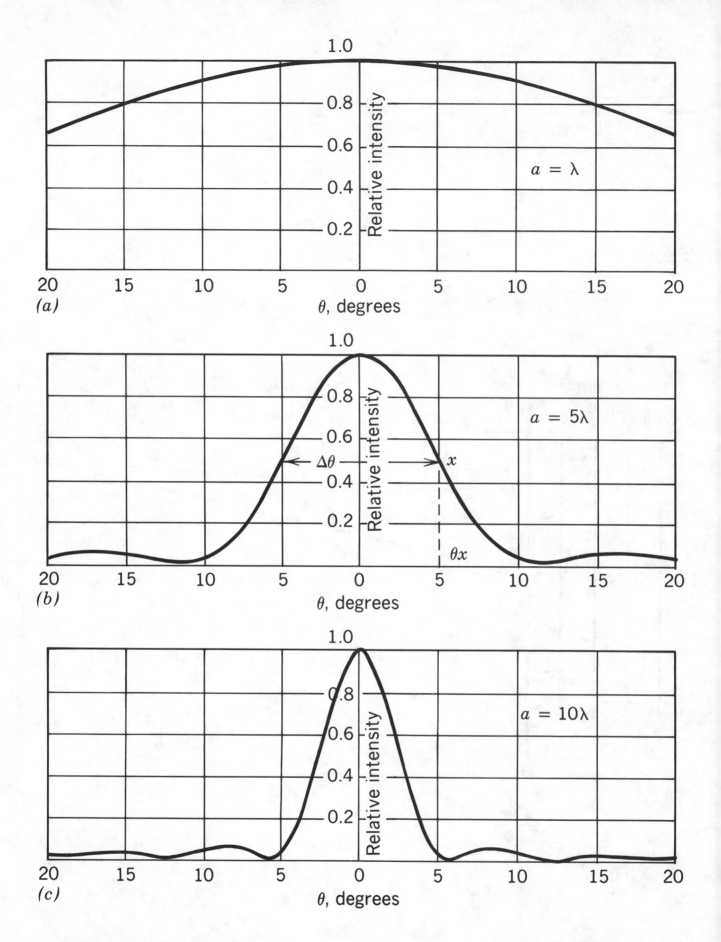

Single-slit diffraction pattern.

FIG. 41.11

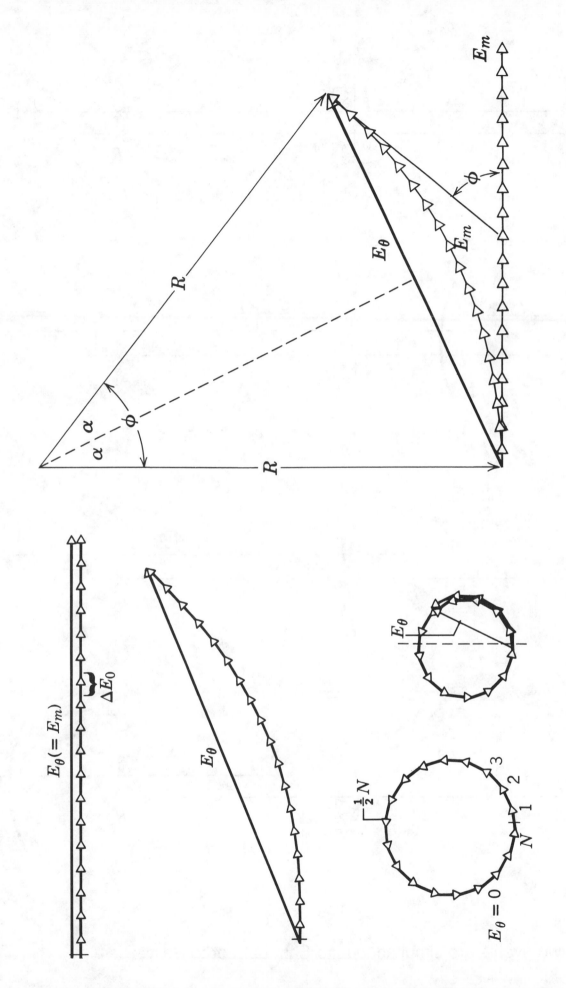

Addition of phasors corresponding to wavelets from a single slit.

FIG. 41.10 & 41.12

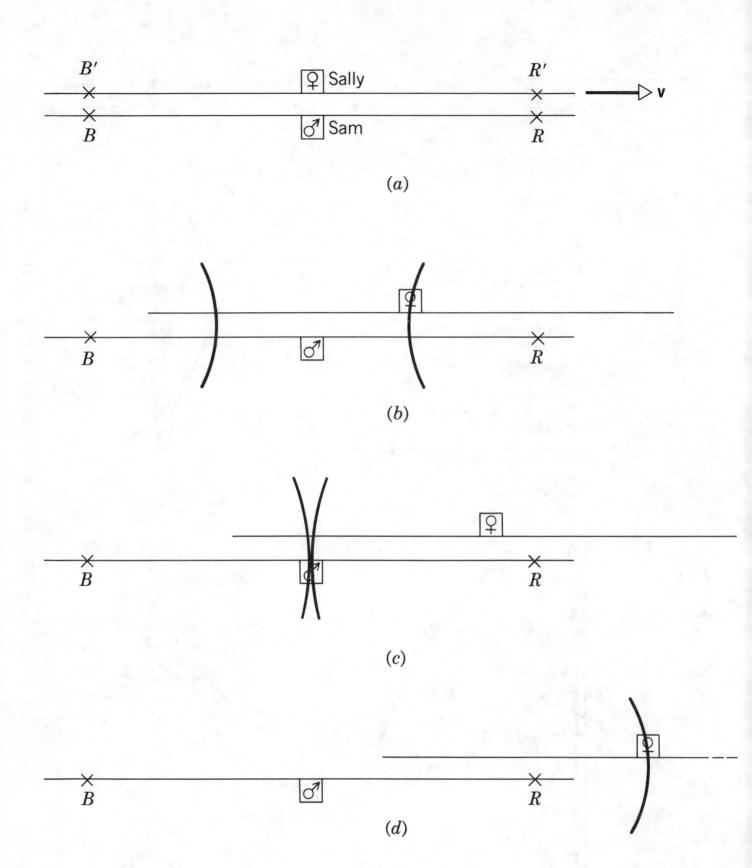

The two events are simultaneous to Sam but not to Sally.

FIG. 42.5

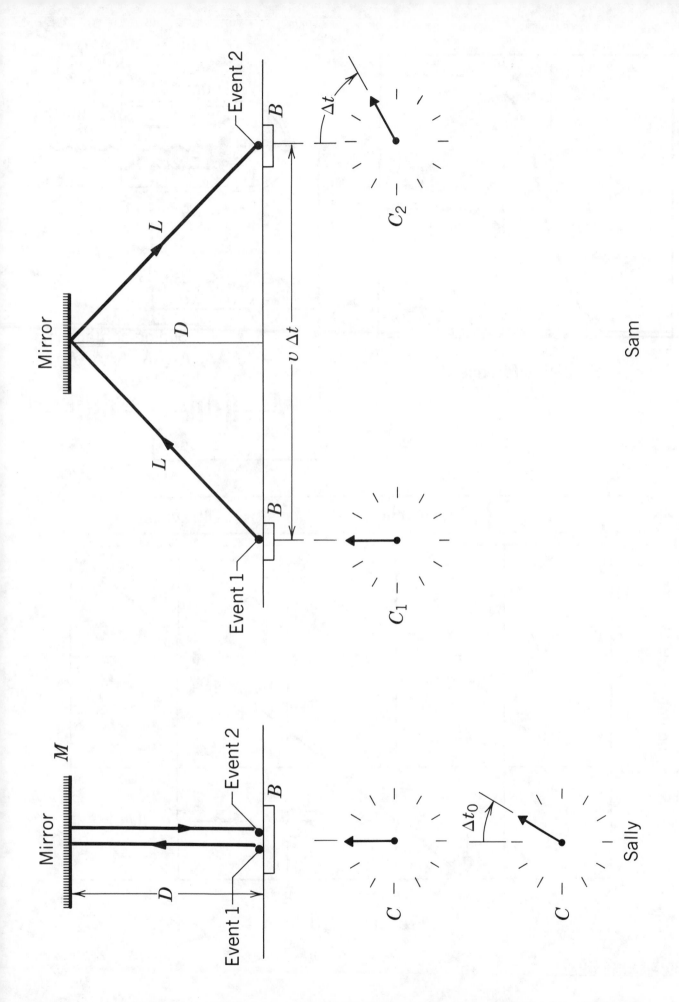

The time interval between two events, measured with clocks in different reference frames.

FIG. 42.6

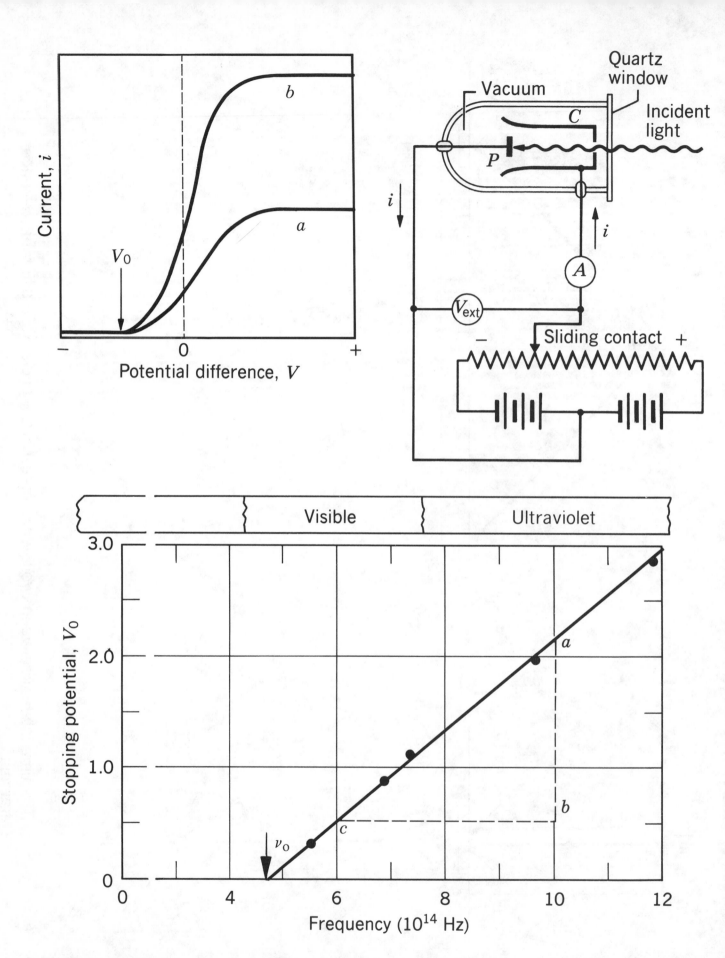

Photoelectric effect.

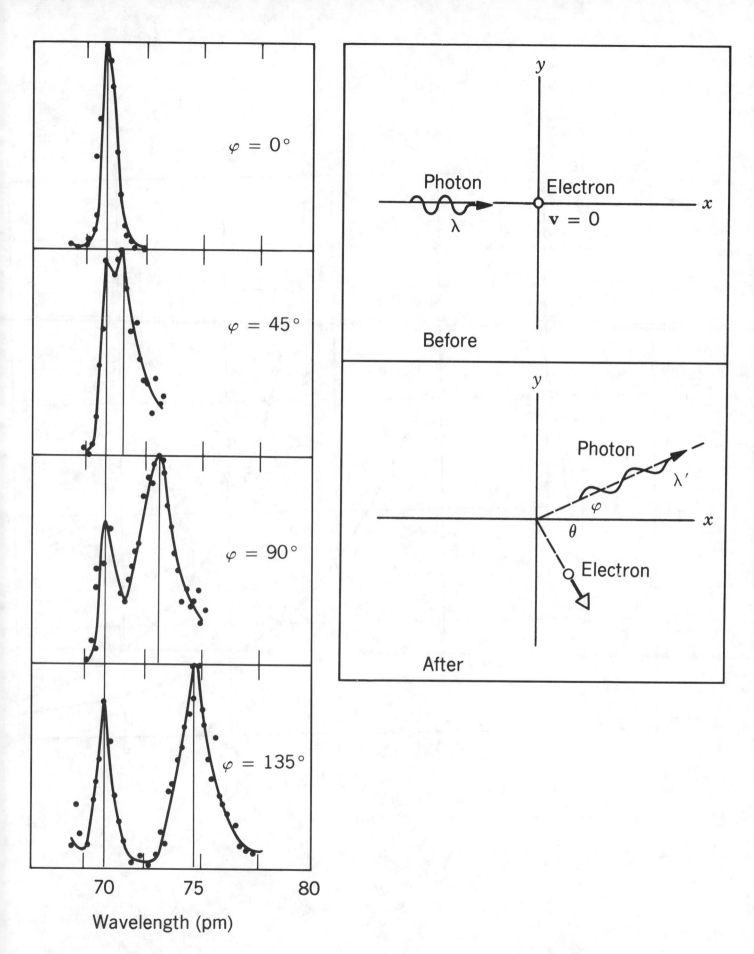

$\varphi = 0°$

$\varphi = 45°$

$\varphi = 90°$

$\varphi = 135°$

Wavelength (pm)

Compton effect.

FIG. 43.5 & 43.6 **161**

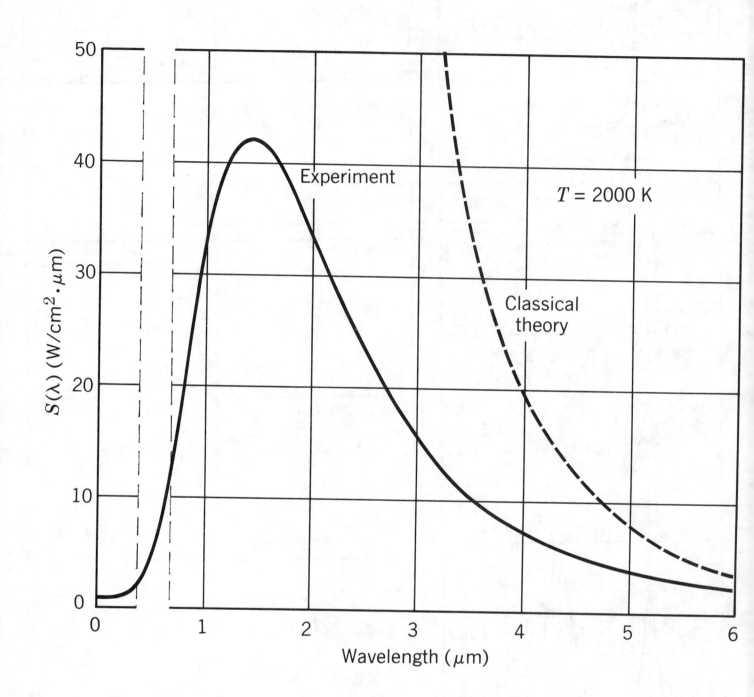

Classically predicted and experimentally determined spectral radiancy curves.

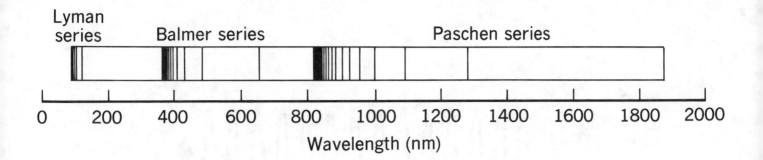

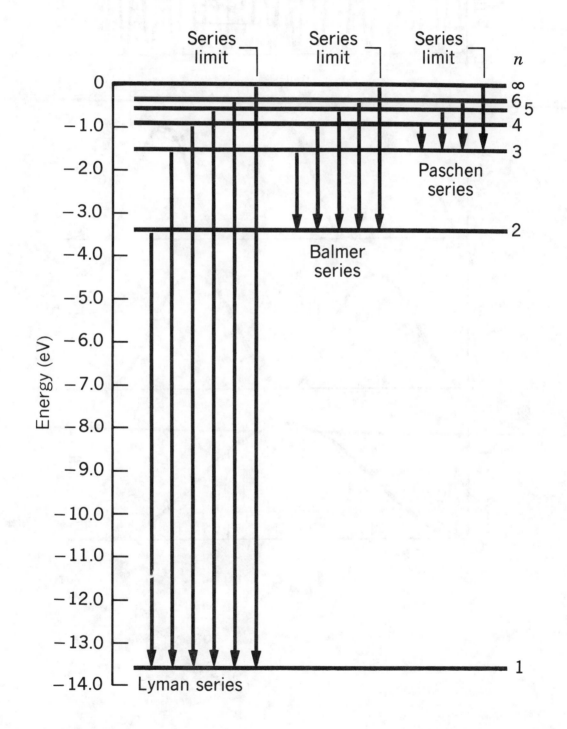

The hydrogen spectrum.

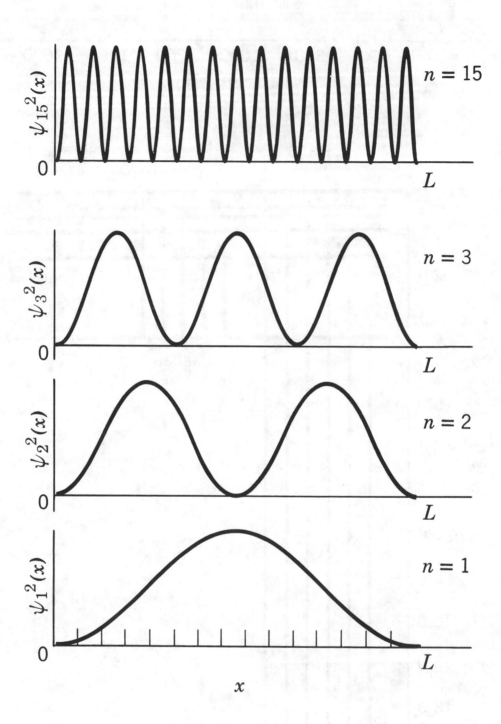

Probability densities for a particle in a one-dimensional square well.

FIG. 44.10   **164**

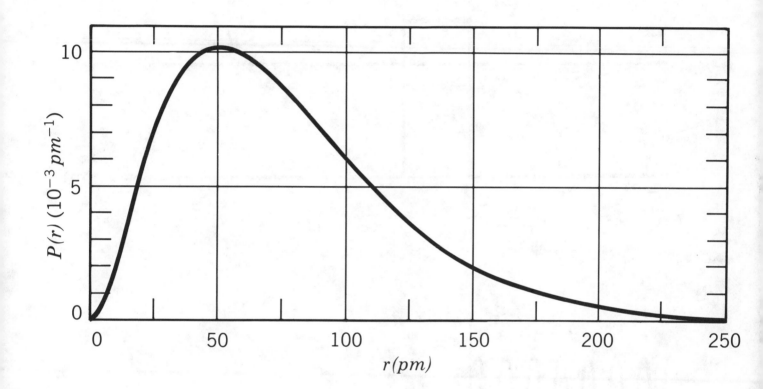

Radial probability density for the ground state of hydrogen.

FIG. 44.12

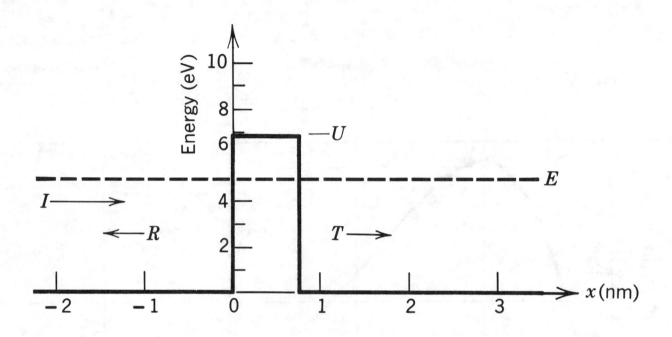

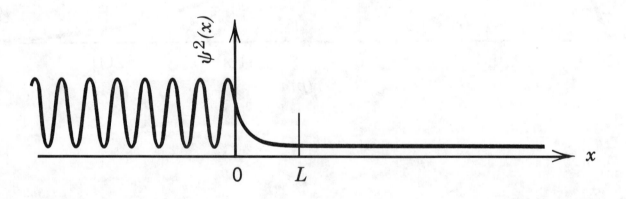

Tunneling.

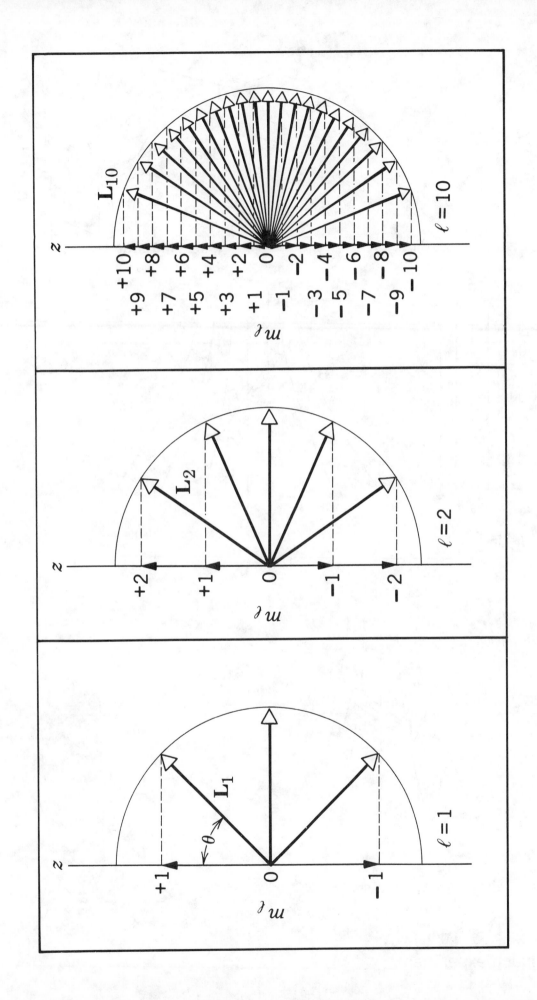

Orientations of angular momentum vectors.

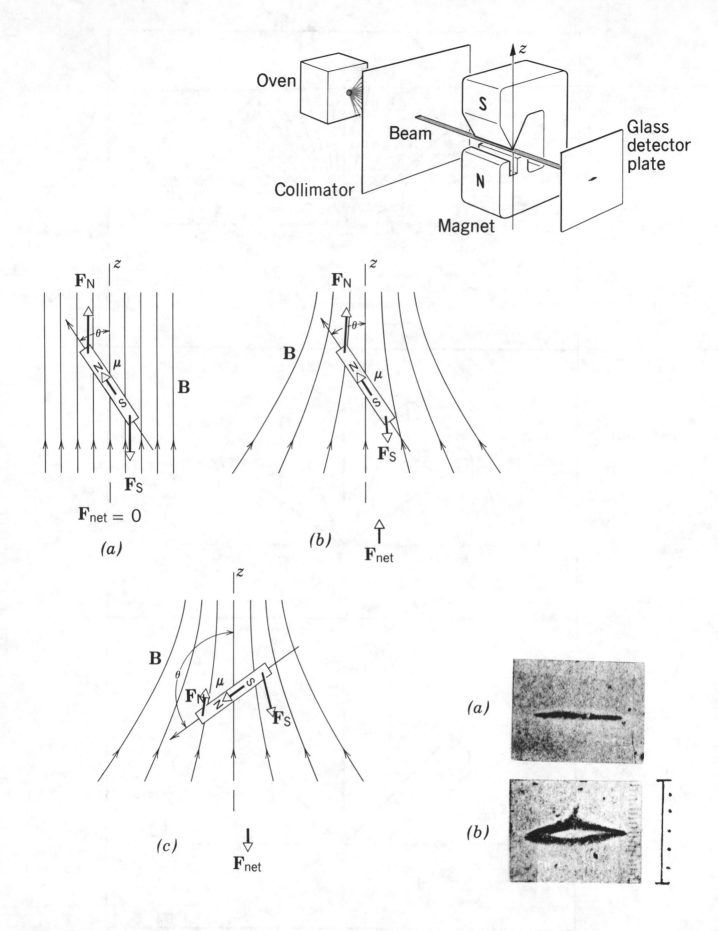

Stern-Gerlach experiment.

FIG. 45.10, 45.11 & 45.12 **168**

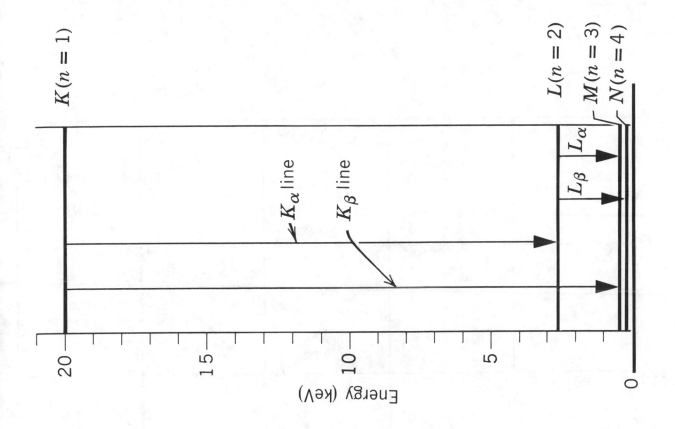

$K(n = 1)$

K_α line

K_β line

$L(n = 2)$
$M(n = 3)$
$N(n = 4)$

L_β L_α

Energy (keV)

20

15

10

5

0

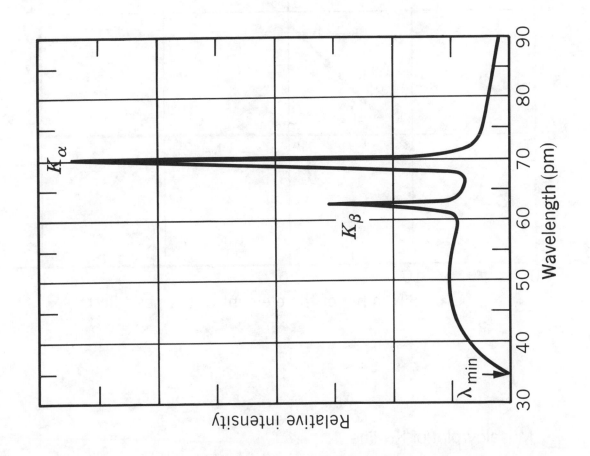

K_α

K_β

λ_{min}

Relative intensity

Wavelength (pm)

30   40   50   60   70   80   90

X-ray spectrum of molybdenum.

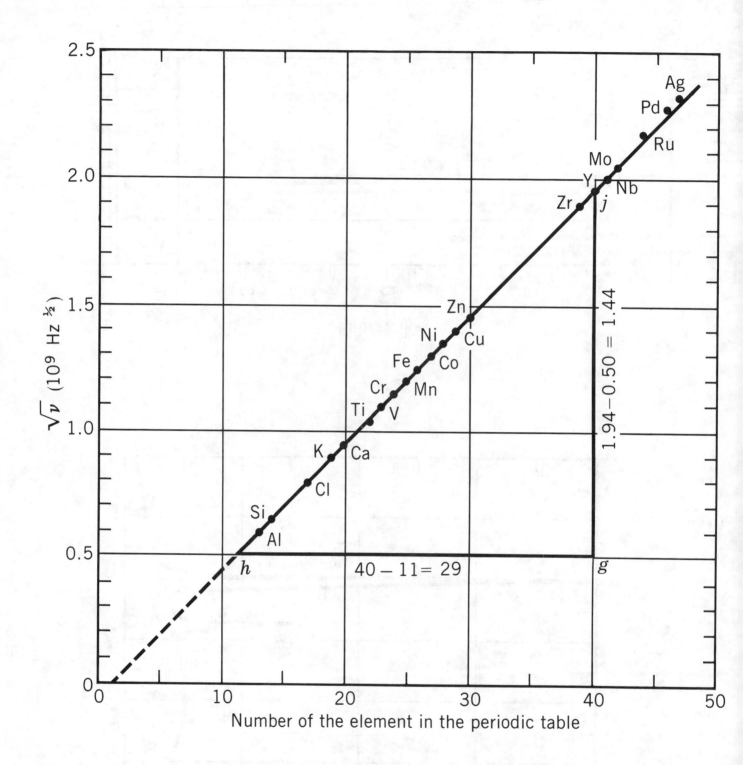

Moseley plot of Kα line.

FIG. 45.21

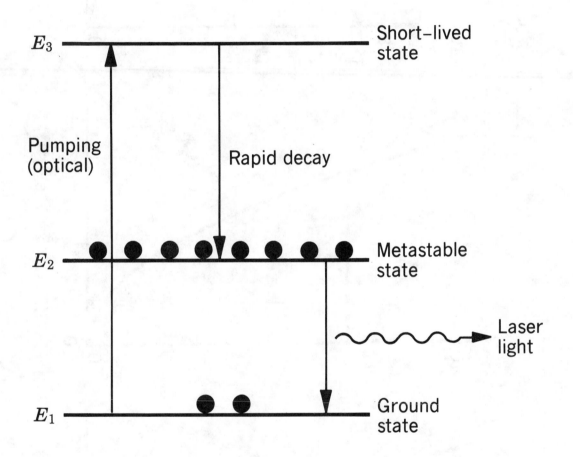

E_3 — Short–lived state

Pumping (optical)

Rapid decay

E_2 — Metastable state

Laser light

E_1 — Ground state

3-level laser.

FIG. 45.27

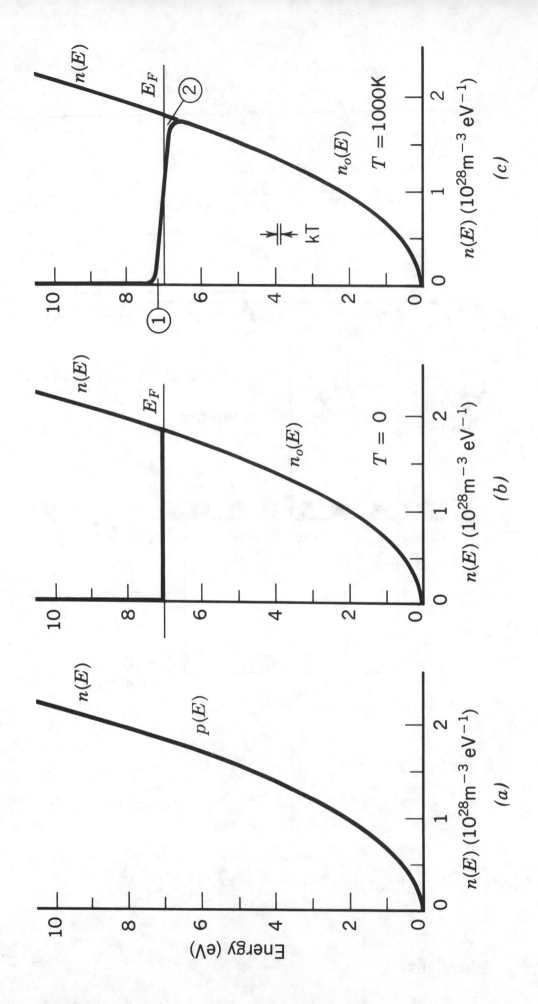

The filling of electron states in a conductor.

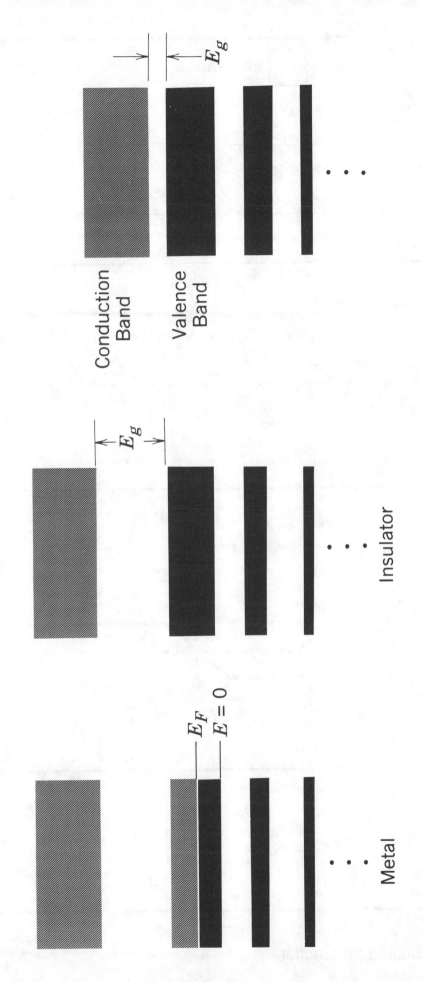

Conduction
Band

Valence
Band

E_g

E_g

E_F
$E = 0$

Insulator

Metal

Filling of states in a conductor, insulator, and semiconductor.

FIGS. 46.4, 46.5 & 46.8 **173**

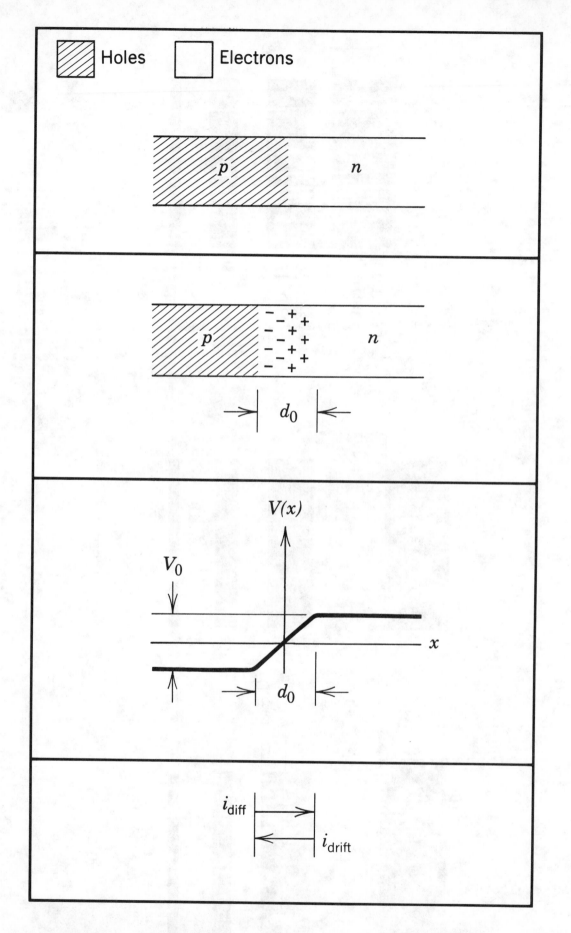

An unbiased pn junction.

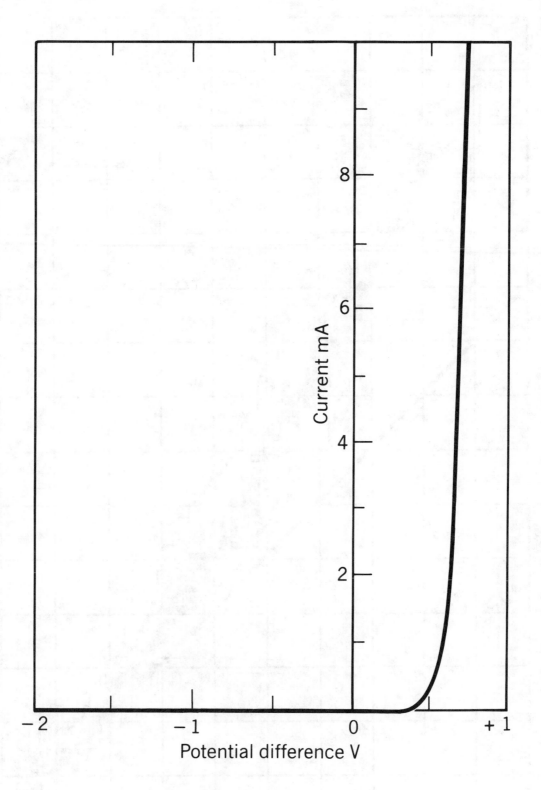

Current vs applied potential difference for a pn junction.

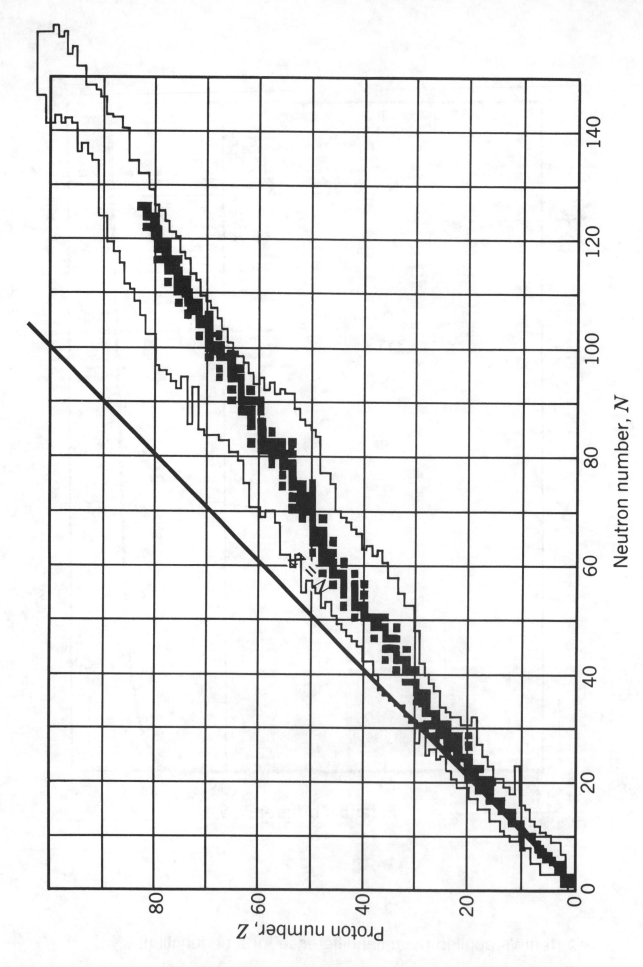

Neutron number, N

Proton number, Z

Proton and neutron numbers for the known nuclides. Stable nuclides are represented by black squares.

FIG. 47.4

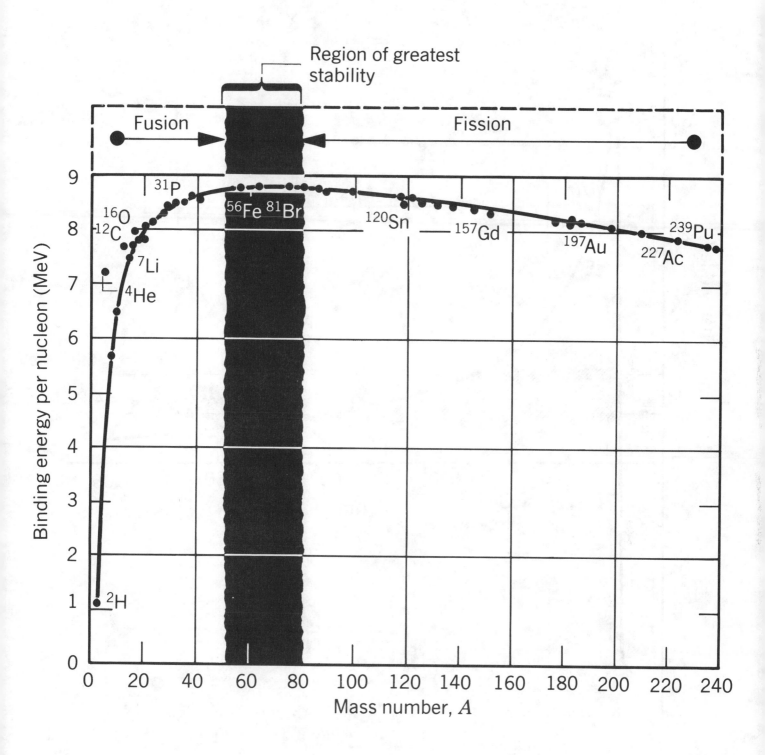

Binding energy per nucleon as a function of mass number.

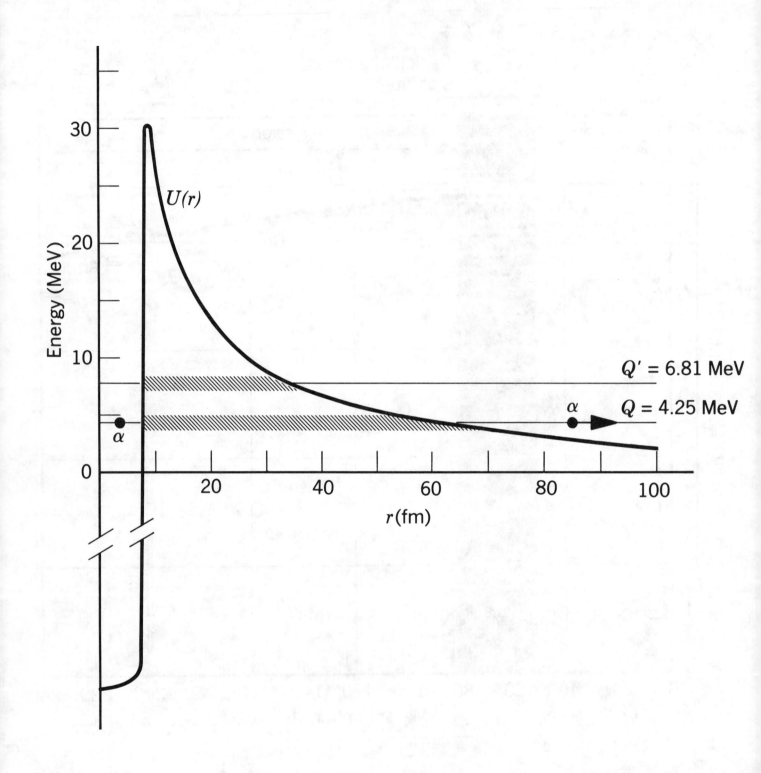

The barrier to alpha emission.

FIG. 47.9

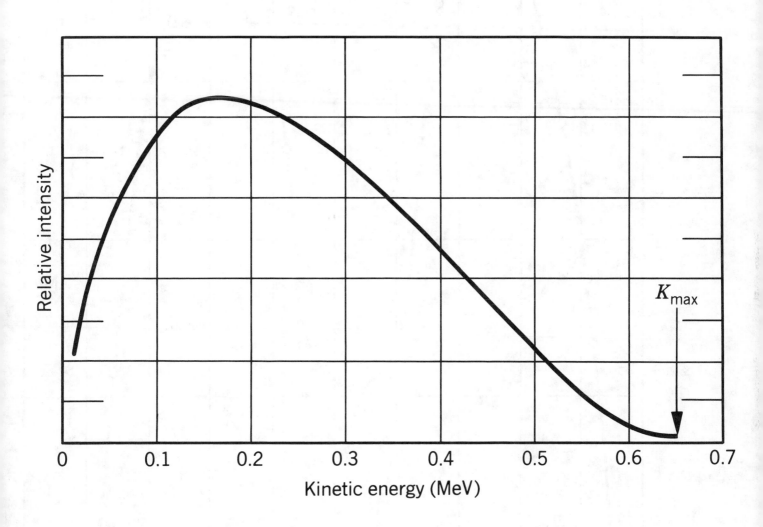

Distribution of positron kinetic energies for beta decay of ^{64}Cu.

FIG. 47.10

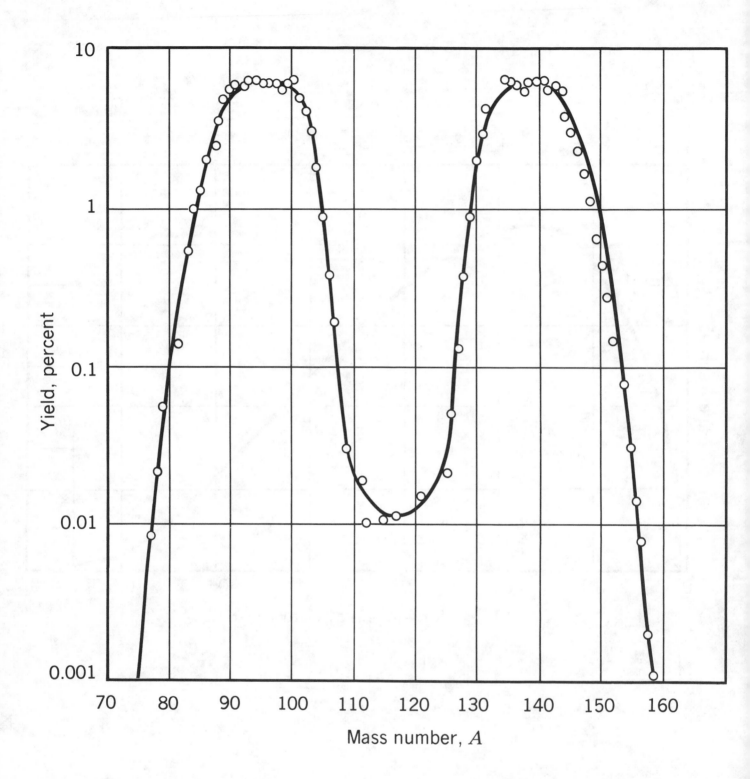

Distribution of fission fragments of ^{236}U.

(a) Neutron — A ^{235}U nucleus absorbs a thermal neutron

(b) It forms a ^{236}U nucleus, with excess energy; it oscillates violently

(c) The motion may produce a neck

(d) Coulomb forces stretch it out

(e) Fission occurs

(f) The fragments separate; prompt neutrons boil off

Typical fission process.

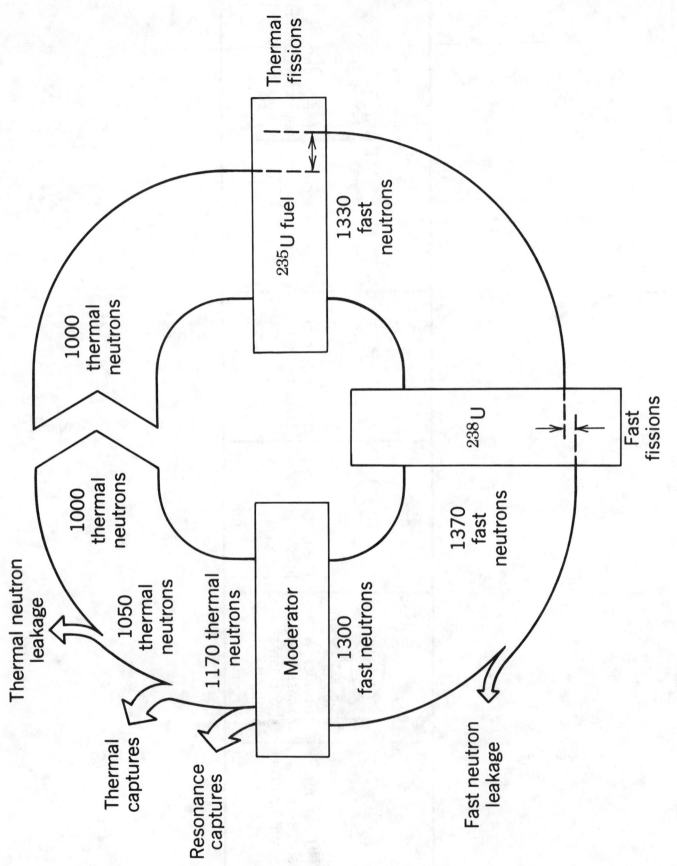

The fate of neutrons in a nuclear reactor.

FIG. 48.5

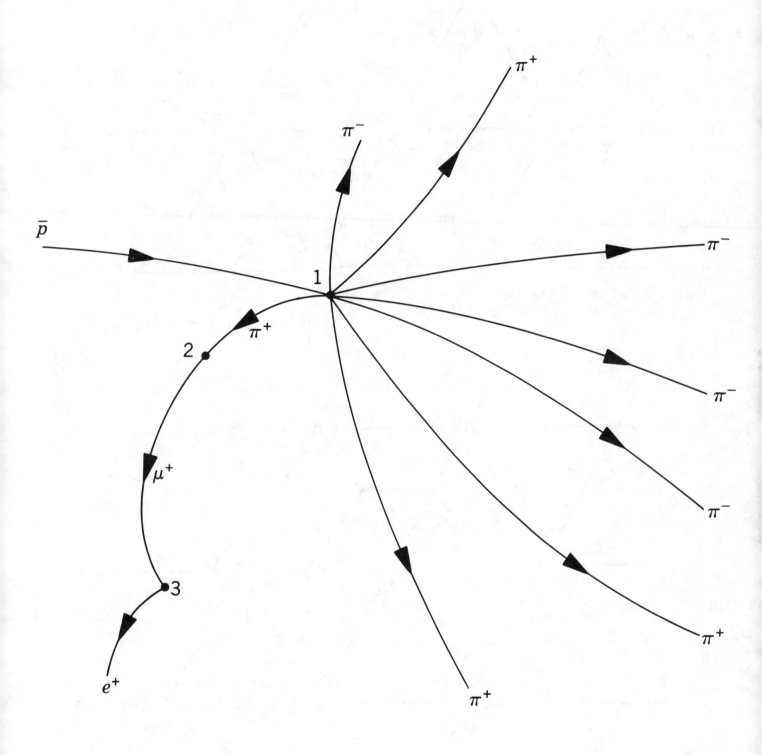

Bubble chamber tracks.

FIG. 49.3

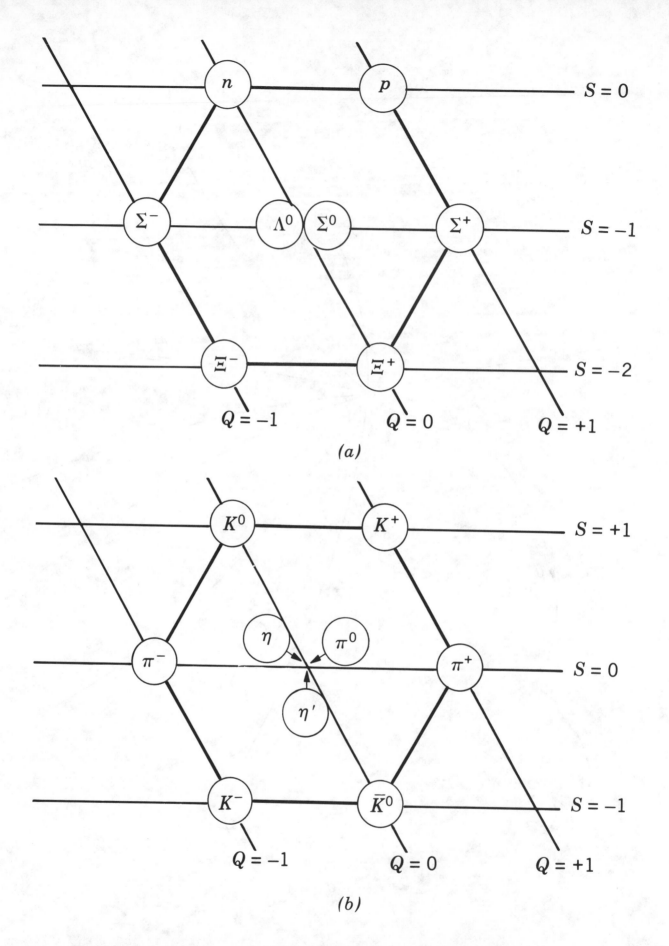

Eightfold pattern for spin-½ baryons and spin-0 mesons.

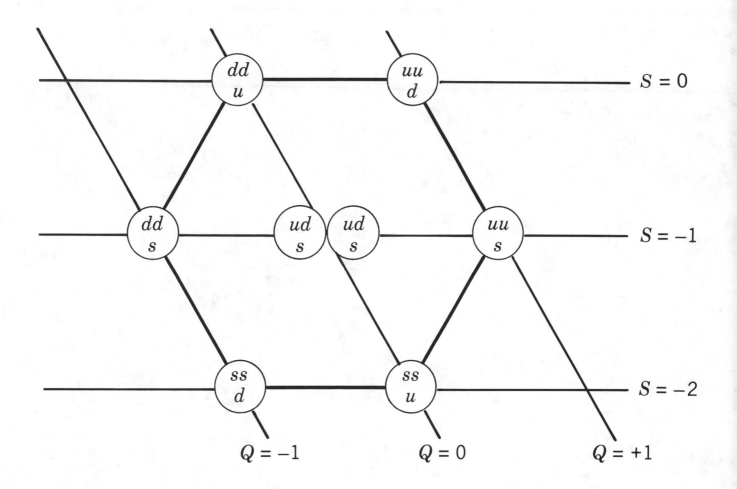

Quark content of spin-½ baryons.